Preface For the *Old School Calculus* Series

If we turn to the problems to which the calculus owes its origin, we find that not merely, not even primarily, geometry, but every other branch of mathematical physics—astronomy, mechanics, hydrodynamics, elasticity, gravitation, and later electricity and magnetism—in its fundamental concepts and basal laws contributed to its development and that the new science became the direct product of these influences.

— William Osgood

The birth of calculus in the 18th century may constitute the single most significant intellectual advance in the history of mankind since the development of classical geometry. Indeed, with the possible exceptions of fire, agriculture, spoken languages, writing and the counting numbers, it may be the single most important creation of homo sapiens in terms of the sheer magnitude of its consequences.

The quote by the old Harvard University master above-who knew quite a bit about both the content and teaching of calculus- underscores both the importance of the subject and its deep connection to the study of the empirical world. Before calculus, the most basic calculations of the properties of natural phenomena-such as the area under curves, volumes under surfaces and the velocity of moving objects- required significant ingenuity utilizing exhaustively detailed geometric and algebraic computations. For example, Galileo attempted to compute the area under the curve generated by a rolling circle in 1599 by an elaborate procedure of tracing and cutting out the curve and its' generating circle in sheet metal and weighing them to produce proportional areas. Interestingly, it gave erroneous results. The problem was officially solved nearly 40 years later by Galileo's students Torricelli and Viviana, using the tangent methods of analytic geometry that anteceded the differential calculus. (It should be noted that the correct area had been produced prior in unpublished works by several others.)

Of course, using the methods of calculus, this area can be very easily calculated now directly from the formula for the cycloid. Indeed, without the calculus, so many ideas in both the physical and social sciences we take as obvious today-such as work, acceleration, compound interest and temperature-were incapable of being defined precisely, let alone determined with any accuracy. For that reason alone, the importance of at least an intuitive understanding of calculus by most educated people is of paramount importance.

A common comment by mathematicians is that every standard introductory calculus textbook is a version of the 18th century calculus textbooks by Leanord Euler, *Institutiones calculi differentialis* and *Institutionum calculi integralis* . While this statement is ludicrous if taken literally, it's intent is quite truthful, namely that the basic substance of a calculus class for a general audience hasn't changed dramatically in topic selection or presentation since the great German master wrote his texts over 2 centuries ago. A basic calculus course has traditionally focused on the applications of the subject to geometry and physical/social sciences rather than the rigorous theory, which is now usually reserved for later courses in analysis. Since the firm foundation of calculus is only 120 years old, only the teaching of the applied aspects of the subject have existed throughout its entire existence.

After World War II with the birth of the Space Program, there was a Renaissance in America of mathematical and scientific training at universities and high schools, leading to a dramatic upgrading of undergraduate programs in mathematics and the hard sciences. The college calculus courses in those days were introductions to modern mathematics for beginners, brimming with proofs of important theorems, careful definitions and inequality computations as well as numerous applications to the sciences, primarily mechanics. This was tragically short lived. Over the last 40 years, the training of students at these levels has degraded gradually but steadily, particularly in elementary calculus courses.

Today's calculus courses are-at most universities, particularly American ones- a fading shadow of what they once were. Once, not so long ago, calculus was a true mathematics course-aimed at serious students who, while not necessarily going on to careers in mathematics, nevertheless, were eager to pit themselves against this significantly difficult course and learn some important skills, namely abstract logical thinking and problem solving, that they would use the rest of their lives. Indeed, many colleges in the mid-20th century onwards imposed a requirement of at least a semester of calculus upon pre-med and business students to act as a sieve to limit the number of applicants to the professional schools.

(Alright, this is not a "fact" in the sense I can provide a ton of evidence that this was the intent of the requirement. But this has been testified to by many in academia, far too many to quote here. I suggest going to your favorite search engine and begin finding testimonies.)

My point is that the calculus courses of yesteryear, unlike today, required actual mathematical skills that most of today's students do not possess. When these courses were standard calculus classes, average entering freshmen at most American universities had solid backgrounds in Euclidean geometry, algebra and trigonometry from high school and they were expected to be able to use this background with minimal review. In fact, not only do they not have such skills entering from high school, but they are actively discouraged from developing in their formative years, when it would be easiest for them to acquire them.

The big advantage of the rigorous calculus texts of the past was that it was almost impossible for such students to con their way to a good grade-the fact that rigorous mathematics was an **essential** part of the structure of the course ensured they actually had to learn something to do reasonably well. And the course acted to ensure that students with impure motives who didn't even try **didn't** get good grades.

I remember as a premed sitting around with a number of students taking calculus using Stewart and the discussion of the exam was like they were talking about a football game and how they were going to "beat" the exam. They came up with codes, mnemonics, word games-not a single theorem or concept or proof. I made the idiotic mistake of asking if anyone actually learned the material and the whole table erupted with laughter. The President of the Student Medical Association smiled at me like The Grinch.

"Winning is about APPEARING to know what you're doing, not actually doing it. Don't worry- you can always work taking out the trash in my office on 5th avenue."

Our society rewards this kind of behavior. Why? Because letting these monsters use Stewart and get their A's without learning anything is good for business, that's why. The university gets to pack the classes with 200 **paying** students by making this a required course, the students get their A's which the college can use to improve its' ranking standing so that administrators get promoted for making so much money and helping public relations and off they go to Ivy League medical schools thinking urea is made in the kidney-and worse, not giving a shit.

And 5 years later they're killing and crippling patients left and right and being acquitted at malpractice trials because the only one in the room who's a better liar then they are is the son of a bitch defending them.

Today, pre-med, pre-law and business students encourage each other to take calculus as an "easy A"-and *not* because they're mathematical geniuses. It's a consequence of the purely pragmatic, plug and chug nature of most university calculus courses-which of course lends itself easily to having programmable calculators and IPhones doing all the work. Worse, not only do students not have to do any real thinking in these courses, they're *encouraged* not to because the final grade is all that matters to both students and universities.

In other words, if you're teaching basic calculus given in the mathematics department of a university and your students leave the course thinking it was an "easy A"-*you're doing it wrong.*

Many mathematicians and scientists in other fields of a mathematical bent often mourn this state of affairs. They often complain about the guts of calculus being removed from present day textbooks in order to make the book as easy and accessible as possible for the maximum number of students, many of whom don't give a damn about the theory of calculus. They perceive most students that come into their calculus courses as not only poorly prepared, but completely indifferent to why calculus is important or why it works. Poor preparation is an inconvenience and a headache for both teacher and student, but if the student cares and is willing to work hard, this is not an insurmountable difficulty. But if the students see calculus as an annoying and meaningless hurdle they have to scam their way through before they can go out into the Real World and foreclose on old people's homes for fun-well, there's no book or course in the world that's going to make teaching those students anything but a chore.

I have great sympathy for the pessimism of most practicing mathematicians in regards to the subject. I've discussed my perceptions of the current state of mathematical academia in the mission statement of my company at the main website for our books. What I said there, I think, bears paraphrasing here to encapsulate the current status of calculus courses and textbooks and provide context for the myriad of calculus textbooks I'm preparing for publication under this series banner in addition to the one you're holding in your hand.

To me, there are 2 major factors ruining calculus today for students and professors alike at most universities: The first is the perceived academic psychology of students taking today's calculus courses and the second is the financial incentivizing of the resulting structure. These factors are by no means independent of each other-indeed, focusing on these 2 factors vastly oversimplifies the situation. But since these at the 2 factors that are most determinate in the direct planning of calculus courses each semester by faculty and staff, we'll focus on these. Since they are closely intertwined, we need to analyze them together.

We begin with a quote from one of my favorite sources:

The two biggest obstacles to the success of the Moore method (or, for that matter, of teaching of any kind) are students who don't want to be there and students who want to be somewhere else. The two are not the same thing; let me explain. By students who don't want to be there, I refer to required courses. If a student comes to me and asks my help to learn something that I already know, I am overjoyed. I am sure I can teach it to him, whatever it is, and I expect the process to be pleasant for both of us. If, however, he comes to me and says "I don't really want to know this stuff, but it's required that I get a C in it before I can go out and make a lot of money", then I'm unhappy……….. I dream of the ideal university, full of students who are full of intellectual curiosity. The subset of those among them who take a mathematics course do so because they want to know mathematics. They may be future doctors or chemists or executives in a shirt factory, but, for whatever reason, they want to find out what this mathematics stuff is about, and they come to me free willing and ask me to teach it to them. Oh, joy!-Paul Halmos, **I Want To Be A Mathematician**

Ok, there's a lot to unpack here that reflects directly on the teaching of calculus today compared with those of yesteryear that this book represents. To Halmos then, there are essentially 3 kinds of students in any university mathematics class:

- \# 1) Geeks who absolutely live for understanding mathematics and its related subjects.
- \# 2) Students in other disciplines who, while not mathematics or physical science students, have the same curiosity and passion about learning in general. These students are willing to work hard and try and learn something new, especially if you can convince them it'll be helpful in whatever field they chose.
- \# 3) Sociopathic scam-artists who break into the professors' office to get the final exam the night before, program the entire textbook in code into their calculators and know a hundred other ways to cheat. .

Clearly, in the quote above, Halmos is referring to students of types (2) and (3). "Students that don't want to be there" is type (2) and "students that would rather be somewhere else" are of type (3).

Granted, I'm not describing these students in anywhere near as polite a manner as Halmos is. But having been on the receiving end of such students' mocking and watching them be rewarded so heavily for their evil-well, you can forgive me for being somewhat tactless.

I would argue that the American academic system over the last 20 or so years has virtually eliminated students of type (2) and heavily encouraged students of type (3).Calculus courses today are geared specifically on these assumptions in order to maximize the profit generated by making these courses required.

It would take an entire book-which I hope to write someday-to fully explain this statement and all its implications. But Halmos almost certainly hit the nail on the head about the fundamental difference between students of type (2) and type (3): Students of type (2) are true students and have that most critical requirements for the mastery of any academic discipline: *curiosity*. They want to learn. They enjoy learning, even if it's in an area unrelated to their chosen pursuit and they aren't initially passionate about.

Students of type (3) aren't real students. They aren't really curious about anything. To them, learning anything is for suckers. To them, this is a game they need to play to win at any cost to get to their real goals: ***Money and temporal power.***

Because America is a society which doesn't encourage honesty or values-it encourages people to devour each other-such students are rewarded for cheating any way they can. They mockingly laugh at students that actually study and try and learn things to get high marks while they get straight A's while never understanding a thing. And sure enough, most get into Ivy League medical, dental, law or business schools and eventually kill someone through either cruelty, incompetence or indifference before calling their own 1000 dollar an hour lawyers

And most universities-especially those at the bottom of the academic food chain-couldn't be happier to hand those kids A's and look the other way about cheating because it's great for their bottom line. They get to say they have kids with great GPAs and get to double tuition accordingly.

Let's face it-**that's** the "academic" reason most universities buy non-mathematical 4 pound calculus books with new editions every year. It's so the premeds, accounting students, actuaries, pre-law and all the rest of the master cheaters that form the vast majority of bodies filling the enormous lecture halls of the average 200 student registration-per-semester of the required calculus course can program the solutions of all their exams into their programmable calculators.

And this is where the financial factor of student debt, which is a lifetime burden now for most American students considering going to higher education that aren't wealthy-has its impact differently on each of the student types above.

Because of the college debt deliberately shackling an entire generation, most students who aren't wealthy simply won't be able to stay in college no matter how much they want to learn if they can't succeed. In the academic paths to most lucrative careers, truly superior grades are required to even have a chance at entering these fields. Of most non-wealthy students have to work nearly full time in addition to going to school to be able to live-and that's the best case scenario, assuming they don't have any additional ongoing personal problems. Therefore, when put in an impossible position to compete for their life's dreams, they simply either dispose of their ethics and join group (3) or drop out. And those that manage to stay can only do so by taking the absolute minimum of coursework they can afford to borrow enough to support. Most financial support today for such students is loan-based-actual free federal and state student financial aid has been slashed in most states to the point its' insignificant.

In addition, in the case of calculus courses, the impact of this limiting debt and its resulting pressures has been magnified by the ludicrous prices of calculus textbooks. While nearly all standard American college textbooks have reached astronomical prices-the 400 dollar textbook is now a reality as of this writing in summer 2018-calculus textbooks have become notorious in this regard. This is not only for their substantial price-a complete new 8th edition of the very commonly used *Calculus* by James Stewart is now running over 300 USD at Amazon, with a used set running a still-painful 168 USD-these books are constantly "updated" with "new" editions where insignificant changes are routinely made and accompanied by a significant bump in price.

Again, a complete discussion of the mechanics of the textbook monopoly in America and how its' exacerbating the student debt crisis would take us too far afield here. But in the particular circumstance of calculus, the steadily climbing prices of relatively indistinguishable new editions is impossible to justify beyond simple greed. The situation was very well summed up by an anonymous calculus professor at UCLA in 2015:

The subject of calculus did not change much in the last hundred years….[T]here are no reasons why the textbooks have to be updated every 5 years or even more frequently." (For the record, the sixth edition of Stewart's "Calculus" was published in 2007, and the seventh in 2012.)
….New illustrations are sometimes added, exercises are shuffled and so on, but these do not substantially affect teaching/learning. Textbook publishers produce new editions solely as a means to sell more books and make more profit.(https://www.collegian.psu.edu/opinion/columnists/article_5d3d65b4-9b8d-11e4-8abd-5f63ba6c5556.html *)*

Given the enormous financial and academic pressures on non-wealthy students in today's sink or swim American university environment, it's not hard to understand why the selective pressures at work favor only the extreme student types of (1) and (3). Group (1) students will always be there and sadly, they'll always be in the minority. These students will always register for the hardest classes and buy, beg for and borrow mathematics books to help them master as much as time allows. Unless they suffer a personal tragedy that literally makes it impossible for them to perform at 100 percent, they'll find a way to persevere. Equally sadly, group (3) students will always be there as well and they'll always find a way to game the system. And again, because this is America, they'll be rewarded for it because both they and the people that rely on them profit from their amoral behavior. Its' why while they probably (hopefully?) won't be the majority of students taking calculus, they'll always be far more numerous than the math/science geeks.

The disturbing trend is how the current system actively discourages bright, curious students who are not mathematics or hard science majors-basically Halmos' type (2) students, "students who don't want to be there'-and who would benefit enormously from a meaty introduction to the calculus. It would not only teach them an enormous number of basic skills that are applicable in any number of careers, it would teach them the value of thinking mathematically. Indeed, with the right textbook and/or teacher, it might inspire some of them who never considered it before to pursue a career in mathematics or the hard sciences. For many students in today's universities, calculus is their first and only real exposure to mathematics. Since this exposure most of the curious non-science students gets is not only mediocre, pragmatic and uninspiring, in content, but extremely stressful, both financially and in terms of the student's precious time. In the end, most students see this course as a boring, mechanical, expensive, unpleasant and time consuming chore they resent being forced by their department to muddle through and leave without the slightest clue what the hell they wasted all that time and effort for. Why would any of these students ever take another mathematics course again after that unless it was at academic knifepoint? And if that's the only impression they ever get about mathematics, why would any of them ever see it in a positive light?

****

The Old School Calculus Series will attempt to rectify this situation by making available inexpensively a number of full calculus textbooks and supplementary sources which it is hoped will inspire both teachers and students to take calculus seriously again. The *Old School* series banner doesn't refer so much to the fact many of these books are republished out of print texts from decades ago as the fact all books in this series will attempt to recapture the spirit, comprehensiveness and rigor of these books of yesteryear to inspire a new generation of students. As with many of Blue Collar Scholar initial publications, the first few will be classic out of print texts brought into the 21st century with new original supplementary material. Eventually, we hope to publish original works for all mathematical subjects and calculus will be foremost among them.

These books will range considerably in academic difficulty level. From run of the mill, "plug and chug" non-rigorous textbooks for average beginners, which we hope will appeal to talented high school and weak freshman college students to accessible but somewhat-higher-than-today's-standard-level calculus textbooks, which we hope will raise the bar on commonly used calculus texts and put "the guts" back into ordinary undergraduate calculus classes-to completely rigorous, no-nonsense honors calculus texts for strong, advanced students beginning study to eventually enter top graduate programs in both pure and applied mathematics as well as the physical sciences. Most of all, we hope the inexpensive prices of these books will encourage self-study in this most critical and important of sciences, fostering a rebirth of general interest in both mathematics and the hard sciences in students of all social and financial backgrounds.

Karo Maestro

Founder, Editor, Blue Collar Scholar (LLC pending)

New York City

Summer 2018

CALCULUS FOR BEGINNERS

By ROBERT D. CARMICHAEL
Professor of Mathematics, University of Illinois

JAMES H. WEAVER
Professor of Mathematics, Ohio State University

and LINCOLN LAPAZ
*Associate Professor of Mathematics,
Ohio State University*

New Prefaces And References by Karo Maestro

The Old School Calculus Series

Blue Collar Scholar/Createspace

COPYRIGHT, 1937, BY ROBERT D. CARMICHAEL, JAMES H. WEAVER, AND
LINCOLN LAPAZ
COPYRIGHT, 1927, BY ROBERT D. CARMICHAEL AND JAMES H. WEAVER
ALL RIGHTS RESERVED

937.9

New Edition@2018 Karo Maestro

Preface to The Old School Calculus Series@ 2018 Karo Maestro

PREFACE

In approaching the calculus, as presented in this text, the student will find frequent use for the more elementary parts of algebra, trigonometry, and analytic geometry. These will furnish the basis on which he will build up the conceptions and the processes of the calculus. Some of the ideas to be encountered will at first appear novel and strange to him, and he will find it impossible to master them merely by receiving instruction. He cannot remain passive; he must become active, consciously seeking to initiate himself into the subject. The study of the calculus constitutes a sort of intellectual adventure. To pursue it properly and with the greatest pleasure requires a certain spirit of exploration.

After the preliminary stages it will be useful to the student to devise (when he can) his own methods for dealing with each new problem treated in the text. This will help him to develop that initiative without which he cannot master the new ideas and processes with joy. Before studying a new topic he should think over the main ideas of the preceding ones, making a special effort to recall clearly the more recent developments. When a new process is learned he should formulate clearly its meaning and significance as far as he is able to understand them. As he proceeds, the meaning and significance of the more fundamental processes should become progressively richer and fuller to him. A conscious effort to bring about this enrichment of the subject, as understood by the student, will economize his intellectual energy and add greatly to the comfort of his study and increase in a remarkable manner the measure of his mastery.

To facilitate the understanding of the course as a whole it is desirable that the student shall frequently run over in his mind a brief outline of the work up to the point that he has reached. The summaries given at the ends of several chapters are intended primarily as an aid in this review.

When a topic is difficult and the ideas are not clear after the first reading,* a good plan of procedure is to start again at the beginning of the section and to read with careful attention to the question as to where the first thought is encountered which is not entirely clear. When this initial difficulty has been determined it should usually be resolved before proceeding further. If a careful reading of the earlier part of the section does not clear it up, it is well to inquire whether some preceding ideas or principles are involved which the learner has momentarily forgotten. If the authors have succeeded in making their exposition clear, the student should in this way find it possible to master the text for himself, with only that help from the instructor which is involved in the analysis of the more fundamental and novel ideas of the subject. If difficulties persist, make a careful note of them (mentally or otherwise) and try to formulate them into definite questions. This will clarify the nature of the difficulty and make for efficiency in the discussions of questions in the class.

Every example solved in the text is given to elucidate some point. In many cases it will be desirable to read a section for the second time with the solutions of the illustrative examples in mind. These examples will also assist the student in quizzing himself; that is, in putting to himself such questions as will test his knowledge of the topic under consideration. To think of these questions and answer them for oneself affords one of the best means of making sure of a full mastery, step by step.

In the calculus the student will find a definite way of thinking in numerical terms which he has not encountered in any earlier part of his work. He will approach it with his previous work as a basis, but he must extend his conceptions and learn new processes. The subject has been developed primarily to enable us to describe and understand and interpret the natural phenomena of motion and change as observed in common experience. It affords the most powerful mathematical tool that exists. The principles to be developed are themselves abstract, but they

---

* Sometimes it is desirable to give a section a rapid first reading before setting out to master it.

have many concrete interpretations and applications. It tends to economy in learning if we develop the abstract processes before many of the concrete applications are presented. Consequently we shall usually indicate the nature of the applications briefly by means of one or two concrete instances and then develop the abstract processes; when one of these is suitably mastered we shall proceed to some of its more important elementary applications. In this way the abstract processes will be enriched in significance. The student is urged to observe, from the beginning, the way in which abstract processes have concrete connections, so as to afford himself a growing incentive to master the abstract processes themselves as they arise.

Before entering upon a study of the main topics it is desirable to introduce certain general preliminary considerations. These will serve to bridge the gap between the earlier courses and this one and to clear the field of action for the development of the topics of the calculus proper. To these preliminary considerations the first chapter is devoted.

The material of the course as a whole is so organized that topic grows into topic in a natural way. Things which the student should think together are carefully brought in together in the order and with the emphasis which contributes best to facilitate mastery on the part of the learner. The explanations are always addressed to the learner and never to the teacher, since it is the former (and not the latter) who must acquire from the book a knowledge of the subject. If the teacher misses some statement or method to which he is accustomed, let him remember that the authors have omitted this from reasons growing out of their experience in teaching.

The teacher will observe that the two main parts of the calculus (differential and integral) are carefully interwoven in such a way as to meet the needs of the learner. Throughout the book a continued review of the more important principles is effected by means of carefully chosen exercises so selected as to assist the student in integrating the earlier work with the subject under immediate consideration. The problems set for the student to solve have been selected and graded and arranged with great care so as to lead the student naturally and effectively from step to step and to give him confidence in the handling of the new principles which he is learning.

The total material given is probably sufficient for a year's work for a class that meets five times per week. It is arranged with special reference to the needs of a class that meets four times per week for a year, the material being sufficient in amount to leave some choice to the teacher in the selection of the course. It is believed that material for a three-hour course for a year can well be selected from the book. For even the shortest course the first fifteen chapters would probably be taken almost or quite entire. Selection from the remaining seven chapters would then be made to meet the needs of the particular course being given. The material has been so organized that there would be proper continuity in a course of any one of the three sorts indicated.

R. D. C.
J. H. W.

PREFACE TO REVISED EDITION

This revision is a definite attempt to present the topics of the calculus from the standpoint of their correlation with courses in physics, chemistry, mechanics, and technical subjects in engineering curricula. It is especially valuable for use in colleges where these courses are given simultaneously with the calculus. In order to accomplish this end the concepts of antiderivative and the definite integral are developed along with the concept of derivative and are used to develop the laws of rectilinear motion. The sine and cosine functions are differentiated as examples of the Four-Step Rule so as to be available for early use in the solution of problems involving simple harmonic motion. A chapter on differential equations with applications to problems of physics and engineering has been added. A number of topics, including a discussion of the properties of hyperbolic functions, have been added, and numerous exercises have been included. Other topics have been rewritten.

During the work of revision several departments at the Ohio State University have coöperated, particularly the departments of physics, mechanics, and electrical engineering. Members of the staffs of these departments have taught classes in the calculus and have met regularly for discussion of what topics to include, the arrangement of the topics that were to be included, and the best methods to use in presenting these topics.

For this coöperative help the authors wish to thank those who participated. Their suggestions have been valuable and their encouragement helpful. The authors also wish to thank the publishers for their willingness to undertake the revision and for their painstaking care while the book was in press.

R. D. C.
J. H. W.
L. L.

Preface For the 2018 Edition

Every one who understands the subject will agree that even the basis on which the scientific explanation of nature rests is intelligible only to those who have learned at least the elements of the differential and integral calculus, as well as analytical geometry.

-Felix Klein

The sheer importance of mathematical analysis in the development of the physical and social sciences cannot be understated. Its importance in the training of students who seek to pursue the study of these sciences to any real degree is arguably even more important. The German master whose quote begins this preface understood this as well as perhaps anyone who has born and died since calculus' birth in the late 18th century.

Unfortunately, the study of introductory calculus is no longer taken seriously in America by either students or the universities they attend. Indeed, Klein- whose own world famous lectures on analysis at the University of Gottingen eventually became the basis for what many consider the greatest calculus textbook of all time, Richard Courant's *Differential And Integral Calculus*-probably would have been absolutely aghast with horror at what passes for basic calculus at most colleges in America today. As I already detailed in the Preface to the *OSC* Series, most calculus courses have become simply a profit making machine they fill with as many bodies as they can, with neither concern nor commitment to course content beyond the mechanical calculations that students can program into their IPhones without any real thought or understanding. This felony's been compounded by the ridiculous cost of standard calculus textbooks: Not only by the average price of these texts, but the ongoing scam of new editions that differ insignificantly in content from previous editions, but differ dramatically in price.

The book you hold in your hand is one of a series of new calculus textbooks from Blue Collar Scholar hoping to break the current trends, offering quality courses in calculus at various levels for an audience ranging from honors high school students to honors freshmen at the top universities.

This book was published in 1937 and is the third edition of a standard calculus textbook originally written by Robert. D. Carmichael and then revised by Carmicheal, James Weaver and Lincoln Lapaz. This text was written long before the internet, laptops, desktop computers, IPhones, IPads, tablets, smart phones, programmable calculators or any of the computer science marvels today's students take for granted and without which many of them literally wouldn't know how to live their lives. Calculus existed centuries before any of these things did and it exists, prospers and thrives just fine in their complete absence, thank you very much. That's not to say these things don't have their place in a mathematics class-**they do.** It really hard to justify making students construct complicated curves and surfaces of functions of 2 and 3 variables with pencil, paper and ruler and compass the way they did when this book was new. Just as it's equally hard to justify attempting numerical approximation of functions at given points to n decimal places by forcing students to toil over multiple iterative approximations using up a stack of scratch paper. Both these tasks are not only far easier and quicker by either programmable calculators or a computer algebra systems like *Mathematica* or *Maple,* they can be done far more accurately and extensively as well. Approximating # to a hundred decimal places or more using Newton's method would be insane to do by hand-even if the students didn't drop out of the class the second the professor suggested it, right?

All the standard topics of a single and multivariable calculus course are covered here: Functions, limits, derivatives, integrals, linear approximations and derivatives of higher orders, approximation, infinite series, partial derivatives and multiple integrals. Most standard applications to geometry and physics are covered as well: velocity, speed and acceleration in one and two dimensions, plane curves and arc length, finding local extrema of functions and their resulting graphs, surface areas, differential equations, force, work and much, much more. Since the book is pre-technological, students will have to learn to analyze problems using basic pre-calculus tools such as geometric diagrams and solving inequalities. As a result, they will leave the course with a much greater command of problem solving then they would receive in a modern course. Teachers, of course, should feel free to supplement the book with computer and calculator assignments, but these of course, would be strictly as educational enhancements and not as replacements for learning mathematics.

As a student or teacher (or both if you're a PhD student) looking over this textbook, most of you may be surprised how similar this 8 decade old textbook is to any of the calculus books available today. You might be particularly surprised after the preceding preface for the *Old School Calculus* series, where I droned on at length about the lack of theory and understanding in today's calculus courses and how this mindless approach to calculus is encouraged. And now looking at this book, it doesn't look any harder than today's books. Well, you're right, overall this is a pretty standard, "grind it out" kind of calculus book without many proofs, lots of calculations and a whole lot of applications, particularly to physics. It's not a rigorous calculus course with a lot of hard analysis theorems on limits. So why republish it, what's different about it? And what happened to that speech bemoaning the loss of theory in calculus?

Well, first of all, I meant what I said there. BCS will be publishing other books by year's end under the *OSC* banner that will be far more challenging and mathematically precise in presentation. These books will be intended as introductions to calculus for strong mathematics or science majors beyond this level .This is a book for everyone else, for beginning students that just want to learn some calculus and haven't decided yet how serious they want to get with it. As much as I want to encourage a deeper, more precise approach to calculus,all students have to begin somewhere and this book will give them a nice, gentle place to begin. In fact, the ideal students I'd love to have using this book are honors high school students! If we can get many high school students interested in calculus using this book,that would be a definite victory here.

Secondly, if you look closer at it, you'll notice this book requires a bit more care then most "plain" calculus textbooks do. For example, there are lots of applications of geometry to calculus, both in the text and in the exercises. These kinds of problems need to be thought through, there's no mindless putting them into a calculator to solve-at least, not without a few minutes thinking it through and setting it up first.

Thirdly, while it's not rigorous in the sense of having a lot of proofs, it is careful in the sense of defining things like limits and derivatives. There's not a lot of handwaving here, everything is defined pretty carefully. This approach will make the real mathematics courses that hopefully students will take after this easier to learn

Lastly and most important for the purposes of a standard calculus course like this one, is the cost of the book. One of the reasons universities have watered down calculus over the last few decades is so calculus textbooks will be easy, pleasant and above all, *marketable*. James Stewart's ubiquitous book is notorious for this. At this writing-summer 2018-a brand new 8th edition of Stewart will cost you is now running over 300 USD at Amazon, with a used set running a still-painful 168 USD. Worse, these books are constantly "updated" with "new" editions where insignificant changes are routinely made and accompanied by a significant bump in price.

This book will cost you a total of 25 USD.

Yes, you're not misreading that. 25 bucks for a full blown calculus course that can be used for a full 2 years of basic calculus in a nice little paperback you can hold in your hand.

An ebook Kindle version is also available which is even cheaper-if you're into that. But to be honest, I don't understand why students wouldn't want a real book they can hold in their hands if it's this inexpensive.

Some words on the pedigree of this work. One the key publishing strategies of BCS-at least in the early stages of our existence-will be to deliberately republish out of print works by significant mathematicians when we deem the author and his/her works, once considered important benchmarks, have both been largely forgotten. This is done in the hope of reawakening interest in the author's works and career, which we both believe are still quite valuable to learn from, in students and teachers of a new generation. We've done this so far with once-standard textbooks by Phillip Franklin and Henri Cartan-the former author being nearly forgotten by any but historians of mathematics. This standard calculus text brings up another once famous, now nearly unremembered name from the first half of the last century: Robert D. Carmicheal.

R.D.Carmicheal's name, once a household one in American mathematics, hasn't been entirely forgotten by number theorists. Indeed, his classic textbook on the subject is now in the public domain and has been republished by a number of independent publishers. Originally trained as a minister, he showed a gift for mathematics in his published papers and earned a fellowship for graduate study at Princeton University, earning a PhD at the (then) ripe old age of 32. At a time when most American university mathematics programs were dedicated to teaching rather than research, Carmicheal was one of the most prolific paper producers, mostly in number theory and differential equations. He was also a gifted teacher. It was the latter skill that he would most be remembered for. Not only did he produce a number of textbooks on subjects ranging from calculus to relativity theory, he also produced an enormous number of PhD and successful undergraduate students in mathematics, most when he was the Chair and later Dean of Graduate Studies at the University of Illinois. Paul Halmos, one of those graduate students in the 1930's, described Carmicheal's lectures on number theory as "supremely organized, clearly delivered, inspiring". A wonderful and far more detailed biography of Carmicheal can be found at http://www-groups.dcs.st-and.ac.uk/history/Biographies/Carmichael.html .

Reading through the text you hold in your hands, it's difficult to argue with Halmos' assessment. Carmicheal himself taught calculus at various levels at both Indiana University and The University of Illinois for nearly 20 years. The original version of the book developed from his lecture notes. Both of his co-authors, Weaver and Lapaz, were experienced in teaching calculus from an applied perspective: Weaver as a longtime teacher of calculus at Ohio State and Lapaz as a practicing astronomer. This is clearly a book written by several experienced teachers, with an enormous amount of careful thought put to not only the content, but the organization. While that content is very customary with no real innovations, it is very detailed in its explanations and examples. It is also clearly formulated for a general audience of beginning students, many of which were students in other disciplines. That lengthy experience and the importance of teaching pedagogy by the authors shows on every page.

I hereby commend this book to future generations of student as part of Carmicheal and his co-authors' legacy. I hope it will serve those generations well as an inexpensive and accessible textbook for beginners of all backgrounds in this incredibly vital subject.

Karo Maestro

New York City

July 2018

CONTENTS

Preface To The Old School Calculus Series by Karo Maestro
Preface To The First Edition
Preface To The Revised Edition
Preface To The 2018 Blue Collar Scholar Edition by Karo Maestro

CHAPTER I. INTRODUCTION TO THE CALCULUS

SECTION
1. Formulas for reference
2. Variables and constants
3. Functions
4. Functional notation
5. Graphs
6. Slope of a curve
7. Some problems which arise
8. Limit of a variable; infinitesimals
9. Limiting value of a function
10. Fundamental theorems on limits
11. The concept of infinity (∞)
12. Special limiting values
13. Continuous and discontinuous functions
14. Continuity and discontinuity illustrated by graphs
15. The indeterminate forms $0/0$ and ∞/∞
16. To show that $\lim_{x=0} \frac{\sin x}{x} = 1$
17. The number e
18. Summary

CHAPTER II. GENERAL PRINCIPLES

19. Increments
20. Definition of derivative
21. Symbols for the derivative of a function
22. Differentiable functions
23. The Four-Step Rule for differentiation
24. The slope of a curve
25. The derivative of $\sin x$ and $\cos x$
26. The derivative of a function of a function
27. The derivative of an area
28. The antiderivative
29. Computation of areas
30. The second fundamental limiting process. Preliminaries
31. Areas considered as limits of sums
32. The definite integral and its properties
33. Evaluation of definite integrals. Special methods

SECTION	PAGE
34. Evaluation of definite integrals by means of antiderivatives	60
35. The indefinite integral and its derivative	62
36. Summary	65

CHAPTER III. DIFFERENTIATION

37. General theorems on differentiation	66
38. Use of the general theorems in finding derivatives	70
39. How to find antiderivatives	75
40. Derivatives of inverse functions	77
41. Derivatives of functions represented parametrically	78
42. Implicit and explicit functions	80
43. Derivatives of higher order	81
44. Historical note on the origin of the calculus	82
45. Summary	83

CHAPTER IV. APPLICATIONS OF DIFFERENTIATION AND INTEGRATION

46. Velocity in rectilinear motion	84
47. Acceleration in rectilinear motion	86
48. Rectilinear motion with assigned acceleration	87
49. Speed in curvilinear motion	90
50. Component velocities	90
51. Acceleration in curvilinear motion	92
52. Simple harmonic motion	94
53. Rotation	97
54. Equations of tangent and normal to a curve	98
55. Lengths of tangent, normal, subtangent, subnormal	99
56. Solutions of equations having multiple roots	101
57. The derivative interpreted as a rate	102
58. Summary	105

CHAPTER V. THE DIFFERENTIAL NOTATION. FURTHER APPLICATIONS

59. Differentials	108
60. Approximate computation by means of differentials	109
61. Arc length	111
62. Differential of arc length in rectangular coördinates	113
63. Differential of arc length in polar coördinates	114
64. Formulas for computing differentials	116
65. Successive differentials	118
66. Summary	118

CHAPTER VI. MAXIMA AND MINIMA OF ALGEBRAIC FUNCTIONS

SECTION	PAGE
67. Illustrative example	119
68. Increasing and decreasing functions	120
69. Maximum and minimum of a function	121
70. Summary	126

CHAPTER VII. DIFFERENTIATION OF THE TRIGONOMETRIC FUNCTIONS

71. Introduction	130
72. Differentiation of tan u, cot u, sec u, csc u	131
73. Illustrative examples	132
74. Summary	134

CHAPTER VIII. DIFFERENTIATION OF THE INVERSE TRIGONOMETRIC FUNCTIONS

75. Differentiation of arc sin u and arc cos u	139
76. Differentiation of arc tan u and arc cot u	140
77. Differentiation of arc sec u and arc csc u	140
78. Differentiation of arc vers u	141
79. Summary	141

CHAPTER IX. DIFFERENTIATION OF THE LOGARITHMIC AND EXPONENTIAL FUNCTIONS

80. Differentiation of a logarithm	146
81. Differentiation of the simple exponential function	147
82. Differentiation of the general exponential function	148
83. Logarithmic differentiation	149
84. Hyperbolic functions	152
85. Relations between the hyperbolic functions	154
86. The inverse hyperbolic functions	155
87. Differentiation of the hyperbolic functions and their inverses	156
88. Applications involving exponential functions	157
89. Summary	162

CHAPTER X. INTEGRATION

90. Introduction	163
91. Standard elementary forms of integrals	163
92. Proof of formulas (1) to (9)	165
93. Proof of formulas (10) to (23)	171
94. Proof of formulas (24) to (29)	174
95. Integration by parts; proof of formulas (30) and (31)	177

SECTION	PAGE
96. Miscellaneous methods of integration	181
97. Infinite limits of integration	185
98. Definite integrals of discontinuous functions	186

CHAPTER XI. THE SIMPLER APPLICATIONS OF INTEGRATION

99. Plane areas in rectangular coördinates	189
100. Length in rectangular coördinates	190
101. Areas by the summation of thin strips	192
102. Volumes by the summation of thin slices	194
103. Mean value	198
104. The constant of indefinite integration in geometrical problems	201
105. The constant of indefinite integration in physical problems	204

CHAPTER XII. DIFFERENTIAL EQUATIONS

106. Definitions	208
107. Differential equations in which the variables are separable	209
108. Homogeneous differential equations	211
109. Linear differential equations of the first order	212
110. Linear differential equations of the second order	214
111. Method for finding the solution of a homogeneous linear differential equation of the second order with constant coefficients	218
112. Physical applications	220
113. Method of undetermined coefficients for finding a particular integral, Y, of the nonhomogeneous linear differential equation	222

CHAPTER XIII. SUCCESSIVE DIFFERENTIATION AND INTEGRATION

114. Successive differentiation	225
115. Successive integration	227
116. Rolle's theorem and the law of the mean	228
117. Taylor's theorem	231
118. Maxima and minima	232
119. Indeterminate forms	234

CHAPTER XIV. INFINITE SERIES

120. Introduction	240
121. Tests for convergence	245
122. Power series	250

CHAPTER XV. EXPANSIONS OF FUNCTIONS

123. Expansions in power series	253
124. Taylor's series	254
125. Maclaurin's series	254
126. Taylor's theorem and Taylor's expansion of a given function	255
127. Computation by means of series	258
128. Approximate formulas	261

CHAPTER XVI. PROPERTIES OF PLANE CURVES

SECTION	PAGE
129. Concavity	263
130. Points of inflection	263
131. Singular points	265
132. Asymptotes	269
133. Curve tracing	274
134. Curvature	275
135. Roulettes, involutes, and evolutes	278

CHAPTER XVII. APPLICATIONS TO GEOMETRY AND MECHANICS

136. Plane areas in polar coördinates	283
137. Lengths of curves in polar coördinates	285
138. Volumes of solids of revolution	286
139. Surfaces of revolution, rectangular coördinates	288
140. Surfaces of revolution, polar coördinates	290
141. Force	292
142. Work	294
143. Summary	296

CHAPTER XVIII. INTEGRATION OF SPECIAL CLASSES OF FUNCTIONS

144. Introduction	298
145. Integration of rational fractions. Case I	299
146. Integration of rational fractions. Case II	301
147. Integration of rational fractions. Case III	302
148. Integration of rational fractions. Case IV	304
149. Integrals containing $(ax+b)^{\frac{p}{q}}$	307
150. Integrals containing the radical $\sqrt{x^2 + px + q}$	308
151. Integrals containing the radical $\sqrt{-x^2 + px + q}$	308
152. Integration of $\int \sin^m x \cos^n x\, dx$	311
153. Reduction formulas	312
154. Integration of $\int \tan^n x\, dx$ or of $\int \cot^n x\, dx$	316
155. Integration of $\int \tan^n x \sec^m x\, dx$ or of $\int \cot^n x \csc^m x\, dx$	317
156. The substitution $\tan x/2 = t$	318
157. Use of a table of integrals	319
158. Approximate integration	320
159. Trapezoidal rule	320
160. Simpson's rule	321
161. Integration in series	322
162. Integrating machines	323
163. Summary	323

CHAPTER XIX. FUNCTIONS OF TWO OR MORE VARIABLES. DIFFERENTIATION

SECTION	PAGE
164. Continuous functions of two variables	326
165. Partial derivatives	326
166. Higher partial derivatives	327
167. Partial derivatives interpreted geometrically	328
168. Total derivatives	329
169. Differentiation of implicit functions	331
170. Differentials	332
171. Exact differentials	335
172. Envelopes	337
173. Method of determining the envelope	338
174. The evolute as the envelope of the normals	340
175. Tangent plane to a surface	343
176. Line normal to a surface	344
177. Summary	346

CHAPTER XX. FUNCTIONS OF TWO OR MORE VARIABLES. INTEGRATION

178. Definitions and processes	347

CHAPTER XXI. FURTHER APPLICATIONS TO GEOMETRY

179. Plane area by means of double integration; rectangular coördinates	350
180. Plane area by means of double integration; polar coördinates	353
181. Volume by double integration	356
182. Volumes by triple integration	359
183. Area of a surface	361
184. Summary	362

CHAPTER XXII. FURTHER APPLICATIONS TO MECHANICS

185. Heterogeneous masses; density	364
186. Moments; centroids	366
187. Centroids of curves	368
188. Centroids of areas	370
189. Cylindrical coördinates	371
190. Spherical coördinates	373
191. Centroids of solids	374
192. Theorems of Pappus	376
193. Moments of inertia	379
194. Summary	380
INDEX	381

THE CALCULUS

CHAPTER I

INTRODUCTION TO THE CALCULUS

1. Formulas for reference. The student will find it convenient to have for reference the following formulas from geometry, algebra, trigonometry, and analytic geometry.

I. FROM GEOMETRY

Using r for radius, a for altitude, s for slant height, we have the following:

1. Circumference of a circle $= 2\pi r$.
2. Area of a circle $= \pi r^2$.
3. Surface of a sphere $= 4\pi r^2$.
4. Volume of a sphere $= \frac{4}{3}\pi r^3$.
5. Volume of a right circular cylinder $= \pi r^2 a$.
6. Lateral surface of a right circular cylinder $= 2\pi r a$.
7. Volume of a right circular cone $= \frac{1}{3}\pi r^2 a$.
8. Lateral surface of a right circular cone $= \pi r s$.

II. FROM ALGEBRA

9. Binomial theorem ($n =$ positive integer):

$$(a+b)^n = a^n + na^{n-1}b + \frac{n(n-1)}{2!}a^{n-2}b^2 + \frac{n(n-1)(n-2)}{3!}a^{n-3}b^3 + \cdots + nab^{n-1} + b^n.$$

10. $n! = \lfloor n = 1 \cdot 2 \cdot 3 \cdot 4 \cdots (n-1) \cdot n.$
11. The roots of the equation $ax^2 + bx + c = 0$ are

$$\frac{-b \pm \sqrt{b^2 - 4ac}}{2a}.$$

They are real and unequal when $b^2 - 4ac > 0$, real and equal when $b^2 - 4ac = 0$, complex when $b^2 - 4ac < 0$.

12. In the arithmetical series $a, a+d, a+2d, \cdots, l$ of n terms, where s is the sum of the terms, we have

$$l = a + (n-1)d, \quad s = \frac{n}{2}(a+l) = \frac{n}{2}[2a + (n-1)d].$$

13. In the geometrical series $a, ar, ar^2, \cdots, l$ of n terms, where s is the sum of the terms, we have

$$l = ar^{n-1}, \quad s = \frac{rl-a}{r-1} = \frac{a(r^n-1)}{r-1} = \frac{a}{1-r} - \frac{ar^n}{1-r}.$$

14. $\log ab = \log a + \log b.$

15. $\log \dfrac{a}{b} = \log a - \log b.$

16. $\log a^n = n \log a.$

17. $\log \sqrt[n]{a} = \dfrac{1}{n} \log a.$

18. $\log \dfrac{1}{a} = - \log a.$

19. $\log_a a = 1.$

20. $\log_a 1 = 0.$

III. FROM TRIGONOMETRY

21. $\sin x \csc x = 1, \cos x \sec x = 1, \tan x \cot x = 1.$

22. $\tan x = \dfrac{\sin x}{\cos x}, \cot x = \dfrac{\cos x}{\sin x}.$

23. $\sin^2 x + \cos^2 x = 1, 1 + \tan^2 x = \sec^2 x, 1 + \cot^2 x = \csc^2 x.$

24. $\sin\left(\dfrac{\pi}{2} - x\right) = \cos x, \cos\left(\dfrac{\pi}{2} - x\right) = \sin x, \tan\left(\dfrac{\pi}{2} - x\right) = \cot x.$

25. $\sin(\pi - x) = \sin x, \cos(\pi - x) = -\cos x, \tan(\pi - x) = -\tan x.$

26. $\sin(-x) = -\sin x, \cos(-x) = \cos x, \tan(-x) = -\tan x.$

27. $\sin(x \pm y) = \sin x \cos y \pm \cos x \sin y.$

28. $\cos(x \pm y) = \cos x \cos y \mp \sin x \sin y.$

29. $\tan(x \pm y) = \dfrac{\tan x \pm \tan y}{1 \mp \tan x \tan y}.$

30. $\sin 2x = 2 \sin x \cos x, \cos 2x = \cos^2 x - \sin^2 x, \tan 2x = \dfrac{2 \tan x}{1 - \tan^2 x}.$

31. $\cos^2 x = \tfrac{1}{2}(1 + \cos 2x), \sin^2 x = \tfrac{1}{2}(1 - \cos 2x).$

32. $\sin x + \sin y = 2 \sin \tfrac{1}{2}(x+y) \cos \tfrac{1}{2}(x-y).$

33. $\sin x - \sin y = 2 \cos \tfrac{1}{2}(x+y) \sin \tfrac{1}{2}(x-y).$

34. $\cos x + \cos y = 2 \cos \tfrac{1}{2}(x+y) \cos \tfrac{1}{2}(x-y).$

35. $\cos x - \cos y = -2 \sin \tfrac{1}{2}(x+y) \sin \tfrac{1}{2}(x-y).$

36. $\dfrac{a}{\sin A} = \dfrac{b}{\sin B} = \dfrac{c}{\sin C}.$ (Law of sines for triangles.)

37. $a^2 = b^2 + c^2 - 2bc \cos A.$ (Law of cosines for triangles.)

38. Area of a triangle $= \tfrac{1}{2} ab \sin C.$

IV. From Analytic Geometry

39. The distance from (x_1, y_1) to the line $Ax + By + C = 0$ is given by the formula $\dfrac{Ax_1 + By_1 + C}{\pm \sqrt{A^2 + B^2}}$.

40. The distance between the points (x_1, y_1) and (x_2, y_2) is given by the formula $\sqrt{(x_1 - x_2)^2 + (y_1 - y_2)^2}$.

41. To transform from rectangular to polar coördinates set
$$x = \rho \cos \theta, \quad y = \rho \sin \theta.$$

42. To transform from polar to rectangular coördinates set
$$\rho = \sqrt{x^2 + y^2}, \quad \theta = \arctan \frac{y}{x}.$$

43. As equations of straight lines we have the following:

General form: $\quad Ax + By + C = 0$.

Two-point form: $\quad y - y_1 = \dfrac{y_2 - y_1}{x_2 - x_1} (x - x_1)$.

Intercept form: $\quad \dfrac{x}{a} + \dfrac{y}{b} = 1$.

Slope-point form: $\quad y - y_1 = m(x - x_1)$.

Slope-intercept form: $\quad y = mx + b$.

44. When θ is the angle from a line with slope m_2 to a line with slope m_1, we have
$$\tan \theta = \frac{m_1 - m_2}{1 + m_1 m_2};$$
the lines are parallel when $\quad m_1 = m_2$,

the lines are perpendicular when $\quad m_1 = -\dfrac{1}{m_2}$.

45. $x^2 + y^2 = r^2$, equation of a circle with center $(0, 0)$ and radius r.

46. $(x - a)^2 + (y - b)^2 = r^2$, equation of a circle with center (a, b) and radius r.

47. $y^2 = 4 px$, standard equation of a parabola.

48. $\dfrac{x^2}{a^2} + \dfrac{y^2}{b^2} = 1$, standard equation of an ellipse.

49. $\dfrac{x^2}{a^2} - \dfrac{y^2}{b^2} = 1$, standard equation of a hyperbola.

2. Variables and constants. From the point of view of change in value the quantities which enter into mathematical analysis are of two kinds. A quantity which does not change its value in a given problem is called a *constant*. A *variable* is a quantity which may assume an unlimited number of values in a given problem. The set of values which it may assume is called the *range* of the variable.

A constant which may have any one of an infinitude of values is called an arbitrary constant, or a parameter.

The equation of a circle whose center is the origin of rectangular coördinates x and y and whose radius is a is $x^2 + y^2 = a^2$. Here x and y are variables which change as the point (x, y) moves along the circle; but a is a constant, being the radius of the circle. The range of x, and also of y, is from $-a$ to $+a$. Since a may have any positive value, it is an arbitrary constant, or parameter.

Any given numerical magnitude, as 2 or π, is a constant.

Three types of variable quantities will be of frequent occurrence: (1) quantities which vary with the time, as the temperature at a given point; (2) quantities which vary with the position in space, as the temperature at various points of space at a given instant of time; (3) quantities which vary in an arbitrary manner without reference to space and time.

The earlier letters of the alphabet are usually employed to denote constants, and the later letters to denote variables.

3. Functions. Both in natural phenomena and in mathematical investigations variable quantities usually appear in intimate relation one to another. Thus in the equation $x^2 + y^2 = a^2$ of a circle, the variables x and y are connected by the fact that a point (x, y) on the circle must be at a distance a from the origin of coördinates. The area of a variable square is determined by means of the length of the variable side. The volume of a sphere depends on the length of the radius. The velocity of a freely falling body depends on the time during which it has been falling. The pressure in a given volume of gas depends on the temperature and increases as the temperature rises. The number of bacteria in a developing culture depends on the time during which the development has proceeded and on the temperature and on various other factors. The most fundamental problems in the study of phenomena are connected with the mutual dependence of related variables. When such a dependence exists among variables they are said to be connected by a functional relation.

If two variables are so related that to every value of one of them (taken at will in a certain range) there corresponds a determined value of the other, then the second variable is said to be a function of the first. In this text the range of the first variable is restricted so that the second variable assumes only real values.

Thus, if y depends on x so that to every value of x there corresponds a determined value of y, then y is a function of x.

The area of a square is a function of the side of the square; the area of a circle is a function of its radius; if $y = x^2$ then y is a function of x; if $y = \sin x$ then y is a function of x; if t denotes the time during which a freely falling body has been falling and s denotes the distance traversed, then s is a function of t; the pressure of the gas in a given closed vessel is a function of the temperature.

The mathematical meaning of the word *function* is very different from its meaning outside of mathematics; and it is important that the student shall come to understand its significance as early as possible. In the definition itself there is no reference to any form of expression by means of which a function y of x may be determined in terms of x; but in practice we shall usually have such expressions. Ordinarily we cannot proceed far with the study of a function until we have some sort of mathematical expression for it. For the area A of a circle in terms of its radius r we have $A = \pi r^2$. This gives the expression of A as a function of r. When r is given, A is determined and is also readily computed.

The variable to which values may be assigned at will is called the *independent variable*, or the *argument*. The variable whose value is determined when the value of the independent variable is assigned is called the *dependent variable*.

Frequently we may look upon a given functional relation between two variables from either of two points of view, taking either the one or the other variable to be the independent variable. Thus in the equation $A = \pi r^2$ for the area of a circle in terms of its radius, we may consider r the independent variable and A the dependent variable; or we may reverse our point of view and consider A the independent variable and r the dependent variable. Which shall be considered the independent variable is determined in a particular case by the nature of the problem in hand.

The notion of function may be extended. The dependent variable may be a function of two or more independent variables. The area A of a triangle is a function of its base b and its altitude h, being determined in terms of these by means of the relation $A = \frac{1}{2} bh$. The volume of a rectangular solid is a function of its three dimensions. If the value of a variable y is determined when values are assigned to certain other variables, then y is said to be a function of these other variables. In such a relation y is the dependent variable and the others are all independent variables.

It may happen that the dependent variable is defined only for a limited range of values of the independent variable. In the relation $y = \sqrt{x}$ we have y defined only when x is positive or zero. Likewise in the equation $x^2 + y^2 = a^2$ of a circle the dependent variable y is defined only for those values of x which lie between $-a$ and $+a$. This example also illustrates an additional point. When the value of x is assigned, the value of y is in general either one of *two* values; for when x is assigned, there are *two* corresponding values of y. For this reason y is said to be a double-valued or a two-valued function of x. A function may be three-valued or four-valued, and so on. Usually, however, we shall be dealing with one-valued (or single-valued) functions; that is, functions such that there is a single value of the dependent variable determined by a given assignment of the independent variable or variables.

Let us consider certain other examples involving the range of the independent variable. Each of the functions

$$x^2 - 7x + 22, \quad \sin x, \quad \sqrt[3]{x},$$

is defined for every value of x. We may say, in each case, that the range of x is from $-\infty$ to $+\infty$. But the functions

$$\sqrt{x}, \quad \sqrt[4]{x}$$

are defined only for non-negative values of x; in this case the range of x is from 0 to $+\infty$. The function

$$\log_a x \quad \text{for} \quad a > 0 \quad a \neq 1$$

is defined only for positive values of x, so that the range of x in this case is from 0 to $+\infty$ exclusive of 0. The function

$$\frac{2}{x-3}$$

is not defined for $x = 3$; otherwise x may range from $-\infty$ to $+\infty$. Since the value of $\sin y$ or $\cos y$ lies between -1 and $+1$ inclusive, it follows that the functions

$$\text{arc sin } x \quad \text{and} \quad \text{arc cos } x$$

are defined only for values of x ranging from -1 to $+1$ inclusive.

4. Functional notation. We shall so frequently have to deal with the properties of functions without naming any particular function that it is desirable to have general symbols to represent functions. We shall use the symbol $f(x)$ to denote a function of x. This symbol $f(x)$ is read *f of x*, or, sometimes, *the f function of x*. In order to distinguish two or more functions simultaneously considered it is often desirable to use also the symbols $F(x)$, $\phi(x)$, $f'(x)$, etc., to denote different functions. Thus if y is a function of x, we may express this fact by writing any one of the following relations:

$$y = f(x), \quad y = F(x), \quad y = \phi(x), \quad \text{etc.}$$

In the case of functions of more than one variable we employ the natural extension of this notation. Thus if z is a function of x and y, we may employ such notations as the following:

$$z = f(x, y), \quad z = F(x, y), \quad z = \phi(x, y).$$

Similarly, for functions of three variables we have

$$t = f(x, y, z), \text{ etc.},$$

with a like notation for functions of a greater number of variables.

If a functional symbol occurs more than once in a given problem or discussion, it is of course understood to represent the same function throughout that problem or discussion. In most particular cases the symbol will represent a mathematical expression. In such a case the symbol will denote the same expression throughout, though it may be written in terms of one variable or another. Thus, if

then we have
$$f(x) = x^2 - 3x + 2,$$
$$f(y) = y^2 - 3y + 2,$$
$$f(a) = a^2 - 3a + 2,$$
$$f(t) = t^2 - 3t + 2,$$
$$f(4) = 4^2 - 3 \cdot 4 + 2 = 6,$$
$$f(1) = 1^2 - 3 \cdot 1 + 2 = 0,$$
$$f(-3) = (-3)^2 - 3(-3) + 2 = 20.$$

Similarly, if $\phi(x, y) = x^2 + xy - y^2$,
we have
$$\phi(a, b) = a^2 + ab - b^2,$$
$$\phi(b, a) = b^2 + ba - a^2,$$
$$\phi(y, x) = y^2 + yx - x^2,$$
$$\phi(1, 2) = 1^2 + 1 \cdot 2 - 2^2 = -1.$$

Example 1. If $y = f(x) = \dfrac{1+x}{1-x}$ and $z = f(y) = f[f(x)]$, find the expression for z as a function of x.

We have
$$z = f(y) = \frac{1+y}{1-y} = \frac{1 + \dfrac{1+x}{1-x}}{1 - \dfrac{1+x}{1-x}} = \frac{1-x+1+x}{1-x-1-x} = -\frac{1}{x}.$$

Hence $-1/x$ is the expression for z as a function of x.

Example 2. If $f(x) = \sqrt{1-x^2}$, find $f(\sin\theta)$ and $f(\cos\theta)$.
We have $f(\sin\theta) = \sqrt{1-\sin^2\theta} = \sqrt{\cos^2\theta} = \cos\theta,$
$f(\cos\theta) = \sqrt{1-\cos^2\theta} = \sqrt{\sin^2\theta} = \sin\theta.$

Example 3. Given $\phi(x) = \log\dfrac{1-x}{1+x}$, prove that
$$\phi(x) + \phi(y) = \phi\left(\frac{x+y}{1+xy}\right).$$

We have $\phi(x) + \phi(y) = \log\dfrac{1-x}{1+x} + \log\dfrac{1-y}{1+y}$

$= \log\left(\dfrac{1-x}{1+x} \cdot \dfrac{1-y}{1+y}\right)$

$= \log\dfrac{1+xy-(x+y)}{1+xy+(x+y)}$

$= \log\dfrac{1 - \dfrac{x+y}{1+xy}}{1 + \dfrac{x+y}{1+xy}}$

$= \phi\left(\dfrac{x+y}{1+xy}\right).$

EXERCISES

1. If $f(x) = x^3 - 7x + 2$, find $f(0), f(1), f(2), f(-\tfrac{1}{2}), f(-\tfrac{4}{7})$.
2. If $f(x) = x^6 - 8x^4 + 19x^2 - 27$, show that $f(-x) = f(x)$.
3. If $f(x) = x^7 - 8x^5 + 19x^3 - 27x$, show that $f(-x) = -f(x)$.
4. If $f(x) = 2^x$, find $f(a), f(t), f(2), f(9), f(\tfrac{1}{3}), f(-2)$.
5. If $f(x) = x(x-1)(x+3)(3x-7)$, show that
$$f(0) = f(1) = f(-3) = f(\tfrac{7}{3}) = 0.$$
6. If $f(x) = \dfrac{x-1}{x+1}$, show that
$$\dfrac{f(x) - f(y)}{1 + f(x)f(y)} = \dfrac{x-y}{1+xy}.$$
7. If $F(x) = x^2$, show that $F(x+y) - F(x) = 2xy + y^2$.
8. If $f(t) = 2t^2 - 1$, show that $f(\cos\theta) = \cos 2\theta$.
9. If $f(x) = \sin x$ and $\phi(x) = \cos x$, show that
$$f(x)\phi(y) \pm \phi(x)f(y) = f(x \pm y).$$
10. If $f(\theta) = \tan\theta$, show that
$$\dfrac{f(\theta) - f(t)}{1 + f(\theta)f(t)} = \tan(\theta - t).$$
11. If $y = \dfrac{x+1}{x-1}$, find x in terms of y; that is, find x as a function of y. What is the range of x in the given function? What is the range of y in the computed function?
Ans. $x = \dfrac{y+1}{y-1}$.

12. If $y = \dfrac{ax+b}{cx-a}$, find x as a function of y. What are the ranges of x and y respectively? $\quad$ Ans. $x = \dfrac{ay+b}{cy-a}$.

13. Let Δx (read *delta x*) be a symbol (not the product of Δ and x) denoting a certain increment to be added to the variable x. If $f(x) = x^2$, show that $f(x + \Delta x) = x^2 + 2\,x \cdot \Delta x + (\Delta x)^2$.

14. With the same notation as in problem 13, show that
$$\frac{f(x + \Delta x) - f(x)}{\Delta x} = 2\,x + \Delta x.$$

15. If, in the foregoing problem, x is held fixed and Δx is allowed to vary and to approach the value 0, show that
$$\frac{f(x + \Delta x) - f(x)}{\Delta x}$$
varies and approaches the value $2\,x$.

16. If $f(x) = x^3$ and x is held fixed while Δx is allowed to vary and to approach the value zero, show that
$$\frac{f(x + \Delta x) - f(x)}{\Delta x}$$
varies and approaches the value $3\,x^2$.

5. Graphs. Important properties of a function may often be rendered apparent to the eye by means of a graph. The principal notions involved are already familiar to the student from his study of analytic geometry. We shall review them briefly.

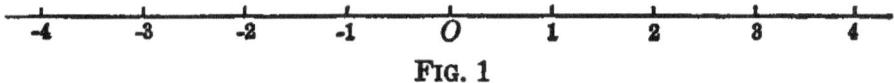

FIG. 1

In order to set up a scale on a horizontal line one proceeds as follows: Choose some convenient point on the line and mark it with O; lay off successive units to the right and mark the extremities with $1, 2, 3, \cdots$; lay off successive units to the left and mark the extremities with $-1, -2, -3, \cdots$. A point a units to the right denotes $+a$ (whether a is integral or fractional or irrational), while a point a units to the left denotes $-a$. Then every real number corresponds to one and just one point on the line, and every point on the line corresponds to one and just one real number. In this book we employ only real numbers unless the contrary is explicitly stated.

If the line lies in any other direction we proceed in the same way, choosing the origin O and one direction from O as positive, the other being negative, and marking the points in the same way.

A variable number x is denoted by a variable point P on the line. If the variable point P ranges between two fixed points A and B corresponding respectively to the numbers a and b and a is less than b, then x is said to range over the interval from a to b. Such an interval is denoted by the symbol (a, b). If the point P moves along the line without jumps, or discontinuities, the corresponding variable x is said to vary continuously over the interval (a, b); it is then called a continuous variable.

In order to represent a function graphically we employ scales on two intersecting lines. It is convenient to take these lines at right angles, as in the figure. Their point of intersection is taken to be the origin on each scale. A point P in the plane of these lines is denoted by a pair of numbers obtained by drawing perpendiculars from P to the scale lines and taking the corresponding numbers. Thus the point P in the figure is denoted by the number pair (3, 2). The line OX is called the x-axis, and the line OY is called the y-axis. Then (x, y) denotes the point

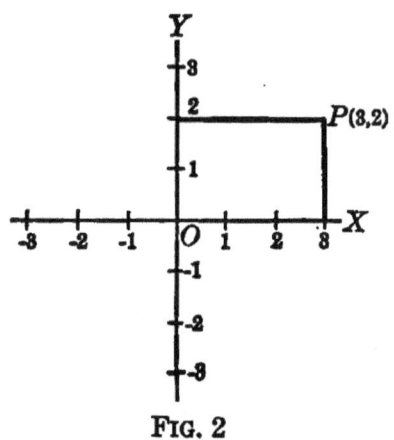

FIG. 2

of intersection of two perpendiculars, one to the x-axis through the point x on that axis and one to the y-axis through the point y on that axis. In the symbol (r, s) for a point the first number r always refers to the scale on the x-axis and the second number s to the scale on the y-axis.

The graph of the function $f(x)$ is then obtained from the equation $y = f(x)$ by assuming convenient values for x, computing the corresponding values for y, plotting the resulting points (x, y), and drawing a smooth curve through these points.

As an example we shall construct the graph of the function x^2. From the equation $y = x^2$ we construct the table of values:

x	-3	$-2\frac{1}{2}$	-2	$-1\frac{1}{2}$	-1	0	1	$1\frac{1}{2}$	2	$2\frac{1}{2}$	3
y	9	$6\frac{1}{4}$	4	$2\frac{1}{4}$	1	0	1	$2\frac{1}{4}$	4	$6\frac{1}{4}$	9

Then we plot the corresponding points, as in the figure, and draw a smooth curve joining them. In this way we obtain the parabola here shown, a figure already familiar to the student.

6. Slope of a curve. Let us consider the function $mx + b$. Its graph is obtained from the equation $y = mx + b$. From analytic geometry the student knows already that this represents a straight line which cuts the y-axis in the point $(0, b)$ and whose slope is m. The slope m is the tangent of the acute angle θ which the line makes with the positive direction of the x-axis. The

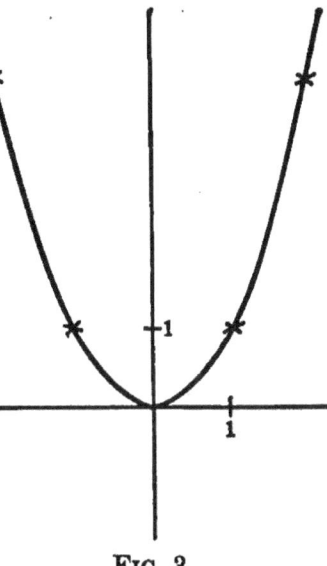

Fig. 3

figure (Fig. 4) shows lines for two values of m, one positive and the other negative.

This conception of the slope of a straight line is to be extended to curves in general. By the *slope of a curve* at a given point we mean the slope of the tangent line to the curve at that point.

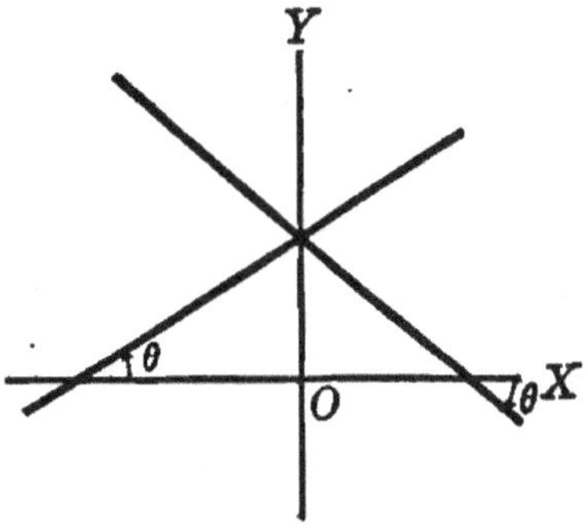

To illustrate the notion of slope let us consider the graph of the function x^2 obtained by means of the equation $y = x^2$. Let (x, y) denote the point P at which the slope is to be found. By definition, the slope of the curve at this point is the slope of the tangent PT to the curve at P. To obtain an

Fig. 4

expression for the slope we shall first find the slope of the secant line PQ and find what value this slope assumes when the secant line is allowed to rotate on the point P until it comes into coincidence with the tangent line PT.

Draw PM and QM parallel to the x-axis and the y-axis respectively. Then, if the point Q is denoted by $(x + \Delta x, y + \Delta y)$, the length of PM is Δx and the length of QM is Δy. Here Δx (read *delta x*) and Δy are certain quantities added to x and y respectively in passing from P to Q. We have

$$y = x^2,$$
$$y + \Delta y = (x + \Delta x)^2$$
$$= x^2 + 2x \cdot \Delta x + (\Delta x)^2.$$

Subtracting the first of these equations from the second, member by member, we have

$$\Delta y = 2x \cdot \Delta x + (\Delta x)^2.$$

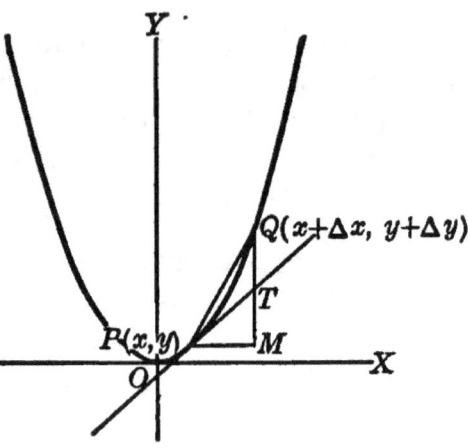

Fig. 5

Now the slope of PQ is equal to tan MPQ and hence is $\Delta y/\Delta x$. But from the last equation we have

$$\frac{\Delta y}{\Delta x} = 2x + \Delta x.$$

If PQ rotates on the fixed point P until it comes into coincidence with the tangent PT, the point Q moves along the curve until it comes into coincidence with P, and at the same time the *increments* Δx and Δy decrease and approach (or come to have) the value 0. But as Δx approaches 0 the second member of the last equation approaches $2x$. But this second member is all the while equal to the slope $\Delta y/\Delta x$ of the secant line. When the secant line comes into coincidence with the tangent line PT, this slope therefore comes to have the value $2x$. Hence the expression for the slope of the graph of the equation $y = x^2$ at the point (x, y) is $2x$, the expression which was being sought.

This expression may be employed to find the slope at any particular point on the graph. For instance, the slope at the point $(1, 1)$ is 2, this being the value of $2x$ at the point $(1, 1)$.

At the point $(-1, 1)$ the slope is -2; at $(2, 4)$ it is 4; at $(3, 9)$ it is 6; and so on.

In order to find the expression for the slope of the graph of any equation $y = f(x)$ we proceed by the method employed in the preceding example. We form the equations

$$y = f(x),$$
$$y + \Delta y = f(x + \Delta x).$$

Subtracting the former from the latter, member by member, we have
$$\Delta y = f(x + \Delta x) - f(x).$$

Then we divide both members by Δx, thus obtaining the relation

$$\frac{\Delta y}{\Delta x} = \frac{f(x + \Delta x) - f(x)}{\Delta x}.$$

In each particular case the second member of this equation is simplified in a suitable manner. In the problem already treated the simplification consisted in dividing the numerator by the denominator. Then we let Δx vary and come to have the value zero, x itself remaining fixed in the meantime. Then we determined the form which the expression in the second member of the last equation comes to have as Δx comes to have the value zero. This is the required *expression* for the slope at the point (x, y). To find the slope at a particular point substitute in this expression the value for x at that point.

EXERCISES

1. Find the expression for the slope of the graph of the equation $y = x^3$ at the point (x, y). *Ans.* $3x^2$.

2. What is the slope of the graph of the equation $y = x^3$ at the point $(1, 1)$? at the point $(0, 0)$? at the point $(3, 27)$? at the point $(-3, -27)$? at the point for which x has the value $7/\sqrt{3}$?
Ans. 3, 0, etc.

3. Find the expression for the slope of the graph of the equation $y = 1/x$ at the point (x, y). *Ans.* $-1/x^2$.

SUGGESTION. When the value of Δy is found in the form

$$\frac{1}{x + \Delta x} - \frac{1}{x},$$

simplify (by subtraction) into the form

$$-\frac{\Delta x}{x(x+\Delta x)}$$

before dividing by Δx.

Find the expression for the slope of the graph of each of the following equations at the point (x, y):

4. $y = x^4$. \hfill Ans. $4x^3$.

5. $y = \dfrac{1}{x^2}$. \hfill Ans. $-\dfrac{2}{x^3}$.

6. $y = x^3 + 3x^2 - 7x + 9$. \hfill Ans. $3x^2 + 6x - 7$.

7. Some problems which arise. Later it will be seen that the process employed in deriving the expression for the slope of the graph involves one of the most important essential ideas of the calculus. For the present we shall merely indicate some of the problems which arise.

We have several times spoken of a quantity as varying and coming to have a certain value. In doing so we have tried to use terms of ordinary discourse and to avoid those which have a technical mathematical meaning. But this process of variation is so frequent in the calculus and of such central importance that it is imperative to have technical terms for referring to it and to study it abstractly for itself. This brings us to the problem of *limits*; it is one of the central problems of the calculus.

It is well known to the student that division by zero is not defined as an operation of algebra. Nor do we undertake to introduce such an operation in the calculus. Such division is impossible. But we frequently have to consider the value which a fraction approaches when its numerator and denominator both approach zero. Such a problem has already arisen in determining the slope of a curve. The slope of the secant line was found to have the value $\Delta y/\Delta x$. As the secant is revolved into the position of the tangent both Δx and Δy approach the value zero, and the fraction $\Delta y/\Delta x$ is said to approach the indeterminate form $0/0$. The evaluation of such forms is frequently required in important problems of the calculus.

We may speak of the *rate* with which a varying quantity changes; and this brings us to the problem of rates. By considering the definition of the slope of a curve at a given point and the meaning of the tangent of an angle, it may be seen that the slope of a curve at a point measures the rate of rise of the curve per unit change in the value of the abscissa x. To this problem of rates we shall return in a later chapter. We shall now indicate its nature by means of a more direct example.

If a ball is thrown upward at a velocity of 100 ft. per second, it is known from physics that the height in feet which it attains in t seconds is given by the formula *

$$h = 100\,t - 16\,t^2.$$

Let us consider the question of the rate, or speed, with which it is rising at the end of two seconds. Since the speed is obviously changing all the while, it is clear that we must resort to some special means to define the speed at the precise instant two seconds after the ball started upward. The average speed for the next second will not be the speed which we desire; neither will the average speed for the next half second, nor the average speed for the next quarter second, nor the average speed for the next eighth of a second. But if we consider the average speed for a shorter and shorter interval after two seconds it is clear that we come closer and closer to the speed at the end of two seconds. We may ask what value the average speed approaches as closely as we please when the length of the interval is decreased indefinitely. This limiting value which the average speed would thus approach as closely as we please is called the instantaneous speed at the end of two seconds. This statement is to be taken as the definition of this instantaneous speed.

In order to obtain the value of this instantaneous speed we consider any short interval of time Δt measured in seconds. In Δt seconds after the end of two seconds the ball has been rising $2 + \Delta t$ seconds.

When $t = 2$ we have $h = 100 \cdot 2 - 16 \cdot 2^2 = 136$; when $t = 2 + \Delta t$ we have

$$h = 100(2 + \Delta t) - 16(2 + \Delta t)^2$$
$$= 136 + 36\,\Delta t - 16(\Delta t)^2.$$

* For the sake of simplicity in this formula we take the acceleration g of gravity to be 32 ft. per second per second, whereas a more nearly correct value is 32.2.

Hence the difference Δh in the heights at time $t = 2$ and time $t = 2 + \Delta t$ is
$$\Delta h = 36\, \Delta t - 16(\Delta t)^2.$$

Dividing by Δt, we have
$$\frac{\Delta h}{\Delta t} = 36 - 16\, \Delta t$$

as the average speed for the interval Δt of time. Now if this interval Δt decreases toward zero, it is clear that this average speed approaches the value 36. Hence the instantaneous speed at the end of two seconds is 36.

It will be observed that we have used essentially the same process in finding instantaneous speed as in finding the slope of a graph. In fact, the slope of the distance-time graph at the point where $t = 2$ is the speed at that point. This fact is true in general, for consider a general time t instead of the particular time $t = 2$. Then we have

$$h = 100\, t - 16\, t^2,$$
$$h + \Delta h = 100(t + \Delta t) - 16(t + \Delta t)^2$$
$$= 100\, t + 100\, \Delta t - 16\, t^2 - 32\, t \cdot \Delta t - 16(\Delta t)^2.$$

Subtracting the first equation from the last, member by member, we have
$$\Delta h = 100\, \Delta t - 32\, t \cdot \Delta t - 16(\Delta t)^2.$$

Dividing by Δt, we have
$$\frac{\Delta h}{\Delta t} = 100 - 32\, t - 16\, \Delta t.$$

If t is held fixed and Δt is allowed to approach zero, then this average speed $\Delta h / \Delta t$ approaches the instantaneous speed $100 - 32\, t$. Hence at the end of t seconds the speed is $100 - 32\, t$.

For $t = 2$ this gives the value 36 already obtained. For $t = 3$ it gives the instantaneous speed 4 at the end of three seconds. For $t = 4$ it gives the instantaneous speed -28 at the end of four seconds. Thus at the end of four seconds the body is *falling* at a speed of 28 ft. per second, the fact that it is falling being shown by the negative sign.

The ball continues to rise until its speed becomes zero; that is, till we have $100 - 32\,t = 0$. Solving this equation for t, we have $t = 3\tfrac{1}{8}$ as the number of seconds during which the ball rises. To find the height to which it rises we may substitute this value of t in the original expression for h.

This problem will serve to illustrate the way in which the calculus is to be employed in studying the problems of motion, the most fundamental problems of physical science.

8. Limit of a variable; infinitesimals. If a variable v takes successively a sequence of values that approach more and more nearly to a constant value l in such a way that the *numerical value $|v - l|$ of the difference between v and l* becomes and remains less than any assigned constant however small, then v is said to *approach the limit l* or to *converge to the limit l*. This is written symbolically as follows:

$$\text{limit } v = l, \quad \text{or,} \quad v \doteq l.$$

The abbreviated form *lim* is often employed in place of the word *limit*.

The limit idea is essential to all the main problems of the calculus. It is therefore important that the student's conception of it shall be enriched as rapidly as possible by his careful examination of problems involving it and by his active effort to comprehend its meaning and significance. The following examples will help to illustrate what is meant by it:

1. The slope of a secant line approaches the slope of a tangent line as the former line rotates and comes into coincidence with the latter in the manner indicated in § 6.

2. If the number of sides of a regular polygon inscribed in a given circle is indefinitely increased, then the area of the polygon varies and approaches the area of the circle as a limit. The perimeter of the polygon likewise approaches the circumference as a limit. In each of these cases the variable is always less than its limit.

3. If the number of sides of a regular polygon circumscribed about a given circle is indefinitely increased, then the area of the polygon varies and approaches the area of the circle as a limit. The perimeter of the polygon likewise approaches the

circumference of the circle as a limit. In each of these cases the variable is always greater than its limit.

4. If a coiled wire spring is attached at one end and is extended by a force at the other end and if it is then released, its length will alternately decrease and increase until the spring acquires its normal (unstretched) length. Thus the length varies and approaches the normal length as a limit. In this case the variable is alternately greater than and less than its limit, repeatedly assuming its limiting value as it oscillates back and forth.

5. A more abstract example is afforded by the infinite series
$$1 + \tfrac{1}{2} + \tfrac{1}{4} + \tfrac{1}{8} + \tfrac{1}{16} + \cdots,$$
each of whose terms is half of the preceding term. Let S_n denote the sum of the first n terms of this series. Then we have
$$S_n = 1 + \frac{1}{2} + \frac{1}{4} + \cdots + \frac{1}{2^{n-1}}.$$
Multiplying by $(1 - \tfrac{1}{2})$, we have
$$\left(1 - \frac{1}{2}\right)S_n = 1 + \frac{1}{2} + \frac{1}{4} + \cdots + \frac{1}{2^{n-1}}$$
$$- \frac{1}{2} - \frac{1}{4} - \cdots - \frac{1}{2^{n-1}} - \frac{1}{2^n}$$
$$= 1 - \frac{1}{2^n}.$$
Hence
$$S_n = 2 - \frac{1}{2^{n-1}}.$$

As n increases indefinitely the fraction $1/2^{n-1}$ evidently approaches zero as a limit. Hence, as n increases indefinitely, S_n approaches the limit 2. This limit 2 is by definition the sum of the foregoing infinite series. In general, the limit (if it exists), as n increases indefinitely, of the sum S_n of the first n terms of an infinite series is by definition the sum of the series. As we shall see later, there are cases in which S_n does not approach a limit. In such a case we shall not speak of the sum of the series.

6. More generally, if r is numerically less than 1 and we write
$$S_n = a + ar + ar^2 + \cdots + ar^{n-1}$$
we have $(1-r)S_n = a + ar + ar^2 + \cdots + ar^{n-1}$
$$ - ar - ar^2 - \cdots - ar^{n-1} - ar^n$$
$$ = a - ar^n,$$
or
$$S_n = \frac{a}{1-r} - \frac{ar^n}{1-r};$$
whence it is seen that S_n has the limit $a/(1-r)$ if n increases indefinitely. Hence the sum of the infinite series
$$a + ar + ar^2 + \cdots$$
is $a/(1-r)$.

An *infinitesimal* is a variable which approaches the limit 0. No constant except 0, however small, is an infinitesimal. In example 5, the variable $1/2^{n-1}$ is an infinitesimal as n increases indefinitely. Infinitesimals occur so frequently in the calculus that it has sometimes been called the *infinitesimal calculus*.

9. Limiting value of a function. Let $f(x)$ be a given function of x. Suppose that the independent variable x is allowed to change and to approach the limit a. Then $f(x)$ will assume a series of values, one for each value of the varying x. It may happen that this series of values of $f(x)$ has a limit A. We say then that *the limit of $f(x)$, as x approaches the limit a, is A*; and we write this statement in the abbreviated form
$$\lim_{x=a} f(x) = A.$$
This definition of the limiting value of a function may be stated more formally as follows:

If a constant A exists such that, as x approaches the limit a, the difference $f(x) - A$ becomes and remains numerically less than any assigned constant, however small, then A is said to be the limit of $f(x)$ as x approaches a.

It is not to be assumed that a function of x always has a limiting value when x approaches a limit. To see this let us consider the function $\sin \dfrac{1}{x^2}$ as x approaches zero. It is evident that $1/x^2$ becomes indefinitely large. If x varies continuously, so does $1/x^2$, provided that x remains away from zero. In this case, as x approaches zero continuously, it is evident that $\sin\dfrac{1}{x^2}$ assumes every value between -1 and $+1$; and in fact that it assumes each of these values indefinitely often. We say that it *oscillates* between -1 and $+1$. Evidently $\sin\dfrac{1}{x^2}$ fails to approach a limit. This example shows that it is necessary to examine the question of the existence of a limit before we seek to find the value of the limit. In the cases which arise in this text the limit usually exists, but in each case the question of the existence of the limit must be examined.

10. Fundamental theorems on limits. We shall have occasion to use several fundamental theorems on limits. It is possible to give rigorous formal proofs of them on the basis of suitable postulates. But they are of such a nature that the student will be convinced of their validity by examining them intuitively in the light of the definition. For the beginner this intuitive grasp of their meaning and validity is to be preferred to a formal demonstration. Consequently we shall state them without proof. They are as follows:

I. If a variable u approaches a limit l and c is a constant, then cu approaches the limit cl.

II. If u and v are variables which approach the limits l and m, respectively, then $u + v$ approaches the limit $l + m$.

III. If u and v are variables which approach the limits l and m, respectively, then uv approaches the limit lm.

IV. If u and v are variables which approach the limits l and m, respectively, and if m is different from zero, then u/v approaches the limit l/m.

V. If a variable v steadily increases but never becomes greater than a given constant A, then v approaches a limit l which is not greater than A.

VI. If a variable v steadily decreases but never becomes less than a given constant B, then v approaches a limit l which is not less than B.

In the statement of these theorems we have said nothing of the nature of the variables except what is involved in the existence of the limits. The variables may be independent variables, or they may be functions of an independent variable x. Thus u may be a function $f(x)$ of x and v a function $\phi(x)$ of x and we may be contemplating the limits as x approaches a. The statements in theorems I to IV then give rise to the following:

$$\lim_{x \to a} cf(x) = c \lim_{x \to a} f(x),$$

$$\lim_{x \to a} \{f(x) + \phi(x)\} = \lim_{x \to a} f(x) + \lim_{x \to a} \phi(x),$$

$$\lim_{x \to a} \{f(x) \cdot \phi(x)\} = \lim_{x \to a} f(x) \cdot \lim_{x \to a} \phi(x),$$

$$\lim_{x \to a} \frac{f(x)}{\phi(x)} = \frac{\lim_{x \to a} f(x)}{\lim_{x \to a} \phi(x)};$$

in the last case it is understood that the limit in the denominator is different from zero.

The student is urged to think over the foregoing six theorems until he comes to have a clear intuitive or direct appreciation of their meaning and validity; he should also return to them from time to time and meditate over their significance in order to enrich his conception of them. A discussion on the part of the teacher will be helpful in illuminating them.

11. The concept of infinity (∞). If a variable v changes in such a way as to become and to remain greater than any assigned constant, however large, it is said to *increase without limit*. This is denoted by the symbol

$$\lim v = +\infty, \quad \text{or} \quad v \doteq +\infty.$$

If v becomes and remains less than any assigned negative constant, however large numerically, then v is said to *decrease without limit*. This is denoted by the symbol

$$\lim v = -\infty, \quad \text{or} \quad v \doteq -\infty.$$

If v becomes and remains in numerical value greater than any assigned positive constant, however large, then v is said to *increase in numerical value without limit*. This is denoted by the symbol $\lim |v| = \infty$, or $|v| \doteq \infty$.

It must be clearly understood that infinity (∞) is not a number. The symbol is used to denote a particular kind of variation of a variable, and no meaning is to be attached to it except that given to it by the foregoing definitions.

If the argument x increases without limit it may happen that the difference between $f(x)$ and some constant l becomes and remains less than any assigned positive quantity, however small. We then write
$$\lim_{x \to \infty} f(x) = l.$$

Thus we have 0 as the limit of the function $1/x^2$ as x increases indefinitely.

12. Special limiting values. There are certain types of limits which frequently recur in the calculus. It is convenient to list some of them here. In each case the student should satisfy himself that the limit has the value indicated. The symbols c and a denote constants.

$\lim_{x \to 0} cx = 0.$ $\qquad\qquad \lim_{x \to -\infty} a^x = \infty$ if $0 < a < 1.$

$\lim_{x \to \infty} cx = \infty$ if $c \neq 0.$* $\qquad \lim_{x \to +\infty} a^x = \infty$ if $a > 1.$

$\lim_{x \to 0} c/x = \infty$ if $c \neq 0.$ $\qquad \lim_{x \to 0} \log_a x = +\infty$ if $0 < a < 1.$

$\lim_{x \to \infty} c/x = 0.$ $\qquad\qquad \lim_{x \to 0} \log_a x = -\infty$ if $a > 1.$

$\lim_{x \to +\infty} a^x = 0$ if $0 < a < 1.$ $\quad \lim_{x \to +\infty} \log_a x = -\infty$ if $0 < a < 1.$

$\lim_{x \to -\infty} a^x = 0$ if $a > 1.$ $\qquad \lim_{x \to +\infty} \log_a x = +\infty$ if $a > 1.$

EXERCISES

Find the value of each of the following limits:

1. $\lim_{x \to 0} \left(x + \dfrac{1}{x} \right).$ $\qquad\qquad$ Ans. $\infty.$

2. $\lim_{x \to \infty} \left(x + \dfrac{1}{x} \right).$ $\qquad\qquad$ Ans. $\infty.$

3. $\lim_{x \to +\infty} (a^x + a^{-x})$ if $a > 1.$ $\qquad$ Ans. $+\infty.$

4. $\lim_{x \to -\infty} (a^x + a^{-x})$ if $a > 1.$ $\qquad$ Ans. $+\infty.$

* The sign $\neq$ means *is different from* or *is not equal to.*

5. $\lim_{x \to a} x^n$, $a \neq 0$. Ans. a^n.

6. $\lim_{x \to a} (cx^n + dx^m)$, $a \neq 0$. Ans. $ca^n + da^m$.

7. Show that $\cos \dfrac{1}{x^2}$ oscillates between -1 and $+1$ as x approaches zero.

8. Show that $\tan \dfrac{1}{x^2}$ oscillates between $-\infty$ and $+\infty$ as x approaches zero.

In the case of each of the following series find the sum S_n of n terms and then find the limit of S_n as n increases without limit (remembering that this limit is by definition the sum of the infinite series):

9. $1 + \dfrac{1}{3} + \dfrac{1}{9} + \dfrac{1}{27} + \cdots + \dfrac{1}{3^{n-1}} + \cdots$. Ans. $\tfrac{3}{2}$.

10. $1 - \dfrac{1}{2} + \dfrac{1}{4} - \dfrac{1}{8} + \cdots + (-1)^{n-1} \dfrac{1}{2^{n-1}} + \cdots$. Ans. $\tfrac{2}{3}$.

11. $\dfrac{3}{10} + \dfrac{3}{100} + \dfrac{3}{1000} + \cdots + \dfrac{3}{10^n} + \cdots$. Ans. $\tfrac{1}{3}$.

12. $\dfrac{27}{100} + \dfrac{27}{10000} + \cdots + \dfrac{27}{100^n} + \cdots$. Ans. $\tfrac{3}{11}$.

Note that problems 11 and 12 give $\tfrac{1}{3}$ and $\tfrac{3}{11}$, respectively, as the values of the infinite repeating decimals $.3333\cdots$ and $.272727\cdots$. In a similar manner find the values of each of the following infinite repeating decimals:

13. $.363636\cdots$. Ans. $\tfrac{4}{11}$.
14. $.43333\cdots$. Ans. $\tfrac{4}{10} + \tfrac{1}{10}(\tfrac{1}{3}) = \tfrac{13}{30}$.
15. $32.73636\cdots$. Ans. $32\tfrac{81}{110}$.

16. If a ball is thrown upward at a speed of 128 ft. per second, it is known from physics that the height h in feet which it attains in t seconds is given by the formula

$$h = 128\,t - 16\,t^2.$$

(Here the acceleration of gravity is given the approximate value 32.)

(1) Find the expression for the instantaneous speed at the end of t seconds. Ans. $128 - 32\,t$.

(2) What is the instantaneous speed at the end of 2 sec.? 3 sec.? 4 sec.? 5 sec.?

(3) For how long does the ball continue to rise? Ans. 4 sec.

(4) What is the greatest height attained by the ball? *Ans.* 256 ft.
(5) When is the ball again at the starting-point? *Ans.* When $t = 8$.
(6) What is the relation between the time during which the ball rises and the time required to fall from the highest point to the starting-point?

13. Continuous and discontinuous functions.* Let us consider the function
$$y = x^3 - 3x^2 + 3x.$$

By plotting points the student may show that the graph of this function is that given in the adjoining figure. Let us now consider a point $P(x, y)$ on the curve. If x changes continuously it is clear that the point P moves continuously along the curve. Moreover, the point P moves as little as we please if the change in x is made sufficiently small. Or, what is the same thing, a small change in x is accompanied by a small change in y. There is no break anywhere in the curve. As x changes continuously there is no sudden change in the values of the function. This is the essence of the notion of continuity. We shall now give a formal mathematical definition of continuity.

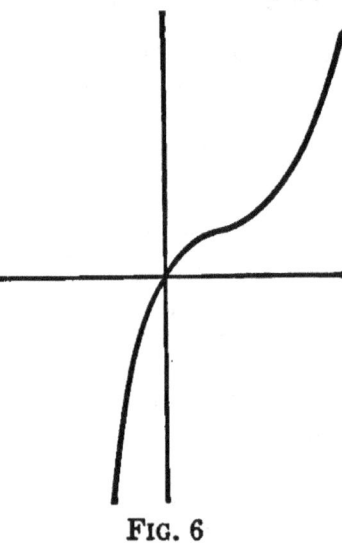

FIG. 6

A function $f(x)$ is said to be *continuous* for $x = a$ if it is defined for this value of x and if, furthermore,
$$\lim_{x \to a} f(x) = f(a).$$

A function $f(x)$ is said to be *continuous in an interval* (a, b) of values of the argument x if it is continuous for each point $x = \alpha$ of the interval.

If a function is not continuous for $x = a$ it is said to be discontinuous for $x = a$.

The functions to be met with in this book are continuous except perhaps for certain isolated values of the argument;

---

* The student is advised to begin the study of this section by reading it through rapidly even though he does not understand all of it.

these values will be excluded from consideration or their relation to the functions will be subjected to a special investigation in each case.

To illustrate continuity and discontinuity let us consider the function
$$f(x) = \frac{x^2 - 4}{x - 2}.$$

As x approaches the limit 3 it is clear that $f(x)$ approaches the limit 5; but $f(3) = 5$; hence $f(x)$ is continuous for $x = 3$. But for $x = 2$ the function $f(x)$ is not defined by the expression given, for the denominator of the fraction is in this case equal to 0. Hence $f(x)$ is discontinuous for $x = 2$, the discontinuity being due to the lack of definition of $f(x)$ for $x = 2$. For x different from 2 it is clear that the fraction used to define $f(x)$ can be simplified to the form $x + 2$. From this it follows that
$$\lim_{x=2} f(x) = 4.$$

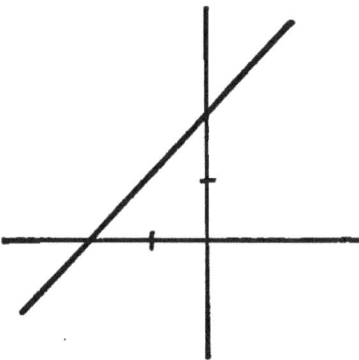

FIG. 7

If we should extend the definition of $f(x)$ by assigning to it the value 4 when $x = 2$ and leaving its definition otherwise the same, we should have a modified function which is continuous for $x = 2$. This is a process of removing the discontinuity by extending the definition of the function. The graph of the function, so extended, is that given in the adjoining figure. The student should verify this by plotting the graph.

The process just indicated can be extended to a wide range of cases. The following theorem, whose truth is evident from the definitions, covers the point:

If $f(x)$ is not defined for the value a of x, and if
$$\lim_{x = a} f(x) = A,$$
then the function $f(x)$ will become continuous for $x = a$ if its value for $x = a$ is defined to be A.

When a discontinuity can be removed in this way we shall generally suppose that it is done. In other words, we shall usually assume in this book that the definition of the function has been extended in accordance with this theorem.

There are discontinuities of functions which cannot be removed by the method indicated in the foregoing theorem. We saw that the function $\sin \dfrac{1}{x^2}$ oscillates between -1 and $+1$ as x approaches zero. Hence the function does not approach a limit as x approaches zero. It is therefore discontinuous in such a way as to prevent the removal of the discontinuity by the method suggested by the theorem.

There is also another case in which that method fails; namely, that in which $f(x)$ increases in numerical value without limit as x approaches a as a limit. Thus the function $1/(x-a)$ not only is undefined by the given expression for $x = a$ but it is also such as to increase in numerical value without limit as x approaches a as a limit.

By means of the definition of continuity for $x = a$, and with the help of the fundamental theorems on limits, the student will find it easy to establish the following theorems:

I. If $f(x)$ is continuous for $x = a$, and c is a constant, then $cf(x)$ is continuous for $x = a$.

II. If $f(x)$ and $\phi(x)$ are both continuous for $x = a$, then $f(x) + \phi(x)$ is continuous for $x = a$.

III. If $f(x)$ and $\phi(x)$ are both continuous for $x = a$, then $f(x) \cdot \phi(x)$ is continuous for $x = a$.

IV. If $f(x)$ and $\phi(x)$ are both continuous for $x = a$, and if $\phi(a)$ is different from zero, then $f(x)/\phi(x)$ is continuous for $x = a$.

After a first rapid reading of this section the student is advised to read the next section; he may then profitably return to this one for a fuller mastery, keeping in mind the geometric interpretations suggested in the following section.

14. Continuity and discontinuity illustrated by graphs. We have already seen that the continuity of a function $f(x)$ for $x = a$ is represented in the graph of $f(x)$ by the smoothness of the curve near the point for which $x = a$. This is illustrated by the graphs in §§ 5 and 13. We have also considered the question as to what is to be done with a function that has a discontinuity due to the function's not being defined for $x = a$. We shall now illustrate other types of discontinuity by means of graphs.

The function $1/x^2$ is represented by the adjoining graph, constructed by means of the relation

$$y = \frac{1}{x^2}.$$

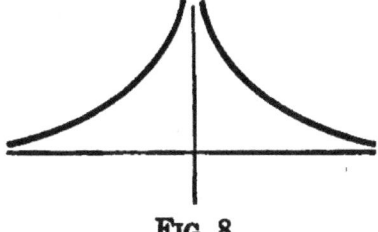

Fig. 8

There is a discontinuity for $x = 0$, illustrated by the fact that the graph for values of x near zero recedes indefinitely far from the x-axis.

A similar discontinuity for the function $1/x$ is shown by the graph (Fig. 9) of the equation

$$y = \frac{1}{x}.$$

Here the two parts of the graph recede to infinity in opposite directions, whereas in the previous case they receded to infinity in the same direction.

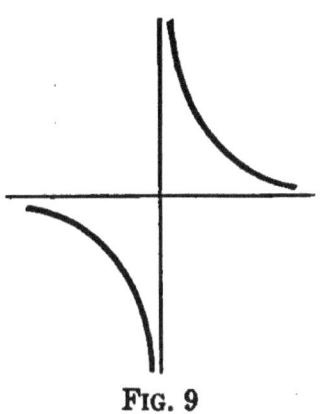

Fig. 9

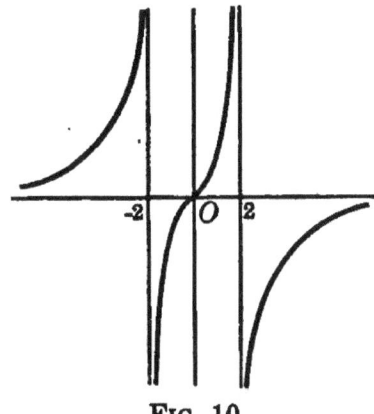

Fig. 10

The graph (Fig. 10) of the equation

$$y = \frac{4x}{4 - x^2}$$

shows to the eye the discontinuities of the function

$$\frac{4x}{4 - x^2}$$

for $x = 2$ and $x = -2$.

The infinitude of discontinuities of the function tan x is represented in the graph of the equation

$$y = \tan x$$

shown in Fig. 11.

The inverse function arc tan x is infinitely many-valued; that is, there is an infinitude of points on the curve for each given value of x. In Fig. 12 we give the graph of the equation

$$y = \text{arc tan } x.$$

Usually, in dealing with such a function, we confine our attention to one branch of the graph. When we do this the restricted function is single-valued and continuous.

The function arc tan $\dfrac{1}{x}$ is also infinitely many-valued. One part of the graph of the equation

$$y = \text{arc tan } \frac{1}{x}$$

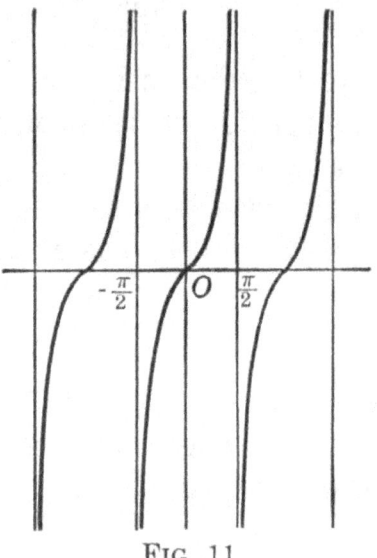

Fig. 11

is shown in Fig. 13, this part being determined by taking for each x the value of y which is least in numerical value. Such a part of the graph of this equation defines a function of x.

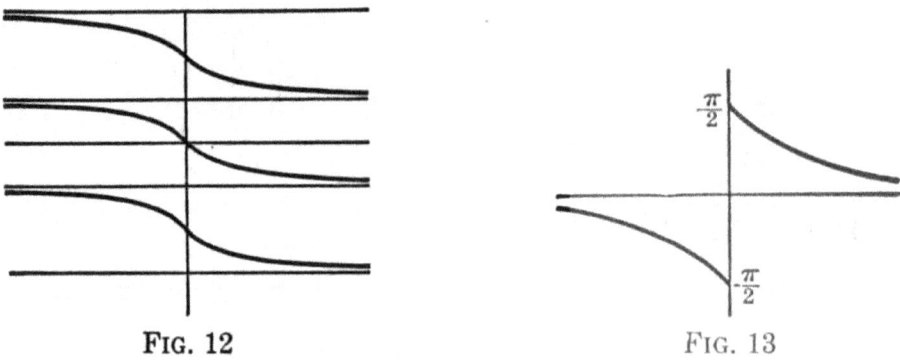

Fig. 12   Fig. 13

From the figure we see that the function so defined approaches the value $-\pi/2$ when x approaches zero from the left, but that it approaches $+\pi/2$ when x approaches zero from the right. As x passes through zero from the left the function value of this function jumps from $-\pi/2$ to $+\pi/2$. The function is undefined for $x = 0$.

Functions have been constructed which are discontinuous for every value of the variable in a given range. In a first course in the calculus we do not meet with such functions. The only discontinuities which we shall encounter are isolated ones of the sorts indicated by the examples in this section and the preceding one.

EXERCISES

1. Draw the graphs of each of the following functions and discuss their discontinuities:

(1) $\dfrac{1}{x^3}$; (2) $\dfrac{1}{x^4}$; (3) $\dfrac{2x}{1-x^2}$; (4) $\dfrac{x-1}{(x+1)(x-2)}$.

2. Draw the graphs of 2^x and $\log_2 x$. Observe that 2^x is continuous for all values of x. For what range of x is the function $\log_2 x$ defined? For what point is $\log_2 x$ discontinuous, and why?

3. Draw the graphs of each of the following functions:

(1) $\cot x$; (2) $\operatorname{arc\,cot} x$; (3) $\operatorname{arc\,cot} \dfrac{1}{x}$.

Discuss these functions from the point of view of continuity and discontinuity.

4. Draw the graph of one part of the function

$$\operatorname{arc\,tan} \frac{1}{x^2}$$

and discuss the function so defined from the point of view of continuity and discontinuity. (Use the part which is obtained by taking for each x the least numerical value of the function.)

15. The indeterminate forms 0/0 and ∞/∞. We have seen that division by 0 is to be excluded as an operation which is not defined. But we have seen also that a fraction *may* approach a limit even when its denominator approaches 0 *provided* that the numerator approaches 0 at the same time. When the numerator and denominator of a fraction simultaneously approach 0 the fraction is said to assume the indeterminate form 0/0.

Some of the most important problems of the calculus require the evaluation of such indeterminate forms; that is, each of

hem requires the finding of the limit approached by a fraction
whose numerator and denominator simultaneously approach 0.
In a later chapter we shall give a systematic method for finding
such limits when they exist. In the meantime, however, we
need to evaluate some such limits. Consequently we shall now
exhibit certain rough-and-ready methods for finding them in
certain rather simple cases. These are best given by means
of examples.

Example 1. Find the limit of $\dfrac{x^2 - 4}{x - 2}$ as x approaches the limit 2.

This takes the form 0/0. But when $x \neq 2$, the given fraction may be simplified to the form $x + 2$. Hence

$$\lim_{x \to 2} \frac{x^2 - 4}{x - 2} = \lim_{x \to 2} (x + 2) = 4.$$

Example 2. Evaluate $\lim\limits_{x \to 0} \dfrac{x^3 - 7x^2 + 3x}{x^2 - 19x}$.

This again takes the form 0/0. But if, for $x \neq 0$, we divide numerator and denominator by x, we have a new fraction which does not take an indeterminate form. Thus we have

$$\lim_{x \to 0} \frac{x^3 - 7x^2 + 3x}{x^2 - 19x} = \lim_{x \to 0} \frac{x^2 - 7x + 3}{x - 19} = -\frac{3}{19}.$$

Example 3. Evaluate $\lim\limits_{x \to 1} \dfrac{x^2 - 3x + 2}{x^2 - 1}$.

We have

$$\lim_{x \to 1} \frac{x^2 - 3x + 2}{x^2 - 1} = \lim_{x \to 1} \frac{(x-1)(x-2)}{(x-1)(x+1)} = \lim_{x \to 1} \frac{x-2}{x+1} = -\frac{1}{2}.$$

Example 4. Evaluate $\lim\limits_{x \to 0} \dfrac{a - \sqrt{a^2 - x^2}}{x^2}$.

We have

$$\lim_{x \to 0} \frac{a - \sqrt{a^2 - x^2}}{x^2} = \lim_{x \to 0} \frac{a - \sqrt{a^2 - x^2}}{x^2} \cdot \frac{a + \sqrt{a^2 - x^2}}{a + \sqrt{a^2 - x^2}}$$

$$= \lim_{x \to 0} \frac{a^2 - (a^2 - x^2)}{x^2(a + \sqrt{a^2 - x^2})}$$

$$= \lim_{x \to 0} \frac{1}{a + \sqrt{a^2 - x^2}}$$

$$= \frac{1}{2a}.$$

It will be observed that the essential element in the solution of each of these problems is the following: The *form* of the fraction whose limiting value is to be found is altered in such a way as *not* to change the value of the fraction for any value of x different from the limiting value and near that limiting value, and this change is effected in such a way that the *new form* of the fraction does *not* approach an indeterminate form when the argument x approaches its limiting value but is of such sort as to be capable of a direct evaluation.

If both numerator and denominator of a fraction increase indefinitely as x approaches a definite limiting value or increases indefinitely, the fraction is said to assume the indeterminate form ∞/∞. In this case neither the numerator nor the denominator approaches a (finite) limit, and yet it may happen that the fraction itself approaches a definite limiting value. For evaluating the indeterminate form ∞/∞ we employ the same general principle as that just named for the form $0/0$.

Example. Evaluate $\lim\limits_{x \to \infty} \dfrac{x^2 + 1}{3x^2 - 7}$.

We have $\lim\limits_{x \to \infty} \dfrac{x^2 + 1}{3x^2 - 7} = \lim\limits_{x \to \infty} \dfrac{1 + \dfrac{1}{x^2}}{3 - \dfrac{7}{x^2}} = \dfrac{1}{3}.$

Here the change in the fraction was effected by dividing both numerator and denominator by x^2, the highest power of x in either the numerator or the denominator.

EXERCISES

Prove the following:

1. $\lim\limits_{x \to 0} \dfrac{x-1}{x+1} = -1.$

2. $\lim\limits_{x \to \infty} \dfrac{x-1}{x+1} = 1.$

3. $\lim\limits_{x \to 0} \dfrac{x^3 + 7x}{x^7 - 9x^2 + x} = 7.$

4. $\lim\limits_{x \to 1} \dfrac{x^3 + x^2 - x - 1}{x^3 - 3x + 2} = -4.$

5. $\lim\limits_{x \to -3} \dfrac{x^2 + 5}{x + 5} = 7.$

6. $\lim\limits_{x \to \infty} \dfrac{ax^2 + bx + c}{dx^2 + ex + f} = \dfrac{a}{d}$ if $d \neq 0.$

7. $\lim\limits_{x \to 1} \dfrac{x^3 - 1}{x - 1} = 3.$

8. $\lim\limits_{h \to 0} \dfrac{(x+h)^n - x^n}{h} = nx^{n-1},\ n = \text{integer}.$

9. $\lim\limits_{x \to \infty} \dfrac{x(x+1)}{(x+2)(x+3)} = 1.$

10. $\lim\limits_{x \to \infty} \dfrac{5x^3 - 3x^2}{x} = \infty.$

11. $\lim\limits_{x \to a} \dfrac{1}{x - a} = -\infty$ if x is increasing as it approaches a.

12. $\lim\limits_{x \to a} \dfrac{1}{x - a} = +\infty$ if x is decreasing as it approaches a.

13. $\lim\limits_{x \to a} \dfrac{x^n - a^n}{x - a} = na^{n-1},\ n = \text{integer},\ a \neq 0.$

14. $\lim\limits_{x \to 0} \dfrac{\sqrt{1+x} - 1}{x} = \dfrac{1}{2}.$

*16. To show that $\lim\limits_{x \to 0} \dfrac{\sin x}{x} = 1$. About the point O as center construct a circle with radius unity. Construct the angles AOP and AOQ each equal to the small positive angle x, *the angle being measured in radians*. Draw PQ. It is perpendicular to OA and is bisected by OA in the point M. At P and Q draw tangents to the circle. These meet the line OA produced in the common point T. From geometry we have

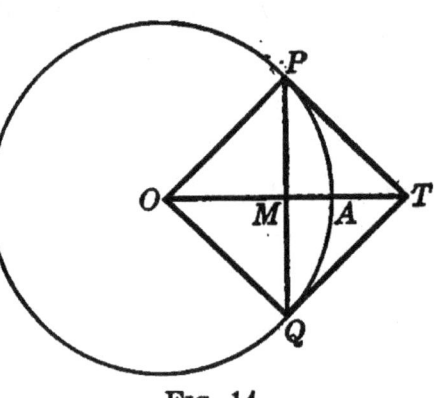

Fig. 14

$$PMQ < PAQ < PTQ;$$

or, $$PM < PA < PT.$$

Hence $$\sin x < x < \tan x.$$

Dividing through by $\sin x$, we have

$$1 < \dfrac{x}{\sin x} < \dfrac{1}{\cos x}.$$

Taking the reciprocal of each member, we get

$$1 > \dfrac{\sin x}{x} > \cos x.$$

---

* The teacher who desires to do so may omit this section until the class is ready to take up Chapter VI.

Now, as x approaches 0 as a limit, $\cos x$ approaches unity. Hence, as x approaches zero as a limit the fraction $(\sin x)/x$ lies between unity and the variable $\cos x$ which approaches unity. Therefore $(\sin x)/x$ itself must approach unity. Hence

If the angle x is measured in radian measure, we have

$$\lim_{x \to 0} \frac{\sin x}{x} = 1.$$

By means of this fundamental limit numerous other limits may be evaluated which have the indeterminate form $0/0$. We give a few examples.

Example 1. Evaluate $\lim\limits_{x \to 0} \dfrac{\tan x}{x}$.

We have
$$\lim_{x \to 0} \frac{\tan x}{x} = \lim_{x \to 0} \frac{\sin x}{x} \cdot \frac{1}{\cos x}$$
$$= \lim_{x \to 0} \frac{\sin x}{x} \cdot \lim_{x \to 0} \frac{1}{\cos x}$$
$$= 1 \cdot 1 = 1.$$

Example 2. Evaluate $\lim\limits_{x \to 0} \dfrac{1 - \cos x}{x^2}$.

We have
$$\lim_{x \to 0} \frac{1 - \cos x}{x^2} = \lim_{x \to 0} \frac{1 - \cos x}{x^2} \cdot \frac{1 + \cos x}{1 + \cos x}$$
$$= \lim_{x \to 0} \frac{1 - \cos^2 x}{x^2} \cdot \frac{1}{1 + \cos x}$$
$$= \lim_{x \to 0} \frac{\sin^2 x}{x^2} \cdot \lim_{x \to 0} \frac{1}{1 + \cos x}$$
$$= 1^2 \cdot \tfrac{1}{2} = \tfrac{1}{2}.$$

Example 3. Evaluate $\lim\limits_{x \to 0} \dfrac{1 - \cos x}{x}$.

We have
$$\lim_{x \to 0} \frac{1 - \cos x}{x} = \lim_{x \to 0} \frac{1 - \cos x}{x^2} \cdot x$$
$$= \lim_{x \to 0} \frac{1 - \cos x}{x^2} \cdot \lim_{x \to 0} x$$
$$= \tfrac{1}{2} \cdot 0 = 0.$$

Example 4. Evaluate $\lim\limits_{x=0} \dfrac{\sin nx}{x}$.

We have
$$\lim_{x=0} \frac{\sin nx}{x} = n \lim_{x=0} \frac{\sin nx}{nx}$$
$$= n \lim_{nx=0} \frac{\sin nx}{nx}$$
$$= n \cdot 1 = n..$$

EXERCISES

Prove the following:

1. $\lim\limits_{x=0} \dfrac{1-\cos x}{\sin x} = 0.$

2. $\lim\limits_{x=0} \dfrac{1-\cos x}{x \sin x} = \dfrac{1}{2}.$

3. $\lim\limits_{x=0} \dfrac{1-\cos x}{\sin^2 x} = \dfrac{1}{2}.$

4. $\lim\limits_{x=0} \dfrac{1}{x} \sin \dfrac{x}{2} = \dfrac{1}{2}.$

5. $\lim\limits_{h=0} \left[\cos(x+h) \dfrac{\sin h}{h} \right] = \cos x.$

6. $\lim\limits_{x=0} \dfrac{\sin x}{\tan x} = 1.$

7. $\lim\limits_{x=0} \dfrac{\sec x - 1}{x^2} = \dfrac{1}{2}.$

8. $\lim\limits_{x=0} \dfrac{\tan x - \sin x}{x^3} = \dfrac{1}{2}.$

*17. **The number e.** A certain constant frequently employed in the calculus is obtained by evaluating the limit
$$\lim_{x=0} (1+x)^{\frac{1}{x}}.$$
A rigorous evaluation of this limit is beyond the scope of this book. We shall be satisfied with a sort of graphical examination of the limit. Let us consider the graph of the equation
$$y = (1+x)^{\frac{1}{x}}.$$
For particular values of x the corresponding values of y may be computed. Thus we have the following table:

x	10	5	2	1	.5	.1	.01	.001
y	1.0096	1.4310	1.7320	2.0000	2.2500	2.5937	2.7048	2.7169

x	$-.5$	$-.1$	$-.01$	$-.001$
y	4.0000	2.8680	2.7320	2.7195

* The teacher who desires to do so may omit this section until the class is ready to take up Chapter VIII.

The graph of the equation is shown in the adjoining figure. From the table we are led to suppose that the value e approached by y when x approaches zero is between 2.7169 and 2.7195. A more exact calculation will show that e has the value $2.71828\cdots$. Hence we may write

$$\lim_{x \to 0}(1+x)^{\frac{1}{x}} = e = 2.71828\cdots.$$

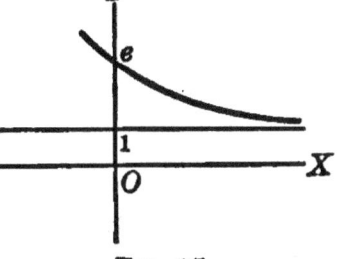

Fig. 15

Later we shall show how to compute e to any named degree of accuracy.

This constant e is the base of the so-called *natural system* of logarithms. Logarithms to this base, the so-called *natural logarithms*, play a very important rôle in mathematics. They are the logarithms usually employed in the calculus. When the base is not indicated we shall in this book always suppose that the base is e.

For *natural logarithms* we have the following important characteristic property:

$$\lim_{x \to 0} \frac{\log(1+x)}{x} = 1.$$

To prove this we observe that

$$\lim_{x \to 0} \frac{1}{x} \log(1+x) = \lim_{x \to 0} \log(1+x)^{\frac{1}{x}}$$
$$= \log \lim_{x \to 0}(1+x)^{\frac{1}{x}}$$
$$= \log e = 1.$$

We observe that $\qquad \log a = \dfrac{1}{\log_a e}.$

For, if $s = \log_a M$, and $t = \log M$, then $a^s = M$ and $e^t = M$. Hence $e^t = a^s$. Therefore $t = s \log a$, or

$$\log M = \log_a M \cdot \log a.$$

Taking $M = e$, we have $1 = \log_a e \cdot \log a$, whence $\log a = \dfrac{1}{\log_a e}.$

EXERCISES

1. Show that $\lim\limits_{y \to 0} \dfrac{\log_a(1+y)}{y} = \log_a e = \dfrac{1}{\log a}.$

2. Show that $\lim\limits_{x \to 0} \dfrac{a^x - 1}{x} = \log a.$

SUGGESTION. Set $y = a^x - 1$, whence $x = \log_a(1 + y)$. Moreover, as x approaches zero so does y, and vice versa. Hence

$$\lim_{x \to 0} \frac{a^x - 1}{x} = \lim_{y \to 0} \frac{y}{\log_a(1 + y)}.$$

3. Show that $\lim\limits_{x \to 0} \dfrac{e^x - 1}{x} = \log e = 1.$

18. Summary. This chapter is preliminary in character. Its purpose has been to clear the way for the main developments of the calculus and to indicate partially the nature of the first problems to be treated. Some of the notions require continued meditation for their full understanding.

The definition of function and the notion of functional relation are to be mastered with thoroughness. A general conception of limit and the main theorems for operating with limits should now be fairly familiar to the student; but he will expect to clarify his understanding of these as he proceeds. In connection with limits it is important to have a clear understanding of the process employed in finding the slope of a curve or the instantaneous speed of a moving body; this *process* is fundamental in much of what follows. It will be developed in more detail in the next chapter.

The notion of continuity, in its abstract form, is elusive. The student should now understand the graphical illustrations of continuity and discontinuity and should expect to return later to the abstract definition in order to come to a fuller mastery of it.

The discussion of indeterminate forms in § 15 is intended as preparatory to Chapter II. In a similar way §§ 16 and 17 are preparatory to Chapters VI and VIII respectively.

CHAPTER II

DIFFERENTIATION. GENERAL PRINCIPLES. ALGEBRAIC FUNCTIONS

19. Increments. If a variable changes from one value to another, the *difference* found by subtracting the first (or initial) value from the second is called an *increment* of the variable. An increment of x is denoted by Δx, read *delta x*; this is not the product of Δ and x, it is a single symbol standing for an increment of x. It is evident that an increment may be either positive or negative according as the variable is increasing or decreasing in value. An increment of y is denoted by Δy, an increment of ϕ by $\Delta \phi$, etc.

If in a functional relation $y = f(x)$ we give to x the increment Δx, we shall understand that Δy denotes the corresponding increment of y, so that $y + \Delta y = f(x + \Delta x)$. The increment Δy is reckoned from the value of y corresponding to the particular value of x from which Δx is reckoned.

We have already seen that certain important results are obtained by considering the limiting value of the ratio $\Delta y/\Delta x$ as Δx approaches zero. This limit, taken for a particular value of x, gives the slope of the graph of the equation $y = f(x)$ at the point corresponding to the initial value of x. We have also seen the intimate connection of a like limit with the problem of the speed of a moving body. We shall later find many other applications for the same limiting process; they arise in many different situations. The process leads to what is known as the derivative of a function, as will be seen from the definition in the next section. (The student is advised to review §§ 6 and 7 before proceeding to the next section.)

20. Definition of derivative. The principal notion of the differential calculus is that of the *derivative of a function* (also called the *differential coefficient* or the *derived function*).

The derivative of a function is the limit of the ratio of the increment of the function to the increment of the independent variable when the initial value of that variable is held fixed and its increment varies and approaches zero.

Thus the derivative of y with respect to x is the limit of $\Delta y/\Delta x$ when x is held fixed and Δx varies and approaches zero. If $y = f(x)$, then we have $y + \Delta y = f(x + \Delta x)$, and therefore

$$\frac{\Delta y}{\Delta x} = \frac{f(x + \Delta x) - f(x)}{\Delta x}.$$

Hence the derivative of $f(x)$ with respect to x is equal to

$$\lim_{\Delta x = 0} \frac{f(x + \Delta x) - f(x)}{\Delta x},$$

x itself being held fixed.

21. Symbols for the derivative of a function. The derivative of y, or $f(x)$, with respect to x is denoted variously by the symbols

$$\frac{dy}{dx},\ \frac{df}{dx},\ D_x y,\ D_x f(x),\ y',\ f'(x).$$

The first and the last are the ones which we shall generally employ. Then we may write

$$f'(x) = \frac{dy}{dx} = \lim_{\Delta x = 0} \frac{\Delta y}{\Delta x} = \lim_{\Delta x = 0} \frac{f(x + \Delta x) - f(x)}{\Delta x}.$$

The symbol dy/dx is read *the derivative of y with respect to x*; the symbol $f'(x)$ is read *f prime of x*.

It is to be carefully noted that dy/dx, as here defined, is not a fraction; in fact, no meaning has as yet been assigned to dy and dx as separate symbols. In Chapter IV we shall define these symbols dy and dx separately. After they are defined we may begin to think of dy/dx as a fraction. For the present it must be treated as one symbol denoting *one* thing, namely, the derivative of y with respect to x.

22. Differentiable functions. If $f(x)$ is continuous for a given value of x, then from the definition of continuity it follows that

$$\lim_{\Delta x = 0} f(x + \Delta x) = f(x).$$

If $f(x)$ is not continuous for the given value of x, then either this limit does not exist or (if it exists) it does not have the value $f(x)$. Hence, when $f(x)$ is not continuous at a given point x, the limit

$$\lim_{\Delta x \to 0} \frac{f(x + \Delta x) - f(x)}{\Delta x}$$

does not exist (as a finite quantity), since the denominator approaches zero and the numerator does not. Therefore $f(x)$ cannot have a derivative for any value of x for which it is discontinuous.

We say that $f(x)$ is differentiable or not differentiable, for a given value of x, according as $f(x)$ has or has not a derivative for this value of x.

We have just seen that a function cannot be differentiable for any value of x for which it is discontinuous. But the converse is not always true. A function may be continuous and yet possess no derivative. But such functions will not occur in this book. We shall be concerned only with such functions as possess derivatives for all values of x for which they are continuous.

23. The Four-Step Rule for differentiation. The process of finding the derivative is called *differentiation*. It consists of the four steps described in the following

FOUR-STEP RULE FOR DIFFERENTIATION

1. *In the function y of x replace x by $x + \Delta x$ and y by the corresponding value $y + \Delta y$.*
2. *Find the value of Δy (the increment of the function) by subtracting the given value of the function from the new value.*
3. *Divide the value of Δy by Δx.*
4. *Find the limit of the quotient $\Delta y/\Delta x$ thus found, when x is held fixed and Δx is allowed to vary and approach zero as a limit. This limit is the derivative of y with respect to x.*

The processes, rules, and theorems relating to differentiation and derivatives make up the *differential calculus*. (We shall later describe another branch of the calculus known as the *integral calculus*.) We shall develop methods of finding the derivatives of many combinations of functions by means of relatively few results established by the help of the Four-Step Rule. But that rule is so important for an understanding of the differential calculus that the student should at once become thoroughly familiar with it. Accordingly we shall now illustrate it by means of examples.

Example 1. Find the derivative of $3x^2 - 4x + 17$.

We have $\quad y = 3x^2 - 4x + 17.$

First Step. $\quad y + \Delta y = 3(x + \Delta x)^2 - 4(x + \Delta x) + 17$
$\quad\quad\quad\quad\quad = 3x^2 + 6x \cdot \Delta x + 3(\Delta x)^2 - 4x - 4\Delta x + 17.$

Second Step. $\quad \Delta y = 6x \cdot \Delta x + 3(\Delta x)^2 - 4\Delta x.$

Third Step. $\quad \dfrac{\Delta y}{\Delta x} = 6x + 3\Delta x - 4.$

Fourth Step. $\quad \dfrac{dy}{dx} = \lim\limits_{\Delta x \to 0} \dfrac{\Delta y}{\Delta x} = 6x - 4.$

Hence the required derivative, or derived function, is $6x - 4$.

If we desire the derivative for a particular value of x we substitute that value for x in the derived function. Thus, for $x = 2$ the derivative has the value 8; for $x = 0$, the value -4; for $x = x_1$, the value $6x_1 - 4$.

(In the next two examples we shall not indicate the steps explicitly as such; but the student is advised to observe the separate steps in the work.)

Example 2. In Boyle's law for gases we have $pv = k$, where p and v denote pressure and volume respectively at a given temperature and k is a constant. Find the derivative of p with respect to v.

We have $\quad p = \dfrac{k}{v},$

$$p + \Delta p = \dfrac{k}{v + \Delta v},$$

$$\Delta p = \dfrac{k}{v + \Delta v} - \dfrac{k}{v} = \dfrac{-k\Delta v}{v(v + \Delta v)},$$

$$\dfrac{\Delta p}{\Delta v} = -\dfrac{k}{v(v + \Delta v)},$$

$$\dfrac{dp}{dv} = \lim\limits_{\Delta v \to 0} \dfrac{\Delta p}{\Delta v} = -\dfrac{k}{v^2}.$$

For $k = 450$ and $v = 15$ we have $dp/dv = -2$.

Example 3. Differentiate $\sqrt{x}$.

We have $y = \sqrt{x}$,
$$y + \Delta y = \sqrt{x + \Delta x},$$
$$\Delta y = \sqrt{x + \Delta x} - \sqrt{x}$$
$$= \frac{(\sqrt{x + \Delta x} - \sqrt{x})(\sqrt{x + \Delta x} + \sqrt{x})}{\sqrt{x + \Delta x} + \sqrt{x}}$$
$$= \frac{\Delta x}{\sqrt{x + \Delta x} + \sqrt{x}}.$$

Therefore $\dfrac{dy}{dx} = \lim\limits_{\Delta x \to 0} \dfrac{\Delta y}{\Delta x} = \lim\limits_{\Delta x \to 0} \dfrac{1}{\sqrt{x + \Delta x} + \sqrt{x}} = \dfrac{1}{2\sqrt{x}}.$

EXERCISES

Use the Four-Step Rule to find the derivatives of the following functions:

1. x^2.   Ans. $2x$.
2. $3x^2$.   Ans. $6x$.
3. $3x^2 - 7$.   Ans. $6x$.
4. $3x^2 + c$.   Ans. $6x$.
5. $5 - 7x$.   Ans. -7.
6. $\dfrac{1}{x}$.   Ans. $-\dfrac{1}{x^2}$.
7. $\dfrac{4}{x^2}$.   Ans. $-\dfrac{8}{x^3}$.
8. $\dfrac{x+2}{x}$.   Ans. $-\dfrac{2}{x^2}$.
9. x^3.   Ans. $3x^2$.
10. x^4.   Ans. $4x^3$.
11. x^5.   Ans. $5x^4$.
12. $\dfrac{3}{x+1}$.   Ans. $-\dfrac{3}{(x+1)^2}$.
13. $1 - 7x^2$.   Ans. $-14x$.
14. $\dfrac{4}{x^2 - 1}$.   Ans. $-\dfrac{8x}{(x^2 - 1)^2}$.
15. $\dfrac{x^2 + 1}{x}$.   Ans. $1 - \dfrac{1}{x^2}$.
16. $\dfrac{x^3 + 7}{x}$.   Ans. $2x - \dfrac{7}{x^2}$.
17. $x^3 - 7x + 22$.   Ans. $3x^2 - 7$.
18. $\dfrac{x^2}{1 + x}$.   Ans. $\dfrac{x^2 + 2x}{(x + 1)^2}$.
19. $x\sqrt{x}$.   Ans. $\tfrac{3}{2}\sqrt{x}$.
20. $x^2 + 7x - 3\sqrt{x}$.   Ans. $2x + 7 - \dfrac{3}{2\sqrt{x}}$.

24. The slope of a curve. One of the fundamental problems in the applications of the differential calculus to geometry is that of finding the slope of a curve whose equation may be taken in the form $y = f(x)$. In § 6 we have discussed this problem in a preliminary way. We shall now treat it with more precision.

Let $y = f(x)$ be the equation of the curve AB in the adjoining figure. Let P and Q be the points (x, y) and $(x + \Delta x, y + \Delta y)$ respectively. Draw the tangent PT to the curve at P. Draw PL parallel to the x-axis; draw PM and QN parallel to the y-axis. Then $\Delta y = QL$ and $\Delta x = PL$.

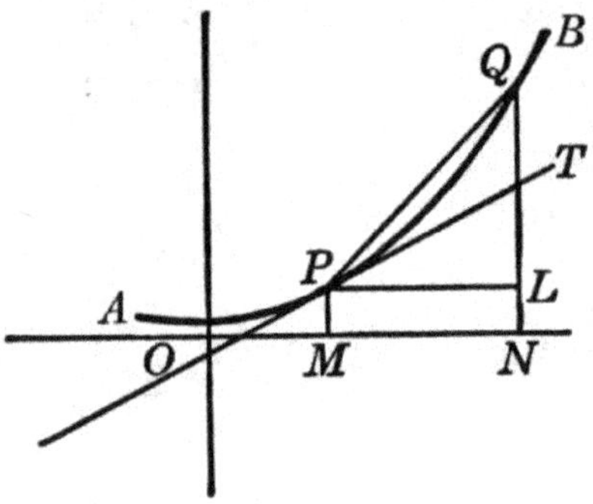

Let us now apply the Four-Step Rule, interpreting the results geometrically. We have

$$y = f(x) = PM.$$

FIG. 16

First Step. $\qquad y + \Delta y = f(x + \Delta x) = QN.$

Second Step. $\qquad \Delta y = f(x + \Delta x) - f(x) = QN - PM = QL.$

Third Step. $\qquad \dfrac{\Delta y}{\Delta x} = \dfrac{f(x + \Delta x) - f(x)}{\Delta x} = \dfrac{QL}{PL}.$

Therefore $\qquad \dfrac{\Delta y}{\Delta x} = \tan LPQ = \text{slope of } PQ.$

As Δx approaches zero the point Q approaches the point P and the secant PQ approaches the tangent PT. Hence

Fourth Step. $\qquad \dfrac{dy}{dx} = \lim\limits_{\Delta x = 0} \tan LPQ = \tan LPT = \text{slope of } PT.$

Therefore $\qquad \dfrac{dy}{dx} = \text{slope of curve at point } P.$

Hence *the derivative of $f(x)$, for a given value of x, is equal to the slope of the curve $y = f(x)$ at the point P which corresponds to the given value of x.*

We have seen that the derivative of x^2 is $2x$. Hence the slope of the parabola $y = x^2$ at the point $(0, 0)$ is zero; that is, the x-axis is tangent to the parabola at the vertex. For $x = \frac{1}{2}$ the derivative $2x$ has

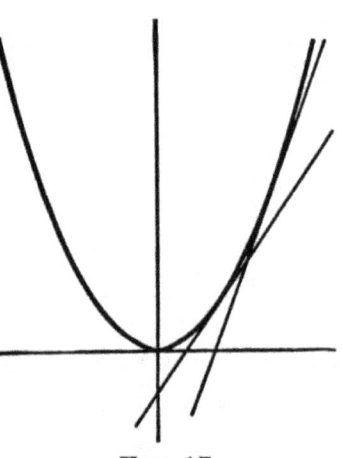

FIG. 17

the value 1. Hence the tangent at the point $(\frac{1}{2}, \frac{1}{4})$ makes an angle of 45° with the x-axis. As x increases the derivative $2x$ increases. Hence the graph in the first quadrant becomes steeper and steeper as we proceed farther to the right.

EXERCISES

Find by differentiation the slopes of the following curves at the points indicated and verify the results graphically:

1. $y = x^2 + 2$, where $x = 1$.   *Ans.* 2.
2. $y = x^2 - 8$, where $x = 1$.   *Ans.* 2.
3. $y = \dfrac{1}{x+1}$, where $x = 3$.   *Ans.* $-\frac{1}{16}$.
4. $y = \dfrac{1}{x+1}$, where $x = 1$.   *Ans.* $-\frac{1}{4}$.
5. $y = x^3$, where $x = 3$.   *Ans.* 27.
6. $y = 7 + 6x - x^2$, where $x = 0$.   *Ans.* 6.

7. If a railway track has the form of a parabola $y = x^2$, with the positive direction of the x-axis extending to the east, and if a train is going east when it passes the vertex of the parabola, in what direction is it going

(1) when half a mile east of the y-axis?   *Ans.* Northeast.
(2) when half a mile west of the y-axis?   *Ans.* Southeast.
(3) when at the point for which $x = \frac{1}{2}\sqrt{3}$?   *Ans.* N 30° E.

25. The derivative of an area. Let us consider the area bounded by a curve $y = f(x)$, a portion AB of the x-axis from the point for which $x = a$ to that for which $x = b$ and the ordinates at these points. It is convenient to treat at the same time the variable area $A(x)$, or $AEFD$, bounded by the curve and the x-axis and the ordinates $x = a$ and $x = x$. (By the ordinate $x = x$ we mean simply the variable ordinate for a general value of x.)

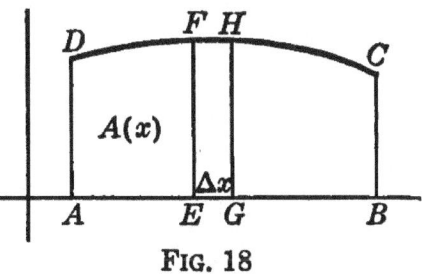

FIG. 18

For this variable area $A(x)$ it is possible to find the derivative by the Four-Step Rule, even though we do not know an expression for $A(x)$ in terms of $f(x)$. We denote the function $A(x)$ by α. Then we have

$$\alpha = A(x) = \text{area } AEFD.$$

First Step. $\quad \alpha + \Delta\alpha = A(x + \Delta x) = \text{area } AGHD.$

Second Step. $\quad \Delta\alpha = A(x + \Delta x) - A(x) = \text{area } EGHF.$

Third Step. $\quad \dfrac{\Delta\alpha}{\Delta x} = \dfrac{A(x + \Delta x) - A(x)}{\Delta x} = \dfrac{\text{area } EGHF}{EG}$

$\qquad\qquad = $ average height of strip $EGHF$.

As Δx approaches zero the strip $EGHF$ becomes more and more narrow and its average height approaches EF. Hence

Fourth Step. $\quad \dfrac{d\alpha}{dx} = \lim\limits_{\Delta x = 0} \dfrac{\Delta\alpha}{\Delta x} = EF = f(x).$

Hence *the derivative of $A(x)$ with respect to x is $f(x)$.*

In order to find the area $ABCD$ it is therefore sufficient to find an expression for the function $A(x)$ and to evaluate this expression for $x = b$. In other words, the area $ABCD$ is equal to $A(b)$. In applying this method it is always desirable to draw a figure to illustrate the work.

26. The antiderivative. The problem of finding the function $A(x)$ of the last foregoing section when $f(x)$ is given is the problem of finding a function when the derivative is known.

An antiderivative of a function $\phi(x)$ is a function whose derivative is $\phi(x)$.

There is no direct method for finding antiderivatives. In a later part of this book we shall systematize the (indirect) methods for finding antiderivatives. The antiderivatives which we shall require in this section can be found by means of the derivatives in § 23 and the exercises which follow it. Thus an antiderivative of $2x$ is x^2; of $3x^2$ is x^3; of $4x^3$ is x^4. From problems 2 and 3 and 4 we see that antiderivatives of $6x$ are $3x^2$ and $3x^2 - 7$ and $3x^2 + c$, where c is a constant.

From the last examples it appears that a function may have more than one antiderivative. In fact, the following theorem is readily proved:

I. If $g(x)$ is an antiderivative of $\phi(x)$, then $g(x) + c$ is also an antiderivative of $\phi(x)$, where c is any constant whatever.

To prove this we write
$$t = g(x) + c.$$
Then $t + \Delta t = g(x + \Delta x) + c,$
$$\Delta t = g(x + \Delta x) - g(x),$$
$$\frac{\Delta t}{\Delta x} = \frac{g(x + \Delta x) - g(x)}{\Delta x},$$
$$\frac{dt}{dx} = \lim_{\Delta x = 0} \frac{g(x + \Delta x) - g(x)}{\Delta x} = \frac{dg}{dx} = \phi(x).$$

Hence the derivative of $g(x) + c$ is $\phi(x)$, as was to be proved.

II. If two functions of x have the same derivative with respect to x, their difference has the derivative 0.

Let $g(x)$ and $h(x)$ have the same derivative $\phi(x)$. Let t denote their difference $g(x) - h(x)$. Then we have
$$t = g(x) - h(x),$$
$$t + \Delta t = g(x + \Delta x) - h(x + \Delta x),$$
$$\Delta t = g(x + \Delta x) - g(x) - [h(x + \Delta x) - h(x)],$$
$$\frac{\Delta t}{\Delta x} = \frac{g(x + \Delta x) - g(x)}{\Delta x} - \frac{h(x + \Delta x) - h(x)}{\Delta x},$$
$$\frac{dt}{dx} = \frac{dg}{dx} - \frac{dh}{dx} = \phi(x) - \phi(x) = 0.$$

Hence the derivative of t with respect to x is 0, as was to be proved.

III. If the derivative of a function of x with respect to x is zero, that function is a constant.

Let the function y of x be $f(x)$. Since the derivative is zero, the slope of the "curve" $y = f(x)$ is zero at every point; that is, the "curve" $y = f(x)$ is everywhere parallel to the x-axis. This can be true only when $f(x)$ is a constant. Hence the theorem.

On combining theorems II and III we have the following theorem:

IV. *If $g(x)$ is an antiderivative of $\phi(x)$, then every antiderivative of $\phi(x)$ is of the form $g(x) + c$, where c is a constant; and every such function $g(x) + c$ is an antiderivative of $\phi(x)$.*

If we know a particular antiderivative of a given function, then every antiderivative may be found by adding to the particular one an arbitrary constant c. This sum is called *the* antiderivative, or the general antiderivative. The constant c is called the arbitrary constant in the antiderivative.

There is one additional elementary principle which will be very helpful to us now in finding antiderivatives. It is embodied in the following theorem:

V. *If $g(x)$ is an antiderivative of $\phi(x)$, then $kg(x)$ is an antiderivative of $k\phi(x)$ if k is a constant.*

Let $t = kg(x)$. Then we have

$$t + \Delta t = kg(x + \Delta x),$$
$$\Delta t = k[g(x + \Delta x) - g(x)],$$
$$\frac{\Delta t}{\Delta x} = k \frac{g(x + \Delta x) - g(x)}{\Delta x},$$
$$\frac{dt}{dx} = k \lim_{\Delta x = 0} \frac{g(x + \Delta x) - g(x)}{\Delta x} = k\phi(x).$$

Hence the derivative of $kg(x)$ is $k\phi(x)$, as was to be proved.

Thus we have seen that x^5 is an antiderivative of $5x^4$. Taking $k = \frac{1}{5}$, we see that an antiderivative of x^4 is $\frac{1}{5} x^5$. Hence the general antiderivative of x^4 is $\frac{1}{5} x^5 + c$, where c is an arbitrary constant.

27. Computation of areas. One of the remarkable applications of the calculus is that involved in finding areas. Areas may be found by means of the principles developed in §§ 25 and 26. We shall illustrate the method by means of some examples.

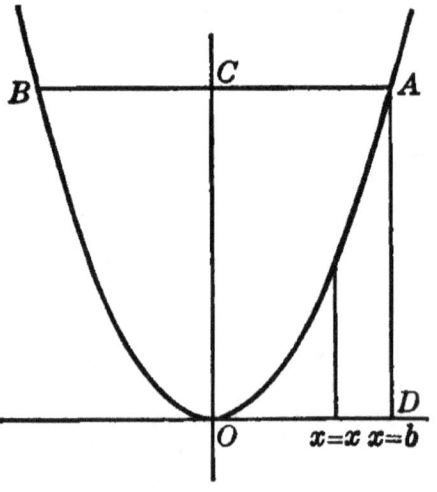

FIG. 19

Example 1. Find the area bounded by the parabola $y = x^2$, the axis of x, and the ordinate $x = b$.

Let $A(x)$ denote the area up to the variable ordinate $x = x$ (Fig. 19). Then $A(0) = 0$, for when x approaches 0 the area $A(x)$ approaches 0. From the result in § 25 it follows that

$$\frac{dA}{dx} = x^2.$$

The antiderivative of x^2 is $\frac{1}{3} x^3 + c$, as one sees from problem 9 of § 23 and theorems IV and V of § 26. Hence

$$A(x) = \tfrac{1}{3} x^3 + c,$$

provided that the constant c is determined so that $A(0) = 0$. This requires that c shall be zero, as is seen by putting zero for x in the expression for $A(x)$. Therefore $A(x) = \frac{1}{3} x^3$. Hence $A(b)$, the required area, is $\frac{1}{3} b^3$.

Example 2. Find the area bounded by the parabola $y = x^2$ and the line $y = b^2$.

In Fig. 19 this is the area OAB. From the symmetry of the figure it follows that OAC is half this area. But from the figure it is seen that the area OAC is equal to the difference between the area of the rectangle $ODAC$ and that of the figure $ODAO$, the area of the latter being $\frac{1}{3} b^3$, as we saw in Example 1. Hence

$$\text{area of } OAC = b^3 - \tfrac{1}{3} b^3 = \tfrac{2}{3} b^3.$$

Therefore   area of $OAB = \tfrac{4}{3} b^3$.

This is the area sought.

EXERCISES

(In these exercises the constants a and b are positive.)

Find the antiderivative of each of the following functions:

1. x. Ans. $\tfrac{1}{2} x^2 + c$.
2. x^2. Ans. $\tfrac{1}{3} x^3 + c$.
3. x^3. Ans. $\tfrac{1}{4} x^4 + c$.
4. x^4. Ans. $\tfrac{1}{5} x^5 + c$.
5. $2 - \dfrac{2}{x^2}$. Ans. $\dfrac{2 x^2 + 2}{x} + c$.
6. $3 x^2 - 7$. Ans. $x^3 - 7 x + c$.

7. Find the area bounded by the straight line $y = mx$ ($m > 0$) and the x-axis and the ordinate $x = b$, and verify the result by elementary geometry.   Ans. $\tfrac{1}{2} mb^2$.

8. Find the area bounded by the curve $y = x^3$ and the x-axis and the ordinate $x = b$.   Ans. $\tfrac{1}{4} b^4$.

9. Find the area bounded by the curve $y = x^3$ and the y-axis and the line $y = b^3$. *Ans.* $\frac{3}{4}b^4$.

10. Find the area bounded by the curve $y = x^4$ and the line $y = b^4$. *Ans.* $\frac{8}{5}b^5$.

11. Find the area bounded by the curve $y = x^4$ and the x-axis and the ordinates $x = a$ and $x = b$, where $b > a$. *Ans.* $\frac{1}{5}(b^5 - a^5)$.

28. General theorems on differentiation. The fundamental theorems on differentiation are proved by means of the Four-Step Rule. We shall now establish the results contained in the following formulas, c and n being constants, and u and v and w being functions of x:

I. $\dfrac{dc}{dx} = 0.$

II. $\dfrac{dx}{dx} = 1.$

III. $\dfrac{d}{dx}(u + v - w) = \dfrac{du}{dx} + \dfrac{dv}{dx} - \dfrac{dw}{dx}.$

IV. (1) $\dfrac{d}{dx}(cv) = c\dfrac{dv}{dx},$ (2) $\dfrac{d}{dx}\left(\dfrac{v}{c}\right) = \dfrac{1}{c}\dfrac{dv}{dx}.$

V. $\dfrac{d}{dx}(uv) = u\dfrac{dv}{dx} + v\dfrac{du}{dx}.$

VI. $\dfrac{d}{dx}(v^n) = nv^{n-1}\dfrac{dv}{dx}.$

VII. $\dfrac{d}{dx}(x^n) = nx^{n-1}.$

VIII. $\dfrac{d}{dx}\left(\dfrac{u}{v}\right) = \dfrac{v\dfrac{du}{dx} - u\dfrac{dv}{dx}}{v^2}.$

IX. $\dfrac{d}{dx}\left(\dfrac{c}{v}\right) = -\dfrac{c\dfrac{dv}{dx}}{v^2}.$

These formulas, together with their proofs, should be thoroughly learned and committed to memory.

1. The first result states that *the derivative of a constant is 0*. If we write $y = c$ and give to x the increment Δx, y remains unchanged, since it is a constant. Hence $\Delta y = 0$. Therefore

$$\frac{\Delta y}{\Delta x} = 0 \quad \text{and} \quad \lim_{\Delta x \to 0} \frac{\Delta y}{\Delta x} = 0.$$

Hence
$$\frac{dc}{dx} = 0.$$

2. *The derivative of a variable with respect to itself is 1.*

Writing $y = x$ we have $y + \Delta y = x + \Delta x$. Therefore

$$\frac{\Delta y}{\Delta x} = 1 \quad \text{and} \quad \lim_{\Delta x \to 0} \frac{\Delta y}{\Delta x} = 1.$$

Hence
$$\frac{dx}{dx} = 1.$$

3. If we write $y = u + v - w,$

we have
$$y + \Delta y = u + \Delta u + v + \Delta v - w - \Delta w,$$
$$\Delta y = \Delta u + \Delta v - \Delta w,$$
$$\frac{\Delta y}{\Delta x} = \frac{\Delta u}{\Delta x} + \frac{\Delta v}{\Delta x} - \frac{\Delta w}{\Delta x}.$$

Taking the limit as Δx approaches 0, we have

$$\frac{dy}{dx} = \frac{du}{dx} + \frac{dv}{dx} - \frac{dw}{dx}.$$

Therefore
$$\frac{d}{dx}(u + v - w) = \frac{du}{dx} + \frac{dv}{dx} - \frac{dw}{dx}.$$

One may proceed similarly for any finite number of functions.

Hence *the derivative of an algebraic sum of a finite number of functions is equal to the corresponding algebraic sum of their derivatives.*

4. If we write $y = cv,$

we have
$$y + \Delta y = c(v + \Delta v) = cv + c\,\Delta v,$$
$$\Delta y = c\,\Delta v,$$
$$\frac{\Delta y}{\Delta x} = c\frac{\Delta v}{\Delta x},$$
$$\frac{dy}{dx} = c\frac{dv}{dx}.$$

[By I, §10]

Therefore
$$\frac{d}{dx}(cv) = c\frac{dv}{dx}.$$

Similarly, we have
$$\frac{d}{dx}\left(\frac{v}{c}\right) = \frac{1}{c}\frac{dv}{dx}.$$

Hence *the derivative of a constant times a function is equal to the product of the constant times the derivative of the function.*

5. If we write $y = uv$,

we have
$$y + \Delta y = (u + \Delta u)(v + \Delta v)$$
$$= uv + u\,\Delta v + v\,\Delta u + \Delta u \cdot \Delta v,$$
$$\Delta y = u\,\Delta v + v\,\Delta u + \Delta u \cdot \Delta v,$$
$$\frac{\Delta y}{\Delta x} = u\frac{\Delta v}{\Delta x} + v\frac{\Delta u}{\Delta x} + \Delta u \cdot \frac{\Delta v}{\Delta x},$$
$$\frac{dy}{dx} = u\frac{dv}{dx} + v\frac{du}{dx} + 0 \cdot \frac{dv}{dx}. \qquad \text{[By II, §10]}$$

Therefore
$$\frac{d}{dx}(uv) = u\frac{dv}{dx} + v\frac{du}{dx}.$$

Hence *the derivative of the product of two functions is equal to the first function times the derivative of the second plus the second function times the derivative of the first.*

This principle may be extended to the product of any finite number of functions. If
$$y = v_1 v_2 v_3 \cdots v_k,$$
we may consider y as the product of v_1 by $v_2 v_3 v_4 \cdots v_k$.

Hence $\dfrac{d}{dx}(v_1 v_2 \cdots v_k) = v_2 v_3 \cdots v_k \dfrac{dv_1}{dx} + v_1 \dfrac{d}{dx}(v_2 v_3 \cdots v_k).$

Therefore
$$\frac{\frac{d}{dx}(v_1 v_2 \cdots v_k)}{v_1 v_2 \cdots v_k} = \frac{\frac{dv_1}{dx}}{v_1} + \frac{\frac{d}{dx}(v_2 v_3 \cdots v_k)}{v_2 v_3 \cdots v_k}$$
$$= \frac{\frac{dv_1}{dx}}{v_1} + \frac{\frac{dv_2}{dx}}{v_2} + \frac{\frac{d}{dx}(v_3 \cdots v_k)}{v_3 \cdots v_k}$$
$$= \frac{\frac{dv_1}{dx}}{v_1} + \frac{\frac{dv_2}{dx}}{v_2} + \frac{\frac{dv_3}{dx}}{v_3} + \cdots + \frac{\frac{dv_k}{dx}}{v_k}.$$

Therefore

$$\frac{d}{dx}(v_1 v_2 v_3 \cdots v_k) = (v_2 v_3 \cdots v_k)\frac{dv_1}{dx} + (v_1 v_3 \cdots v_k)\frac{dv_2}{dx} + \cdots$$
$$+ (v_1 v_2 \cdots v_{k-1})\frac{dv_k}{dx}.$$

Hence *the derivative of the product of a finite number k of factors is the sum of the k products obtained by multiplying the derivative of each factor by the product of the other k − 1 factors.*

6. Let us consider the function $y = v^n$. We may think of v^n as the product of n factors each equal to v. Since the product of any $n - 1$ of these factors is v^{n-1}, it is clear that each of the n products formed in accordance with the foregoing principle is

$$v^{n-1}\frac{dv}{dx}.$$

Since there are n of these, we have, by means of the foregoing principle, the result

$$\frac{d}{dx}(v^n) = nv^{n-1}\frac{dv}{dx}.$$

This formula may also be proved directly by means of the Four-Step Rule and the binomial theorem. Such a proof will furnish a good exercise for the student. The formula yields the following theorem:

The derivative of the nth power of a function is n times the (n − 1)st power of the function times the derivative of the function.

The given method of proof of this result applies only when n is a positive integer. But the result holds for every real constant n, as we shall prove in Chapter VIII. In the meantime we shall employ the theorem in the general form when n is taken to be any real constant whatever (positive or zero or negative).

7. In the important special case in which $v = x$ the result in 6 becomes

$$\frac{d}{dx}(x^n) = nx^{n-1}$$

since dv/dx now has the value 1.

8. If we write
$$y = \frac{u}{v},$$

we have
$$y + \Delta y = \frac{u + \Delta u}{v + \Delta v},$$

$$\Delta y = \frac{u + \Delta u}{v + \Delta v} - \frac{u}{v} = \frac{v \Delta u - u \Delta v}{v(v + \Delta v)},$$

$$\frac{\Delta y}{\Delta x} = \frac{v \frac{\Delta u}{\Delta x} - u \frac{\Delta v}{\Delta x}}{v(v + \Delta v)},$$

$$\frac{dy}{dx} = \frac{v \frac{du}{dx} - u \frac{dv}{dx}}{v^2}. \quad \text{[By II and IV, § 10]}$$

Therefore
$$\frac{d}{dx}\left(\frac{u}{v}\right) = \frac{v \frac{du}{dx} - u \frac{dv}{dx}}{v^2}.$$

Hence *the derivative of a fraction is equal to its denominator times the derivative of its numerator minus its numerator times the derivative of its denominator, all divided by the square of its denominator.*

9. In the special case of the last result in which u is the constant c we have $du/dx = 0$. For this case, therefore, the last foregoing formula takes the following simpler form:

$$\frac{d}{dx}\left(\frac{c}{v}\right) = -\frac{c \frac{dv}{dx}}{v^2}.$$

29. Use of the general theorems in finding derivatives. We shall now show by means of examples how to employ the general theorems in finding the derivatives of algebraic functions.

Example 1. $y = x^{\frac{9}{5}}$.

We have
$$\frac{dy}{dx} = \frac{d}{dx}(x^{\frac{9}{5}}) = \frac{9}{5} x^{\frac{4}{5}}, \qquad \text{[By VII, § 28]}$$

Example 2. $y = ax^m + cx^n$.

We have
$$\frac{dy}{dx} = \frac{d}{dx}(ax^m + cx^n)$$

$$= \frac{d}{dx}(ax^m) + \frac{d}{dx}(cx^n) \qquad \text{[By III, § 28]}$$

$$= a\frac{d}{dx}(x^m) + c\frac{d}{dx}(x^n) \qquad \text{[By IV, § 28]}$$

$$= amx^{m-1} + cnx^{n-1}. \qquad \text{[By VII, § 28]}$$

Example 3. $y = (x^2 + x - 7)^6$.

We have
$$\frac{dy}{dx} = 6(x^2 + x - 7)^5 \frac{d}{dx}(x^2 + x - 7) \qquad \text{[By VI, § 28]}$$

$$= 6(x^2 + x - 7)^5 (2x + 1).$$

Example 4. $y = \sqrt{a^2 - x^2} = (a^2 - x^2)^{\frac{1}{2}}$.

We have
$$\frac{dy}{dx} = \frac{1}{2}(a^2 - x^2)^{-\frac{1}{2}} \frac{d}{dx}(a^2 - x^2).$$
[By VI, § 28, with $v = a^2 - x^2$ and $n = \frac{1}{2}$]

Hence
$$\frac{dy}{dx} = \frac{1}{2}(a^2 - x^2)^{-\frac{1}{2}}(-2x) = -\frac{x}{\sqrt{a^2 - x^2}}.$$

Example 5. $y = (a^2 + x^2)(a^2 - x^2)^{\frac{1}{2}}$.

We have
$$\frac{dy}{dx} = (a^2 + x^2)\frac{d}{dx}(a^2 - x^2)^{\frac{1}{2}} + (a^2 - x^2)^{\frac{1}{2}}\frac{d}{dx}(a^2 + x^2).$$

[By V, § 28, with $u = a^2 + x^2$ and $v = (a^2 - x^2)^{\frac{1}{2}}$]

Hence
$$\frac{dy}{dx} = (a^2 + x^2)\left(-\frac{x}{\sqrt{a^2 - x^2}}\right) + (a^2 - x^2)^{\frac{1}{2}}(2x)$$

[By Example 4]

$$= -\frac{x(a^2 + x^2)}{\sqrt{a^2 - x^2}} + \frac{2x(a^2 - x^2)}{\sqrt{a^2 - x^2}}$$

$$= \frac{a^2 x - 3x^3}{\sqrt{a^2 - x^2}}.$$

Example 6. $y = \dfrac{x^2 - 7x + 1}{x^2 + 1}.$

We have

$$\frac{dy}{dx} = \frac{(x^2+1)\dfrac{d}{dx}(x^2-7x+1) - (x^2-7x+1)\dfrac{d}{dx}(x^2+1)}{(x^2+1)^2}$$ [By VIII, § 28]

$$= \frac{(x^2+1)(2x-7) - (x^2-7x+1)(2x)}{(x^2+1)^2}$$

$$= \frac{7(x^2-1)}{(x^2+1)^2}.$$

Example 7. $y = \dfrac{\sqrt{x^2+1} - x}{\sqrt{x^2+1} + x}.$

In this problem it is convenient to rationalize the denominator before differentiating. We have

$$y = \frac{\sqrt{x^2+1} - x}{\sqrt{x^2+1} + x} \cdot \frac{\sqrt{x^2+1} - x}{\sqrt{x^2+1} - x}$$

$$= \frac{x^2 + 1 - 2x\sqrt{x^2+1} + x^2}{x^2 + 1 - x^2} = 2x^2 + 1 - 2x\sqrt{x^2+1}.$$

Hence $\quad \dfrac{dy}{dx} = \dfrac{d}{dx}(2x^2+1) - 2x\dfrac{d}{dx}(x^2+1)^{\frac{1}{2}} - (x^2+1)^{\frac{1}{2}}\dfrac{d}{dx}(2x)$

$$= 4x - 2x\left(\frac{x}{(x^2+1)^{\frac{1}{2}}}\right) - 2(x^2+1)^{\frac{1}{2}}$$

$$= 4x - \frac{4x^2 + 2}{\sqrt{x^2+1}}.$$

EXERCISES

Prove the following by differentiation:

1. $\dfrac{d}{dx}(6x^3 - 7x^2 + 19) = 18x^2 - 14x.$

2. $\dfrac{d}{dx}(x^m - x^n) = mx^{m-1} - nx^{n-1}.$

3. $\dfrac{d}{dx}(x^k + kx) = k(x^{k-1} + 1).$

4. $\dfrac{d}{dx}(\tfrac{1}{3}x^3 - \tfrac{1}{7}x^7) = x^2 - x^6.$

5. $\dfrac{d}{dx}(ax^2 + bx + c) = 2ax + b.$

6. $\dfrac{d}{dx}(x^{-7}) = -7\,x^{-8} = -\dfrac{7}{x^8}$.

7. $\dfrac{d}{dx}\left(\dfrac{1}{x^6}\right) = -\dfrac{6}{x^7}$.

8. $\dfrac{d}{dx}(\tfrac{2}{3}x^{-3} - \tfrac{4}{7}x^{-2}) = -\dfrac{2}{x^4} + \dfrac{8}{7\,x^3}$.

9. $\dfrac{d}{dt}(t^{\tfrac{4}{7}} - 9\,t^{\tfrac{8}{7}}) = \tfrac{4}{7}t^{-\tfrac{3}{7}} - \tfrac{72}{7}t^{\tfrac{1}{7}}$.

10. $\dfrac{d}{dt}(\sqrt{t}) = \dfrac{d}{dt}(t^{\tfrac{1}{2}}) = \tfrac{1}{2}t^{-\tfrac{1}{2}} = \dfrac{1}{2\,t^{\tfrac{1}{2}}}$.

11. $\dfrac{d}{dt}(\sqrt[8]{t} - \tfrac{2}{3}\sqrt[4]{t}) = \dfrac{1}{3\,t^{\tfrac{2}{3}}} - \dfrac{1}{6\,t^{\tfrac{3}{4}}}$.

12. $\dfrac{d}{ds}(5\,s^{\tfrac{1}{5}} - s^5) = s^{-\tfrac{4}{5}} - 5\,s^4$.

Perform each of the following indicated differentiations:

13. $\dfrac{d}{dx}(4\,x^3 - 7\,x^2 + 92)$.

14. $\dfrac{d}{dx}(3\,x^{\tfrac{1}{3}} - 5\,x^{\tfrac{1}{5}})$.

15. $\dfrac{d}{dt}(\sqrt[5]{t^2} + \sqrt[7]{t^6})$.

16. $\dfrac{d}{ds}(s^{\tfrac{4}{5}} + \sqrt[5]{s^3})$.

17. $\dfrac{d}{d\theta}(a\theta^3 + b\theta^{-3})$.

18. $\dfrac{d}{d\theta}(\theta^6 + \theta^{-6})$.

19. $\dfrac{d}{d\rho}(\rho + \rho^{-1} + 2)$.

20. $\dfrac{d}{d\rho}(\rho^m + m\rho^n)$.

21. $\dfrac{d}{ds}(\tfrac{3}{2}s^{\tfrac{4}{3}} + \tfrac{2}{3}s^{\tfrac{3}{2}})$.

22. $\dfrac{d}{ds}\left(\tfrac{4}{5}s^{\tfrac{m}{n}} - \tfrac{5}{4}s^{\tfrac{n}{m}}\right)$.

(The student should practice the simpler differentiations by performing them mentally, writing only the results. After a little practice many of the foregoing exercises may be solved in this way. A considerable number of the following ones may also be solved mentally.)

Perform the following differentiations involving theorem VI:

23. $y = (x^3 + 3\,x^2 - 7)^3$. $\qquad y' = 9\,x(x+2)(x^3 + 3\,x^2 - 7)^2$.

24. $y = (a^2 - x^2)^{\tfrac{3}{2}}$. $\qquad y' = -3\,x(a^2 - x^2)^{\tfrac{1}{2}}$.

25. $y = (x^2 + 2\,x + 2)^{-\tfrac{1}{2}}$. $\qquad y' = -(x+1)(x^2 + 2\,x + 2)^{-\tfrac{3}{2}}$.

26. $y = (x + \sqrt{x^2 + a^2})^{\frac{1}{2}}$. $y' = \dfrac{(x + \sqrt{x^2 + a^2})^{\frac{1}{2}}}{2\sqrt{x^2 + a^2}}$.

27. $y = \sqrt[3]{x^3 + 7x + 1}$. $y' = \frac{1}{3}(3x^2 + 7)(x^3 + 7x + 1)^{-\frac{2}{3}}$.

Perform the following differentiations involving theorem V:

28. $y = (x^3 + 7)(x^2 + 1)$. $y' = 5x^4 + 3x^2 + 14x$.

29. $y = (a^2 - x^2)(a^2 + x^2)^{\frac{1}{2}}$. $y' = -\dfrac{x(3x^2 + a^2)}{\sqrt{a^2 + x^2}}$.

30. $y = (x + 1)^5 (x^2 + 1)^4$. $y' = (13 x^2 + 8 x + 5)(x + 1)^4 (x^2 + 1)^3$.

31. $y = x^3 (x + 7)^6 (x - 1)^2$. $y' = (11 x^2 + 26 x - 21) x^2 (x + 7)^5 (x - 1)$.

32. $y = (x^{\frac{1}{2}} - 2)\sqrt{x^{\frac{1}{2}} + 1}$. $y' = \dfrac{3}{4(x^{\frac{1}{2}} + 1)^{\frac{1}{2}}}$.

Perform the following differentiations involving theorems VIII and IX:

33. $\dfrac{d}{dx}\left(\dfrac{c-x}{c+x}\right) = -\dfrac{2c}{(c+x)^2}$.

34. $\dfrac{d}{dx}\left(\dfrac{x^3}{1+x^2}\right) = \dfrac{x^4 + 3x^2}{(1+x^2)^2}$.

35. $\dfrac{d}{dx}\left(\dfrac{2x^2 - 1}{x\sqrt{1+x^2}}\right) = \dfrac{1 + 4x^2}{x^2(1+x^2)^{\frac{3}{2}}}$.

36. $\dfrac{d}{dx}\left(\dfrac{x-a}{\sqrt{2ax - x^2}}\right) = \dfrac{a^2}{(2ax - x^2)^{\frac{3}{2}}}$.

37. $\dfrac{d}{dv}\left(\dfrac{c}{v-b} - \dfrac{a}{v^2}\right) = -\dfrac{c}{(v-b)^2} + \dfrac{2a}{v^3}$.

Perform the differentiations indicated in problems 38 to 52.

38. $y = \dfrac{x}{\sqrt{a^2 + x^2}}$. $y' = \dfrac{a^2}{\sqrt{(a^2 + x^2)^3}}$.

39. $y = \dfrac{x}{\sqrt{a^2 - x^2}}$. $y' = \dfrac{a^2}{\sqrt{(a^2 - x^2)^3}}$.

40. $y = \dfrac{1}{\sqrt{a^2 - x^2}}$. $y' = \dfrac{x}{\sqrt{(a^2 - x^2)^3}}$.

41. $y = \dfrac{a}{(a+bx)^2} - \dfrac{2}{a+bx}$. $y' = \dfrac{2b^2 x}{(a+bx)^3}$.

42. $y = \dfrac{1}{x}\sqrt{a^2 + x^2}$. $y' = -\dfrac{a^2}{x^2\sqrt{a^2 + x^2}}$.

43. $y = \dfrac{1}{x}\sqrt{a^2 - x^2}.$ $y' = -\dfrac{a^2}{x^2\sqrt{a^2 - x^2}}.$

44. $y = (x + \sqrt{x^2 - 1})^n.$ $y' = n(x + \sqrt{x^2 - 1})^n (x^2 - 1)^{-\frac{1}{2}}.$

45. $y = \sqrt{\dfrac{x-1}{x+1}}.$ $y' = \dfrac{1}{(x+1)\sqrt{x^2 - 1}}.$

46. $y = \dfrac{\sqrt{x^2 + 1} + \sqrt{x^2 - 1}}{\sqrt{x^2 + 1} - \sqrt{x^2 - 1}}.$ $y' = 2x + \dfrac{2x^3}{\sqrt{x^4 - 1}}.$

47. $y = \dfrac{x}{x + \sqrt{x^2 - 1}}.$ $y' = 2x - \dfrac{2x^2 - 1}{\sqrt{x^2 - 1}}.$

48. $y = \sqrt{2px}.$ $y' = \dfrac{p}{y}.$

49. $y = \dfrac{b}{a}\sqrt{a^2 - x^2}.$ $y' = -\dfrac{b^2 x}{a^2 y}.$

50. $y = (a^{\frac{2}{3}} - x^{\frac{2}{3}})^{\frac{3}{2}}.$ $y' = -\sqrt[3]{\dfrac{y}{x}}.$

51. $y = \sqrt{\dfrac{x^2 - x + 1}{x^2 + x + 1}}.$ $y' = \dfrac{x^2 - 1}{(x^2 + x + 1)\sqrt{x^4 + x^2 + 1}}.$

52. $y = \dfrac{(1 - x)\sqrt{1 - x^2}}{1 + x}.$ $y' = -\dfrac{(2 + x)\sqrt{1 - x^2}}{(1 + x)^2}.$

30. How to find antiderivatives. There is no direct method of finding antiderivatives. They are sought by means of the derivatives already known and the general theorems on differentiation. We shall later reduce this process to a system by the help of which the labor may be facilitated. In the meantime it is desirable that the student shall have some experience in attacking the problem without the aid of a systematized procedure. We shall illustrate the method by means of examples:

Example 1. Find the antiderivative of $x^3 + 7x^2 - 8x + 1$.

The separate terms of this expression have the following antiderivatives:
$$\tfrac{1}{4}x^4,\ \tfrac{7}{3}x^3,\ -4x^2,\ x.$$

Hence the required antiderivative is
$$\tfrac{1}{4}x^4 + \tfrac{7}{3}x^3 - 4x^2 + x + c,$$

where c is the arbitrary constant.

Example 2. Find the antiderivative of $(a^2 + x^2)^{-\frac{3}{2}}$.

By the help of problem 38 of the last section we see that the required antiderivative is
$$\frac{x}{a^2(a^2 + x^2)^{\frac{1}{2}}} + c.$$

Example 3. Find the antiderivative of x^k when $k \neq -1$.

In theorem VII of §28 take $n = k + 1$. Then we see that the required antiderivative is
$$\frac{x^{k+1}}{k + 1} + c.$$

EXERCISES

Find the antiderivatives of the following functions:

1. $(a^2 - x^2)^{-\frac{3}{2}}$.   *Ans.* $\dfrac{x}{a^2}(a^2 - x^2)^{-\frac{1}{2}} + c.$

2. $x(a^2 - x^2)^{-\frac{3}{2}}$.   *Ans.* $(a^2 - x^2)^{-\frac{1}{2}} + c.$

3. $x^{\frac{3}{2}} + ax^{\frac{7}{3}} + b$.   *Ans.* $\dfrac{2}{5}x^{\frac{5}{2}} + \dfrac{3a}{10}x^{\frac{10}{3}} + bx + c.$

4. $x(x^2 + 1)^3$.   *Ans.* $\tfrac{1}{8}(x^2 + 1)^4 + c.$

5. $(3x^2 + 7)(x^3 + 7x - 19)$.   *Ans.* $\tfrac{1}{2}(x^3 + 7x - 19)^2 + c.$

6. $\dfrac{1}{(x + 1)^2}$.   *Ans.* $-\dfrac{1}{x + 1} + c.$

7. Find the area bounded by the curve $y = x(1 - x^2)$ and the x-axis from 0 to 1. Sketch the curve.   *Ans.* $\tfrac{1}{4}$.

8. Show that the area bounded by the curve $y = 1/(x + 1)^2$, the y-axis, the x-axis, and the ordinate $x = k$ is
$$1 - \frac{1}{k + 1}.$$

Note that this approaches 1 as k approaches ∞. On this account the area bounded by the curve and the coördinate axes is said to be 1. Draw this graph and show this area stretching off to infinity along the x-axis.

9. Find the area bounded by the curves $y = x^2$ and $y^2 = x$.   *Ans.* $\tfrac{1}{3}$.

HINT. These curves intersect in the points (0, 0) and (1, 1). The area bounded by the curve $y = x^2$ and the x-axis and the line $x = 1$ may be shown to be $\tfrac{1}{3}$. The area bounded by the curve

$y = \sqrt{x} = x^{\frac{1}{2}}$ and the x-axis and the line $x = 1$ may be shown to be $\frac{2}{3}$. By constructing a figure the student will see that the required area is the difference between the two areas just mentioned.

10. Find the area bounded by the parabola $y^2 = 9x$ and the straight line $y = 3x$.   Ans. $\frac{1}{2}$.

11. Find the area bounded by the parabolas $y^2 = 9x$ and $y = 3x^2$.   Ans. 1.

12. Find the area bounded by the parabolas $y^2 = 2px$ and $x^2 = 2py$.   Ans. $\frac{4}{3} p^2$.

31. Derivative of function of a function. If y is a function of v, and v is a function of x, then y is indirectly a function of x. The following theorem, which will now be proved, will facilitate the labor of finding the derivative of such a function y:

If $y = f(v)$ and $v = \phi(x) \neq$ constant, then the derivative of y with respect to x is equal to the product of the derivative of y with respect to v and the derivative of v with respect to x.

Employing the Four-Step Rule, we proceed as follows:

$$y = f(v),$$
$$y + \Delta y = f(v + \Delta v),$$
$$\Delta y = f(v + \Delta v) - f(v),$$
$$\frac{\Delta y}{\Delta x} = \frac{f(v + \Delta v) - f(v)}{\Delta x}$$
$$= \frac{f(v + \Delta v) - f(v)}{\Delta v} \cdot \frac{\Delta v}{\Delta x}.$$

When Δx approaches zero so does Δv, since v must be continuous if it is to have a derivative. Hence if we employ theorem III of §10 we see that

$$\lim_{\Delta x = 0} \frac{\Delta y}{\Delta x} = \lim_{\Delta v = 0} \frac{f(v + \Delta v) - f(v)}{\Delta v} \cdot \lim_{\Delta x = 0} \frac{\Delta v}{\Delta x}.$$

Hence
$$\frac{dy}{dx} = \frac{df}{dv} \cdot \frac{dv}{dx} = \frac{dy}{dv} \cdot \frac{dv}{dx},$$

or
$$\frac{dy}{dx} = f'(v) \, \phi'(x).$$

This is the result which was to be proved.

Example. If $y = \dfrac{v^2+7}{v+1}$ and $v = x + \dfrac{1}{x}$, what is the derivative of y with respect to x?

We have $\dfrac{dy}{dv} = \dfrac{v^2+2v-7}{(v+1)^2} = 1 - \dfrac{8}{(v+1)^2} = 1 - \dfrac{8x^2}{(x^2+x+1)^2}$,

$\dfrac{dv}{dx} = 1 - \dfrac{1}{x^2} = \dfrac{x^2-1}{x^2}$.

Hence $\dfrac{dy}{dx} = \left[1 - \dfrac{8x^2}{(x^2+x+1)^2}\right]\dfrac{x^2-1}{x^2}$.

EXERCISES

1. Find dy/dx when $y = \sqrt{v}$ and $v = x^2 - 7$.   Ans. $\dfrac{x}{\sqrt{x^2-7}}$.

2. Find dy/dx when $y = \dfrac{v+1}{v-1}$ and $v = \sqrt{x^2+1}$.

     Ans. $\dfrac{-2x}{\sqrt{x^2+1}(\sqrt{x^2+1}-1)^2}$.

3. Find dy/dx when $y = \dfrac{v+1}{v}$ and $v = \sqrt{x+1}$. Ans. $-\dfrac{1}{2(x+1)^{\frac{3}{2}}}$.

32. Derivatives of inverse functions. If y is given as a function of x by means of the relation

$$y = f(x),$$

it may be possible to solve for x in terms of y and obtain a relation of the form

$$x = \phi(y),$$

defining x as a function of y. Thus, if $y = \sin x$, we have $x = \arcsin y$. In the relation $y = f(x)$ we think of x as the independent variable, whereas in the relation $x = \phi(y)$ we think of y as the independent variable. The functions $f(x)$ and $\phi(y)$, thus related, are said to be *inverse functions*. Sometimes it is convenient to speak of the first as the direct function and the second as the inverse function. When two functions are thus connected we have the following important relation between their derivatives:

The derivative of the inverse function and that of the direct function are reciprocals each of the other.

The proof is easy. When Δx and Δy are both different and nonzero, then as they both approach 0:
$$\frac{\Delta y}{\Delta x} = \frac{1}{\frac{\Delta x}{\Delta y}}.$$

When Δx approaches zero so does Δy, since y is a continuous function of x. Hence
$$\lim_{\Delta x \to 0} \frac{\Delta y}{\Delta x} = \lim_{\Delta y \to 0} \frac{1}{\frac{\Delta x}{\Delta y}}.$$

Therefore $\quad \dfrac{dy}{dx} = \dfrac{1}{\frac{dx}{dy}} \quad$ and $\quad \dfrac{dx}{dy} = \dfrac{1}{\frac{dy}{dx}},$

as was to be proved.

Example. Find dy/dx when $y^2 = 2\,px$.

We have $\quad x = \dfrac{y^2}{2p}, \quad \dfrac{dx}{dy} = \dfrac{y}{p}.$

Hence $\quad \dfrac{dy}{dx} = \dfrac{p}{y}.$

EXERCISES

Find the functions which are inverse to the following functions:

1. $y = x^5$.   *Ans.* $x = y^{\frac{1}{5}}$.
2. $y = \sqrt{a^2 - x^2}$.   *Ans.* $x = \sqrt{a^2 - y^2}$.
3. $y = \dfrac{ax + b}{cx + d}$.   *Ans.* $x = \dfrac{-dy + b}{cy - a}$.

4. Find the function which is inverse to y when
$$y = \log_e(x + \sqrt{x^2 + 1}). \quad \textit{Ans.} \ x = \tfrac{1}{2}(e^y - e^{-y}).$$

5. Find dy/dx when $y^5 = x - 2$.   *Ans.* $\dfrac{1}{5\,y^4}.$

6. In the case of problem 2 find dy/dx and dx/dy in terms of y, and by taking their product verify that each of these derivatives is the reciprocal of the other.

33. Derivatives of functions represented parametrically. If
$$y = f(t) \quad \text{and} \quad x = \phi(t),$$

and the latter relation has a solution for t in terms of x, then from the relation $y = f(t)$ we can express y as a function of x. In this case the functional relation between x and y is said to be represented parametrically.

Since the given parametric equations are sufficient to define y in terms of x it is clear that they may be taken as the *parametric equations* of a curve. In a particular case the curve may be plotted by points by assigning values to t and computing the corresponding values of x and y. The points (x, y) are then to be plotted and joined by a smooth curve.

It is sometimes convenient to obtain the derivative of y with respect to x without solving the parametric equations. Now we have

$$\frac{\Delta y}{\Delta x} = \frac{\frac{\Delta y}{\Delta t}}{\frac{\Delta x}{\Delta t}}.$$

Let Δt approach zero. Then Δy and Δx both approach zero when $f(t)$ and $\phi(t)$ are continuous. Hence we see that

$$\frac{dy}{dx} = \frac{\frac{dy}{dt}}{\frac{dx}{dt}}.$$

Example. Find dy/dx when $y = t^2 + t + 1$, $x = t^2 - t + 1$.

We have
$$\frac{dy}{dx} = \frac{\frac{d}{dt}(t^2 + t + 1)}{\frac{d}{dt}(t^2 - t + 1)} = \frac{2t + 1}{2t - 1}.$$

EXERCISES

1. When $y = t^2 + 1$ and $x = t^{\frac{1}{2}}$, show that $dy/dx = 4\, x^3$.
2. When $s = \frac{1}{2} gt^2$ and $v = gt$, show that $ds/dv = v/g$.
3. Work Exs. 1 and 2 by eliminating t before differentiating.

Find the derivative of y with respect to x in the following cases:

4. $y = t^3 + 1$, $x = t^2 + 1$.   Ans. $\frac{3}{2} t$.

5. $y = \sqrt{t^2+1}$, $x = \sqrt{t^2-1}$.   Ans. $\frac{x}{y}$.

6. $y = bt - \frac{1}{2} ct^2$, $x = at$.   Ans. $\frac{1}{a}(b - ct)$.

7. Plot the curves defined by the equations in Exs. 4–6. By eliminating t prove that the second curve is a hyperbola and that the third is a parabola.

34. Implicit and explicit functions. A function is called an *explicit function* when it is expressed directly in terms of the independent variable. Thus the relation

$$y = x^3 + 7 x^2 - 1$$

defines y as an explicit function of x. If a function is defined by means of a relation connecting it with the independent variable but not giving an explicit expression for it, it is said to be defined as an implicit function. The equation

$$xy - 7x + 8y - 1 = 0$$

defines y as an implicit function of x. If this equation is solved for y in the form

$$y = \frac{7x+1}{x+8},$$

we have y as an explicit function of x. But it is sometimes inconvenient to solve an implicit functional relation so as to exhibit the dependent variable explicitly, as the following example will show:

$$y^3 + 3 x^2 y^2 + 4 x^3 y - 7x + 8 = 0.$$

It is not convenient to solve this equation for y.

From such an implicit relation it is possible to find dy/dx readily in terms of x and y without solving the equation for y. The method is to differentiate the terms of the equation throughout, remembering that y is a function of x. Thus for the foregoing equation we have

$$\frac{d}{dx}(y^3) + 3 \frac{d}{dx}(x^2 y^2) + 4 \frac{d}{dx}(x^3 y) - 7 \frac{d}{dx}(x) = 0,$$

or $3y^2 \dfrac{dy}{dx} + 6xy^2 + 6x^2y \dfrac{dy}{dx} + 12 x^2 y + 4 x^3 \dfrac{dy}{dx} - 7 = 0.$

Solving for dy/dx, we have
$$\frac{dy}{dx} = -\frac{6xy^2 + 12 x^2 y - 7}{3 y^2 + 6 x^2 y + 4 x^3}.$$

EXERCISES

In each of the following cases find dy/dx in terms of x and y:

1. $x^2 + y^2 = a^2$. *Ans.* $-\dfrac{x}{y}.$

2. $x^3 + xy + y^3 = 25$. *Ans.* $-\dfrac{3x^2 + y}{3y^2 + x}.$

3. $x^3 + y^3 = 3\,axy$. *Ans.* $\dfrac{x^2 - ay}{ax - y^2}.$

4. What is the slope of the curve $x^3 + y^3 = 3\,axy$ at the point $\left(\dfrac{3\,a}{2}, \dfrac{3\,a}{2}\right)$? *Ans.* $-1.$

5. What is the slope of the circle $x^2 + y^2 = a^2$ at the point $(-\tfrac{1}{2} a\sqrt{2}, \tfrac{1}{2} a\sqrt{2})$? *Ans.* $1.$

35. Derivatives of higher order. The derivative of y with respect to x is itself a function of x. This derivative itself may be differentiated, and this in turn may be differentiated, and so on, the *successive derivatives* of y with respect to x being thus obtained. What we have heretofore called the derivative of y with respect to x may be called the first derivative of y with respect to x. Its derivative is called the second derivative of y with respect to x; it is denoted by the symbol

$$\frac{d^2 y}{dx^2}$$

(read *the second derivative of y with respect to x* or *d second y over dx square*). The derivative of the second derivative is called the third derivative and is denoted by the symbol

$$\frac{d^3 y}{dx^3}.$$

This is continued to derivatives of higher order. Therefore:

$$\frac{d^n y}{dx^n}$$

is called the *n*th derivative of y with respect to x. We sometimes say that it is the derivative of order n.

Other symbols for derivatives of higher orders are

$$y'',\ y''',\ y^{iv},\ \cdots,\quad D_x^2 y,\ D_x^3 y,\ \cdots,\quad f''(x),\ f'''(x),\ \cdots.$$

EXERCISES

Find the higher derivatives of y with respect to x in the following cases:

1. $y = x^3 + 7\, x^2 - 8$.
2. $y = x^4$.
3. $y = x^5$.
4. $y = x^n$, n being a positive integer.
5. Find the *n*th derivative of $y = \dfrac{1}{1-x}$.   *Ans.* $\dfrac{n!}{(1-x)^{n+1}}$.
6. Find the *n*th derivative of $y = \dfrac{1}{x}$.   *Ans.* $\dfrac{(-1)^n \cdot n!}{x^{n+1}}$.

36. Historical note on the origin of the calculus. The infinitesimal calculus consists of the rules, processes, and principles by means of which one may deal mathematically with continuously varying magnitudes. In the early days infinitesimals were vaguely conceived as being neither zero nor finite but in some intermediate state — a confusion which was wholly unnecessary and one which has been removed in more recent times. The founders did not develop the calculus in accordance with strictly logical principles from precisely defined notions. To begin with, it gained its place through the great variety and importance of its applications rather than through any cogency in the development of its principles. But it has now been brought into a secure form and rendered accessible to all those who have need for its powerful processes.

When the student has learned the meaning of integration (in Chapter IX), so that he will have before him the main processes of the calculus, he will find it of interest to consider some of the more important points in the history and development of this science. These are set forth in an interesting way in the first two divisions of the article "Infinitesimal Calculus" in the eleventh edition of the Encyclopædia Britannica. From this article he may see how, even among the ancient Greeks, certain methods were developed which foreshadow those of the calculus. But he will see that the men of that earlier day did not sufficiently realize their importance and the possibility of developing them, so that the matter was not then followed up. Beginning with a work of Kepler's in 1609, more definite progress toward the invention of the calculus was made. Important developments were introduced by Cavalieri and Fermat and Barrow. But it remained for Newton (1642–1727) and Leibniz (1646–1716) to bring the calculus explicitly into existence. A great controversy arose as to priority on the part of these two investigators.

The principal problems lying at the origin of this science were those of determining tangents and of finding areas. We have already seen, in a preliminary way, how our methods come into relation with these two problems. Once they were solved, the results and methods were found to be of great and extended use in widely separated parts of pure and applied mathematics. The creation of the calculus has been and remains one of the chief conquests of the human mind.

37. Summary. In this chapter we have defined increment, derivative, derivatives of higher orders, and antiderivatives. The most important process learned in the chapter is that of differentiation, as described in the Four-Step Rule and extended by means of theorems and formulas. This rule is employed in proving the nine fundamental general theorems on differentiation given in § 28. It is also used to show that the derivative gives the slope of a curve. Again, it is employed in finding the derivative of an area. By aid of the derivative of an area and the notion of antiderivative a means is developed for finding areas. Finally, several additional principles are given relating respectively to a function of a function, inverse functions, functions defined parametrically, and functions defined implicitly.

CHAPTER III

APPLICATIONS OF DIFFERENTIATION

38. Velocity in rectilinear motion.* Let a point P move in the straight line AB. Let s be the distance of the moving point P from the fixed point A, and let t be the time required for the point to traverse the distance s from A. (The quantities s and t may be measured in any convenient units; as foot and second, for instance.) Then s is a function of t and we may write

$$s = f(t).$$

Let t receive the increment Δt; then s receives a corresponding increment Δs. Then

$$\frac{\Delta s}{\Delta t} = \textit{the average velocity}$$

of the moving point during the interval Δt.

If this ratio has the same value for every interval of time, the motion is said to be *uniform*. In this case the average velocity is also the velocity at any instant.

In the general case we define the velocity v at any instant as the limit of $\Delta s/\Delta t$ as Δt approaches zero; thus

$$v = \lim_{\Delta t \to 0} \frac{\Delta s}{\Delta t}.$$

(The naturalness of this definition is brought out by means of the example given in § 7.)

For motion in a straight line the terms *velocity* and *speed* are often employed interchangeably. But, strictly speaking, *speed*

---

* Before taking up this section the student will do well to review the preliminary discussion of the motion of a ball given in § 7.

is used to denote the rate of change of distance with respect to the time, whereas *velocity* denotes this rate together with the direction of the motion. If we make this distinction, then what we have called velocity in the foregoing discussion would be called *speed*. When we are dealing with curvilinear motion we shall make this distinction. When we are speaking of rectilinear motion we shall employ the two terms interchangeably.

EXERCISES

1. If a body is thrown vertically upward with an initial velocity of 322 ft. per second, its height s at the end of t seconds is given by the formula
$$s = 322\, t - 16.1\, t^2.$$

(1) Find the expression for the speed at the end of t seconds.
(2) What is the speed at the end of 2 sec.? 4 sec.?
(3) For how long does the body continue to rise?   *Ans.* 10 sec.
(4) What is the greatest height attained by the body?
     Ans. 1610 ft.
(5) At what time is the body again at the starting-point?
     Ans. At the end of 20 sec.

2. A body was rolled down an inclined plane. Its distance s (in feet) from the top varied with the time in accordance with the law $s = 8\, t^2$. Find the speed at the end of t seconds. What is the speed at the end of 3 sec.?

3. A body slides in a straight line along an inclined plane in accordance with the law
$$s = 8\, t^2 - 32\, t,$$
where s denotes the distance from a fixed point and t denotes the time. Discuss the motion of this body after the manner suggested by the questions in problem 1.

4. Suppose that a train moves along a straight track according to the law
$$s = \tfrac{1}{4} t^4 - \tfrac{14}{3} t^3 + 24\, t^2,$$
where s denotes distance in miles and t is measured in hours.
(1) Find its velocity.   *Ans.* $t^3 - 14\, t^2 + 48\, t$.
(2) At what time does the train reverse its direction?
     Ans. When $t = 6$ and $t = 8$.
(3) Describe the motion during the first twelve hours, telling when the train is going forward and when it is going backward.

39. Speed in curvilinear motion. Let us now consider the motion of a point P on the curved line AB. We let s denote the distance of a moving point P from a fixed point A, this distance being *measured along the curve*.

If t denotes the time required for the point P to move along the curve the distance s, then (as before) we have s a function of t and we may write

$$s = f(t).$$

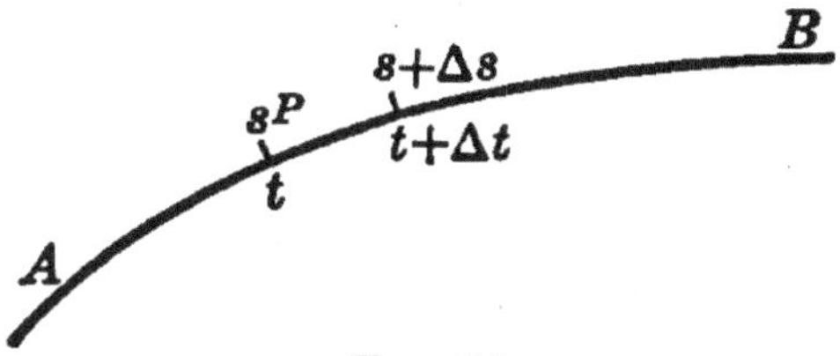

FIG. 21

The *instantaneous speed* v at time t is again defined by means of the relation

$$v = \lim_{\Delta t = 0} \frac{\Delta s}{\Delta t}.$$

Hence we have

$$v = \frac{ds}{dt} = f'(t).$$

The speed is the derivative of the distance (along the curve) with respect to the time.

40. Component velocities. Let us suppose that the curve along which P moves lies in a plane; and let us choose coördinates x and y in this plane. As P moves along AB in accordance with the law $s = f(t)$ the coördinates x and y are changing as functions of t according to laws which may be written in the form

$$x = \phi(t), \quad y = \psi(t).$$

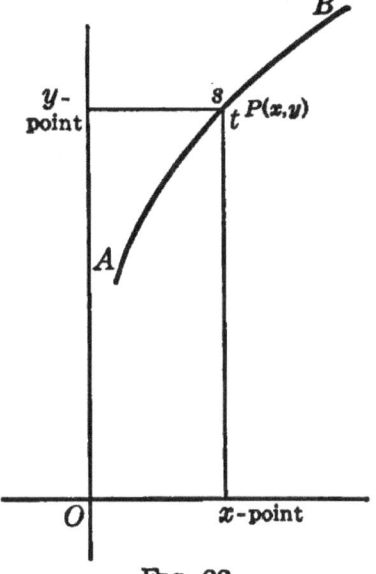

FIG. 22

(These equations may be interpreted as parametric equations of the path.) We may think of a point (the x-point) moving along the x-axis in accordance with the law $x = \phi(t)$, and of a point (the y-point) moving along the y-axis in accordance with the law $y = \psi(t)$.

The velocity of the x-point is called the *x-component* of the velocity of the point P; it is denoted by v_x. The velocity of the

y-point is called the *y-component* of the velocity of the point P; it is denoted by v_y. Since the x-point and the y-point each moves in a straight line we may apply the definition of velocity given in § 38 to show that
$$v_x = \frac{dx}{dt}, \quad v_y = \frac{dy}{dt}.$$

A *vector* is a straight-line segment of definite *length* and *direction*. Vectors are convenient in representing velocities (and also accelerations, forces, etc., as we shall see later). A velocity **v** is denoted by a vector whose length is equal to the speed v and whose direction is that of the given motion.

Let the vector **v**, or CD of length v, denote the velocity of the point P treated in the earlier part of the section. It is supposed that this velocity is considered at a particular instant. If at that instant the motion had become uniform, the point would have

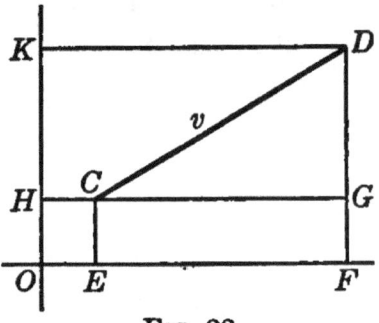

Fig. 23

moved over the line CD in a unit of time. Hence the x-point would have moved over the line EF, so that $v_x = EF = CG$. The y-point would have moved over the line HK, so that $v_y = HK = GD$. But $CD^2 = CG^2 + GD^2$; hence
$$v^2 = v_x^2 + v_y^2,$$
or
$$v = \frac{ds}{dt} = \sqrt{\left(\frac{dx}{dt}\right)^2 + \left(\frac{dy}{dt}\right)^2}.$$

This treatment of component velocities can readily be extended to three dimensions. In this case we have rectangular coördinates x and y and z and three corresponding components of velocity v_x and v_y and v_z. Using the methods employed for the case of the plane, we have
$$v_x = \frac{dx}{dt}, \quad v_y = \frac{dy}{dt}, \quad v_z = \frac{dz}{dt},$$
$$v^2 = v_x^2 + v_y^2 + v_z^2,$$
$$v = \frac{ds}{dt} = \sqrt{\left(\frac{dx}{dt}\right)^2 + \left(\frac{dy}{dt}\right)^2 + \left(\frac{dz}{dt}\right)^2}.$$

41. Acceleration. The speed v is in general a function of the time t. If we give to t the increment Δt, then v will take a corresponding increment Δv. Then

$$\frac{\Delta v}{\Delta t} = \text{the average tangential acceleration}$$

of the moving point during the interval Δt. We define the tangential acceleration a at any instant as the limit of the ratio $\Delta v/\Delta t$ as Δt approaches zero. Thus

$$a = \frac{dv}{dt}.$$

The tangential acceleration is the derivative of the speed with respect to the time.

For motion in a straight line we define the acceleration to be the derivative of the speed with respect to the time.

Since $$v = \frac{ds}{dt} \quad \text{and} \quad a = \frac{dv}{dt},$$

it follows that $$a = \frac{d^2 s}{dt^2};$$

that is, *the tangential acceleration is the second derivative of the distance with respect to the time.* We have obviously the same formula for acceleration in the case of motion in a straight line.

EXERCISES

1. Find the acceleration of the moving body in each of the problems of § 38.   *Ans.* -32.2 ft. per (sec.)^2, 16, 16, $3t^2 - 28t + 48$.

2. Let a projectile P be thrown into the air with an initial velocity v_0 and in a direction making an angle ϕ with the horizontal. If the time t is measured in seconds from the beginning of the flight, and the coördinates x and y (as in the figure) are measured in feet, and if the resistance of the air is neglected, then it is shown in physics that x and y are expressed in terms of t by the equations

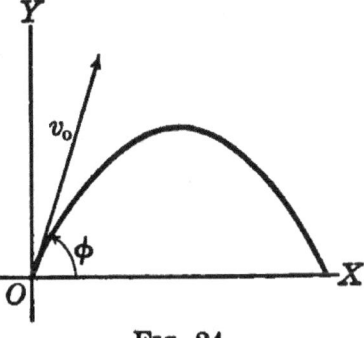

Fig. 24

$$x = v_0 \cos \phi \cdot t, \quad y = v_0 \sin \phi \cdot t - 16.1 \, t^2.$$

Find the velocity, tangential acceleration, component velocities, and component accelerations (1) at time $t = t$, (2) at time $t = 2$, when $v_0 = 200$ and $\phi = 45°$. (3) Find the equation of the path by eliminating t. (4) What is the slope of the path at the point where $t = 2$, when $v_0 = 200$ and $\phi = 45°$?

Ans. (1) $v_x = v_0 \cos \phi$, $v_y = v_0 \sin \phi - 32.2\, t$;

(3) $y = x \tan \phi - \dfrac{16.1\, x^2}{v_0^2 \cos^2 \phi}$.

3. If the projectile of problem 2 is thrown vertically upward, then $\phi = 90°$ and x remains zero, while y is given by the equation

$$y = v_0 t - 16.1\, t^2.$$

If a cannon ball is fired vertically upward with a muzzle velocity of 966 ft. per second, what is its velocity at the end of 20 sec.? How long will it continue to rise? What is the greatest height reached by it?   Ans. 322 ft. per second; 30 sec.; 14,490 ft.

4. If a body moves in a straight line under an attracting force varying as the inverse square of the distance, then $v^2 = k/s$. Show that the acceleration is

$$-\dfrac{k}{2\, s^2}.$$

42. Rotation. Let the point P move on the circle with center O, starting from the initial position X. Let t be the time required to pass from X to P and let θ (measured in radians) be the angle swept out by the revolving radius which connects P to the center O. If $\Delta \theta$ is the increment of angle corresponding to a given increment Δt of time, then the average angular speed for this interval of time is $\Delta \theta / \Delta t$. The instantaneous angular speed ω is the limit of $\Delta \theta / \Delta t$ as Δt approaches zero. Hence

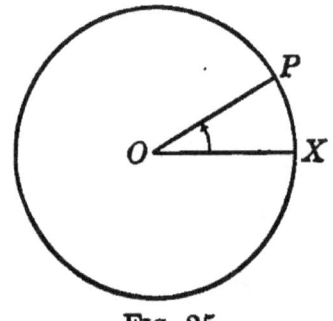

Fig. 25

$$\omega = \dfrac{d\theta}{dt}.$$

Thus *the angular speed is the derivative of the angle swept over with respect to the time.*

In a similar way we define the *angular acceleration* α as the derivative of the angular speed with respect to the time. Thus we have
$$\alpha = \frac{d\omega}{dt} = \frac{d^2\theta}{dt^2}.$$

EXERCISES

1. Find ω and α when

(1) $\theta = at^2 + bt + c$; (2) $\theta = at^{\frac{5}{2}}$; (3) $\theta = t^3 - 6t$.

Ans. (1) $\omega = 2at + b$, $\alpha = 2a$, etc.

2. If $\theta = t^3 - 5t^2$, what is the angular speed at the end of 3 sec.? What is the angular acceleration? At what time is the direction of rotation reversed? *Ans.* $-3, 8$, when $t = 3\frac{1}{3}$.

3. Discuss the angular motion when
$$\theta = \tfrac{1}{4} t^4 - 4 t^3 + 16 t^2,$$
giving angular speed and acceleration, direction of rotation, and times of change of direction of rotation.

43. Equations of tangent and normal to a curve. We have already given two important geometric applications of differentiation; namely, those connected with finding slopes and areas. In this section and the next we give two other geometric applications.

Let us consider the curve
$$y = f(x)$$
and the point (x_1, y_1) on this curve. From analytic geometry it is known that the line of slope m through the point (x_1, y_1) has the equation
$$y - y_1 = m(x - x_1).$$
But the slope m of the tangent at this point has the value $m = f'(x_1)$, the value of the derivative at this point, as we saw in § 24. Hence

The equation of the tangent to the curve $y = f(x)$ at the point (x_1, y_1) is
$$y - y_1 = f'(x_1)(x - x_1).$$

The normal to a curve is defined to be the line which is perpendicular to the tangent at the point of tangency. The condition that two lines shall be perpendicular is that the slope of each shall be the negative reciprocal of the slope of the other. But the slope of the tangent at the point (x_1, y_1) is $f'(x_1)$. Hence the slope of the normal is $-1/f'(x_1)$. Therefore

The equation of the normal to the curve $y = f(x)$ at the point (x_1, y_1) is

$$y - y_1 = -\frac{1}{f'(x_1)}(x - x_1).$$

Example. Find the equations of the tangent and normal to the circle $x^2 + y^2 = a^2$ at the point (x_1, y_1); to the circle $x^2 + y^2 = 25$ at the point $(3, 4)$.

We have $\quad \dfrac{dy}{dx} = -\dfrac{x}{y} = -\dfrac{x}{\sqrt{a^2 - x^2}} = f'(x).$

Hence the equation of the tangent is

$$y - y_1 = -\frac{x_1}{\sqrt{a^2 - x_1^2}}(x - x_1) = -\frac{x_1}{y_1}(x - x_1),$$

or $\quad\quad\quad yy_1 - y_1^2 = -xx_1 + x_1^2,$

or $\quad\quad\quad xx_1 + yy_1 = x_1^2 + y_1^2,$

or $\quad\quad\quad xx_1 + yy_1 = a^2.$

Similarly, the equation of the normal is

$$y - y_1 = \frac{y_1}{x_1}(x - x_1), \quad \text{or} \quad x_1 y - x y_1 = 0.$$

For the particular case the equations of the tangent and normal are, respectively,

$$3x + 4y = 25, \quad 3y - 4x = 0.$$

44. Lengths of tangent, normal, subtangent, subnormal. Certain *line segments* associated with a curve and its tangent and normal are occasionally met with. They may be defined by the aid of the adjacent figure as follows. Let PT be tangent to the curve $y = f(x)$ at the point $P(x_1, y_1)$, and let PN be the corresponding normal. Then PT is called *the tangent* and PN *the* normal, these words being thus used in a partic-

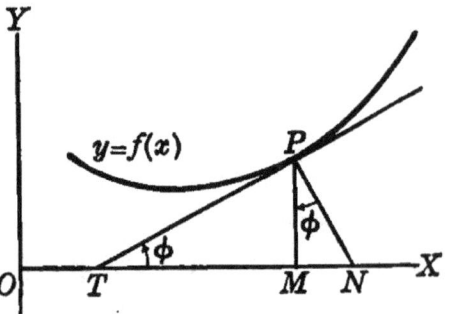

Fig. 26

ular sense different from that employed in the previous section. Also, TM and MN, respectively, are called *the subtangent* and *the subnormal*.

It is easily seen from geometry that the two angles marked ϕ are equal, for each of them is the complement of the angle TPM; moreover, we have seen that $\tan \phi = f'(x_1)$. The line PM is of length y_1. It is therefore possible, by means of trigonometry, to compute the lengths of the tangent, the normal, the subtangent and the subnormal. Thus,

$$\frac{TM}{PM} = \cot \phi = \frac{1}{\tan \phi}.$$

Hence, $$\text{subtangent} = \frac{y_1}{f'(x_1)}.$$

In a similar way we show that

$$\text{subnormal} = y_1 f'(x_1).$$

Example. Find the length of the subtangent to the curve $y = x^2$ at the point (x_1, y_1); at the point $(2, 4)$.

We have $$\text{subtangent} = \frac{y_1}{f'(x_1)} = \frac{x_1^2}{2\,x_1} = \frac{1}{2} x_1.$$

At the point $(2, 4)$ this has the value 1.

EXERCISES

Find the equations of the tangents and normals to the following curves at the points indicated:

1. $y^2 = 2\,px$, at (x_1, y_1).   Ans. $yy_1 = p(x+x_1), y - y_1 = -\frac{y_1}{p}(x - x_1)$.

2. $\frac{x^2}{a^2} + \frac{y^2}{b^2} = 1$, at (x_1, y_1).   Ans. $\frac{xx_1}{a^2} + \frac{yy_1}{b^2} = 1, y - y_1 = \frac{a^2 y_1}{b^2 x_1}(x - x_1)$.

3. $y^2 = \frac{x^3}{2\,a - x}$, at (a, a).   Ans. $y = 2\,x - a, 2\,y + x = 3\,a$.

4. $y = \frac{8\,a^3}{4\,a^2 + x^2}$, at $(2a, a)$.   Ans. $x + 2\,y = 4\,a, y = 2\,x - 3\,a$.

5. $y = x^3 - 3\,x + 10$, at $(2, 12)$.   Ans. $9\,x - y = 6, x + 9\,y = 110$.

Find the lengths of the tangents, the normals, the subtangents, and the subnormals of the following curves at the points indicated:

6. $y = x^3$, at $(1, 1)$. Ans. $\frac{1}{3}\sqrt{10}$, $\sqrt{10}$, $\frac{1}{3}$, 3.

7. $y^2 = 4x$, at $(1, 2)$. Ans. $2\sqrt{2}$, $2\sqrt{2}$, 2, 2.

8. Show that the subnormal to the parabola $y^2 = 2px$ is constant and equal to p.

9. Show that the subtangent to the parabola $y^2 = 2px$ at any point is bisected at the vertex.

10. Show that the equation of the tangent to the hyperbola $xy = c$ at (x_1, y_1) is $x_1 y + x y_1 = 2c$.

11. Show that the length of the normal to the circle $x^2 + y^2 = a^2$ is constant.

12. Show that the equation of the tangent to the curve $x^2(x - y) = a^2(x + y)$ at the origin is $x + y = 0$.

45. Solutions of equations having multiple roots. A root occurring more than once in an equation is called a *multiple* root. Thus the equation $x^3 + x^2 - x - 1 = 0$, which may be written in the form $(x + 1)^2(x - 1) = 0$, has the roots $-1, -1, 1$; -1 occurs as a multiple root.

Let $f(x)$ be a polynomial such that the equation $f(x) = 0$ has a as a multiple root; and let us suppose that a occurs m times as a root so that $f(x)$ may be written in the form

$$f(x) = (x - a)^m \phi(x),$$

where $\phi(x)$ is also a polynomial and is such as not to be divisible by $x - a$. Differentiating both sides of this equation we have

$$f'(x) = m(x - a)^{m-1}\phi(x) + (x - a)^m \phi'(x)$$
$$= (x - a)^{m-1}[m\phi(x) + (x - a)\phi'(x)].$$

From this equation it follows that $f'(x)$ is divisible by $(x - a)^{m-1}$ but by no higher power of $x - a$. Hence

If the polynomial $f(x)$ is divisible by the mth power of $x - a$ and by no higher power, then its derivative $f'(x)$ is divisible by the $(m - 1)$th power of $x - a$ and by no higher power.

Therefore if the equation $f(x) = 0$ has a multiple root occurring m times, then $f'(x) = 0$ has the same root occurring $m - 1$ times.

This gives rise to the following rule for finding the multiple roots of the equation $f(x) = 0$.

Find $f'(x)$. Find the highest common factor of $f(x)$ and $f'(x)$. Find the roots obtained by setting this highest common factor equal to zero. Each of these roots is a multiple root of the equation $f(x) = 0$, occurring once more in this equation than in that obtained from the highest common factor.

Example. Solve the equation $x^3 - 12x + 16 = 0$.

We have $f(x) = x^3 - 12x + 16$, $f'(x) = 3x^2 - 12$. The highest common factor of these is $x - 2$. Hence 2 occurs twice as a root of the given equation. The third root is -4.

EXERCISES

Solve the following equations by the methods of this section:

1. $x^3 - 3x + 2 = 0$.   *Ans.* $1, 1, -2$.
2. $x^4 - 4x^3 + 6x^2 - 8x + 8 = 0$.   *Ans.* $2, 2, \pm \sqrt{-2}$.
3. $x^4 + x^3 - 3x^2 - 5x - 2 = 0$.   *Ans.* $-1, -1, -1, 2$.
4. $x^5 - 15x^3 + 10x^2 + 60x - 72 = 0$.   *Ans.* $2, 2, 2, -3, -3$.

Show that none of the following equations has a multiple root:

5. $x^2 + 4x + 7 = 0$.
6. $x^3 - 6x^2 + 12x - 6 = 0$.
7. $x^n - c = 0$, $c \neq 0$.

8. Show that the equation $x^3 + 3px + q = 0$ has a double root when and only when $4p^3 + q^2 = 0$.

46. Summary. In this chapter we have treated the applications of differentiation to problems dealing with the following subjects: velocity, speed, acceleration, components of speed and acceleration, rotation, equations of tangents and normals, lengths of tangents and normals and subtangents and subnormals, and multiple roots of equations. In the previous chapter we had applications to problems involving slopes and areas.

CHAPTER IV

THE DIFFERENTIAL NOTATION. RATES. FURTHER APPLICATIONS

47. Differentials. In representing the derivative of $y = f(x)$ by the notation
$$\frac{dy}{dx} = f'(x)$$
we emphasized the fact that dy/dx was not defined as a fraction but as a single symbol; but we stated that we would later define the terms dy and dx so that we would be able to treat dy/dx as a fraction. We do this by defining dy and dx separately as differentials.

The differential dx of the independent variable x is any assigned increment of x. The differential dy of the dependent variable y in the relation $y = f(x)$ is the product of the derivative $f'(x)$ of the function by the differential dx of the independent variable, or $dy = f'(x)dx$.

It is in view of this last relation that the derivative $f'(x)$ is sometimes called the differential coefficient; it is the "coefficient" of the differential dx.

In the adjacent graph of the curve $y = f(x)$, where PT is the tangent to the curve at the point $P(x, y)$, we have

$$\frac{dy}{dx} = f'(x) = \tan \phi = \frac{AT}{PA}.$$

Hence if we take PA to be dx we have $AT = dy$. Therefore

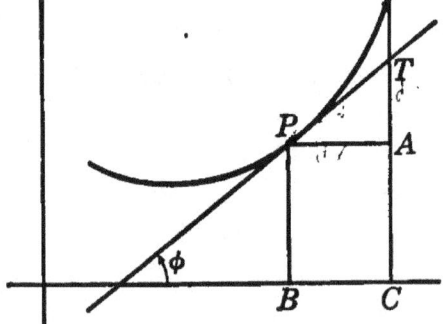

FIG. 27

If dx is any increment of the independent variable x at a point $P(x, y)$ on the curve $y = f(x)$, then in the derivative $dy/dx = f'(x) = \tan \phi$, dy denotes the corresponding increment of y taken, <u>not to the curve</u>, but to the tangent through P.

By means of the definition of differentials we are now able to treat dy/dx as a fraction. We shall often find this convenient. Some of the formulas already developed become, with this interpretation, notationally equivalent to mere algebraic identities. Thus the equations developed in § 32, namely,

$$\frac{dy}{dx} = \frac{1}{\frac{dx}{dy}}, \quad \frac{dx}{dy} = \frac{1}{\frac{dy}{dx}},$$

are in form merely the relations between a fraction and its reciprocal. In § 33 we saw that

$$\frac{dy}{dx} = \frac{\frac{dy}{dt}}{\frac{dx}{dt}},$$

and this is an algebraic identity when interpreted in terms of differentials. The same is true of the relation

$$\frac{dy}{dx} = \frac{dy}{dv} \cdot \frac{dv}{dx}$$

developed in § 31. These observations will aid one in remembering these formulas.

48. Arc length. We have already spoken several times of the length of an arc on the basis of our usual intuitive conception of length. It is necessary to render the notion of length more precise. We begin by assuming a knowledge of the meaning of length along a straight line. What, then, shall we mean by length along a curve?

To obtain the approximate length of an arc AB we begin by separating AB into a large number of shorter arcs.

FIG. 28

Then we draw the chord for each of these arcs, as in the figure. The sum of the lengths of these chords, one will agree, is approximately the length of the whole arc.

Now suppose that we increase the number of smaller arcs indefinitely so that the points of division come indefinitely near together. In each case we take the sum of the lengths of the corresponding chords. As the points of division come indefinitely near together it may happen that this sum of chords approaches a limit. In case it does so, this limit is defined to be the *length* of the arc AB. If it should happen that the limit does not exist we should say that the notion of length is not defined for the curve in consideration. For all the arcs of curves which we shall meet in this book the corresponding limit exists and the notion of length is defined by means of it in accordance with the definition just given.

The student will observe that this definition of length rests on the intuitive conception (see figure in next section) that, as Q approaches P,

$$\lim \left(\frac{\text{chord } PQ}{\text{arc } PQ} \right) = 1.$$

We shall now use this intuitive principle in finding the differential of arc length. In a later chapter we shall show how to find lengths by means of the formula for the differential of arc length.

49. Differential of arc length in rectangular coördinates. From the adjacent figure we have

$$(\text{chord } PQ)^2 = (\Delta x)^2 + (\Delta y)^2,$$

or

$$\frac{\text{chord } PQ}{\Delta x} = \sqrt{1 + \left(\frac{\Delta y}{\Delta x} \right)^2}.$$

Hence

$$\frac{\text{chord } PQ}{\Delta s} \cdot \frac{\Delta s}{\Delta x} = \sqrt{1 + \left(\frac{\Delta y}{\Delta x} \right)^2}.$$

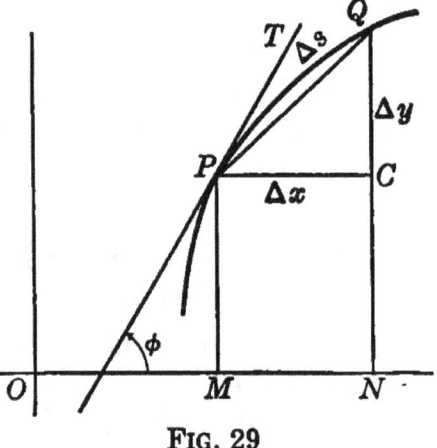

FIG. 29

Taking the limit as Δx approaches zero, and remembering that $(\text{chord } PQ)/\Delta s$ then approaches the limit 1, we see that

$$\frac{ds}{dx} = \sqrt{1 + \left(\frac{dy}{dx} \right)^2}.$$

Treating this as an equation in differentials and multiplying through by dx, we have

$$ds = \sqrt{(dx)^2 + (dy)^2},$$

or

$$ds^2 = dx^2 + dy^2.$$

Either of the last two relations gives us an expression for the differential ds of arc length in terms of dx and dy.

From the last relation we also have

$$\frac{ds}{dy} = \sqrt{\left(\frac{dx}{dy}\right)^2 + 1}.$$

From the formulas for ds/dx and ds/dy we have

$$ds = \sqrt{1 + \left(\frac{dy}{dx}\right)^2}\, dx, \quad ds = \sqrt{\left(\frac{dx}{dy}\right)^2 + 1}\, dy.$$

We have seen that $\quad \tan \phi = \dfrac{dy}{dx},$

where ϕ is the angle of inclination of the tangent at the point P. From the adjacent figure we see then that

$$\sin \phi = \frac{dy}{\sqrt{dx^2 + dy^2}},$$

$$\cos \phi = \frac{dx}{\sqrt{dx^2 + dy^2}}.$$

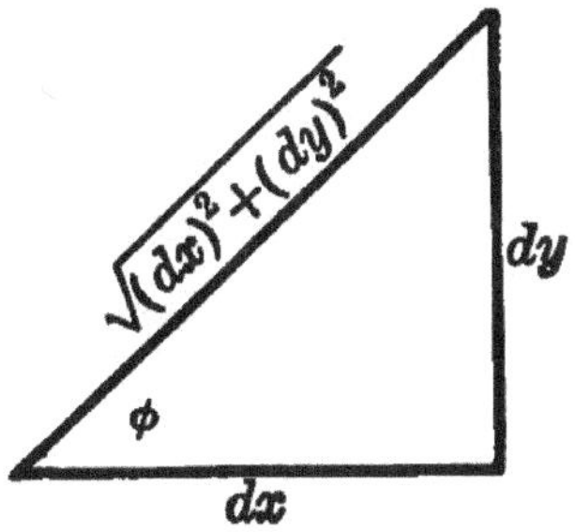

But $\quad ds = (dx^2 + dy^2)^{\frac{1}{2}}.$

Hence $\quad \sin \phi = \dfrac{dy}{ds}, \quad \cos \phi = \dfrac{dx}{ds}.$

Fig. 30

50. Differential of arc length in polar coördinates. From the equations $\quad x = \rho \cos \theta \quad \text{and} \quad y = \rho \sin \theta,$

giving the values of x and y in the polar coördinates ρ and θ, we have

$$\frac{dx}{d\theta} = \frac{d\rho}{d\theta} \cos \theta - \rho \sin \theta,$$

$$\frac{dy}{d\theta} = \frac{d\rho}{d\theta} \sin \theta + \rho \cos \theta.$$

Hence $\quad \left(\dfrac{dx}{d\theta}\right)^2 + \left(\dfrac{dy}{d\theta}\right)^2 = \left(\dfrac{d\rho}{d\theta}\right)^2 + \rho^2.$

Therefore $\quad dx^2 + dy^2 = d\rho^2 + \rho^2 d\theta^2.$

But (§ 49) $\quad ds^2 = dx^2 + dy^2.$

Hence $\quad ds^2 = \rho^2 d\theta^2 + d\rho^2.$

This gives the differential of arc length in polar coördinates. From it we have
$$ds = \sqrt{\rho^2 + \left(\frac{d\rho}{d\theta}\right)^2}\, d\theta,$$
$$ds = \sqrt{1 + \rho^2\left(\frac{d\theta}{d\rho}\right)^2}\, d\rho.$$

Example 1. What is the differential of arc length in the circle $x^2 + y^2 = a^2$?

We have $dy/dx = -x/y$. Hence
$$ds = \sqrt{1 + \left(\frac{dy}{dx}\right)^2}\, dx = \sqrt{1 + \frac{x^2}{y^2}}\, dx$$
$$= \sqrt{\frac{x^2 + y^2}{y^2}}\, dx = \frac{a}{y}\, dx = \frac{a\, dx}{\sqrt{a^2 - x^2}}.$$

In a similar way we find that ds, in terms of y, has the value
$$ds = \frac{a\, dy}{\sqrt{a^2 - y^2}}.$$

Example 2. What is the differential of arc length in the case of the curve $\rho = a\theta$?

We have $d\rho/d\theta = a$. Hence
$$ds = \sqrt{\rho^2 + \left(\frac{d\rho}{d\theta}\right)^2}\, d\theta = \sqrt{a^2\theta^2 + a^2}\, d\theta = a\sqrt{1 + \theta^2}\, d\theta,$$
$$ds = \sqrt{1 + \rho^2\left(\frac{d\theta}{d\rho}\right)^2}\, d\rho = \sqrt{1 + \theta^2}\, d\rho.$$

EXERCISES

Find the differential of arc length in the case of each of the following curves:

1. $y^2 = 2\, px$.   Ans. $ds = \sqrt{\dfrac{p + 2x}{2x}}\, dx$.

2. $x^2 = 2\, py$.   Ans. $ds = \dfrac{\sqrt{x^2 + p^2}}{p}\, dx$.

3. $y = x^3$.   Ans. $ds = \sqrt{1 + 9x^4}\, dx$.

4. $y^3 = x^2$.   Ans. $ds = \frac{1}{2}\sqrt{4 + 9y}\, dy$.

5. $\dfrac{x^2}{a^2} + \dfrac{y^2}{b^2} = 1$.   Ans. $ds = \sqrt{\dfrac{a^2 - e^2 x^2}{a^2 - x^2}}\, dx$, where $e^2 = \dfrac{a^2 - b^2}{a^2}$.

6. $x^{\frac{1}{2}} + y^{\frac{1}{2}} = 1$. Ans. $ds = \sqrt{\dfrac{x+y}{x}}\, dx$.

7. $\rho = a\theta^{\frac{1}{2}}$. Ans. $ds = a\left(\theta + \dfrac{1}{4\,\theta}\right)^{\frac{1}{2}} d\theta$.

8. $\rho = \dfrac{a}{\theta}$. Ans. $ds = \dfrac{a}{\theta^2}(\theta^2 + 1)^{\frac{1}{2}}\, d\theta$.

9. $\rho^2 = a\theta^3$. Ans. $ds = (a\theta^3 + \tfrac{9}{4}\,a\theta)^{\frac{1}{2}} d\theta$.

10. $\rho^3 = a\theta^2$. Ans. $ds = \left(1 + \dfrac{9\,\rho^3}{4\,a}\right)^{\frac{1}{2}} d\rho$.

51. Formulas for computing differentials. The principles for finding derivatives readily yield corresponding principles for finding differentials. Thus the nine theorems in § 28 reduce to the following, when expressed in terms of differentials, as one sees by multiplying the former, member by member, by dx:

I. $d(c) = 0$,

II. $d(x) = dx$,

III. $d(u + v - w) = du + dv - dw$,

IV. (1) $d(cv) = c\, dv$, (2) $d\left(\dfrac{v}{c}\right) = \dfrac{1}{c}\, dv$.

V. $d(uv) = u\, dv + v\, du$,

VI. $d(v^n) = nv^{n-1} dv$,

VII. $d(x^n) = nx^{n-1} dx$,

VIII. $d\left(\dfrac{u}{v}\right) = \dfrac{v\, du - u\, dv}{v^2}$,

IX. $d\left(\dfrac{c}{v}\right) = -c\,\dfrac{dv}{v^2}$.

In finding differentials of explicit functions one may compute the derivative as usual and multiply the result by the differential of the independent variable. Or one may employ differentials from the beginning. The latter method is often preferable. Thus in the case of the function

$$y = \frac{x+1}{x^2+1},$$

if we use theorem VIII, we have

$$dy = \frac{(x^2+1)d(x+1) - (x+1)d(x^2+1)}{(x^2+1)^2}$$

$$= \frac{(x^2+1)\cdot dx - (x+1)\cdot 2x\,dx}{(x^2+1)^2} = \frac{1-2x-x^2}{(x^2+1)^2}dx.$$

In the case of functions connected by an implicit functional relation such as

$$y^2 + x^2y + 7xy - 8x^3 = 0,$$

we may proceed by forming the differentials of the several terms. Thus, for the example given, we have

$$d(y^2) + d(x^2y) + 7\,d(xy) - 8\,d(x^3) = 0;$$

or $2y\,dy + 2xy\,dx + x^2\,dy + 7y\,dx + 7x\,dy - 24x^2\,dx = 0.$

Therefore $\quad dy = \dfrac{24x^2 - 7y - 2xy}{2y + x^2 + 7x}dx.$

EXERCISES

Find the differentials of the functions defined by the following equations:

1. $y = \dfrac{x-7}{x-2}.$ \qquad Ans. $\dfrac{5\,dx}{(x-2)^2}.$

2. $y = x^2(x-a)^{\frac{3}{2}}.$ \qquad Ans. $(\tfrac{7}{2}x^2 - 2ax)(x-a)^{\frac{1}{2}}dx.$

3. $y = (x^4 + 3x - 7)^{\frac{7}{3}}.$ \qquad Ans. $\tfrac{7}{3}(4x^3 + 3)(x^4 + 3x - 7)^{\frac{4}{3}}dx.$

4. $y = (1+x)\sqrt{1-x}.$ \qquad Ans. $\dfrac{1-3x}{2\sqrt{1-x}}dx.$

5. $x^2 + y^{\frac{4}{3}} = c.$ \qquad Ans. $dy = -\tfrac{3}{2}xy^{-\frac{1}{3}}dx.$

6. $x^3 + x^2y - y^3 = c.$ \qquad Ans. $dy = \dfrac{3x^2 + 2xy}{3y^2 - x^2}dx.$

7. $9x^2 + 4y^2 = 36.$ \qquad Ans. $dy = -\dfrac{9x}{4y}dx.$

8. $x^2 + y^2 = a^2.$ \qquad Ans. $dy = -\dfrac{x}{y}dx.$

52. Successive differentials. The differential of a function is in general a function of the independent variable and its increment, these two being independent variables; in fact, it is the product of the derived function and the differential of the independent variable. The differential of the differential dy may be formed; it is denoted by the symbol $d(dy)$, or d^2y. Since $dy = f'(x)dx$ and dx is independent of x, we take $d^2y = f''(x)(dx)^2$. In general the nth differential $d^n y$ of a function y of x is defined to be the nth derivative of y with respect to x multiplied by the nth power of dx. Thus

$$d^n y = f^{(n)}(x)(dx)^n.$$

The second, third, $\cdots$ differentials are called differentials of higher order.

EXERCISES

Find the successive differentials of the following functions:

1. $y = ax^3 + bx^2 + c$. Ans. $dy = (3\,ax^2 + 2\,bx)dx$,
$d^2y = (6\,ax + 2\,b)(dx)^2$,
$d^3y = 6\,a(dx)^3$,
$d^4y = 0$,
$d^5y = 0$,
etc.

2. $y = 4\,x^4 + 3\,x^2 - 7$.
3. $y = (1 - x^2)^3$.

Find the nth differential of each of the following functions:

4. $y = (1 + x)^n$. Ans. $d^n y = n!(dx)^n$.
5. $y = (1 + x)^m$, $m > n$.
   Ans. $d^n y = m(m-1) \cdots (m - n + 1)(1 + x)^{m-n}(dx)^n$.
6. $y = \dfrac{1}{1-x}$. Ans. $d^n y = \dfrac{n!(dx)^n}{(1-x)^{n+1}}$.
7. $y = \dfrac{1}{(1-x)^m}$. Ans. $d^n y = \dfrac{m(m+1) \cdots (m+n-1)(dx)^n}{(1-x)^{m+n}}$.

53. The derivative interpreted as a rate. If a point P describes the curve
$$y = f(x),$$
its position at any instant t is defined by means of certain functions of t to which y and x are respectively equal. From

the point of view of motion along the curve the variables y and x in the preceding equation may be considered as functions of t. Taking this point of view and differentiating with respect to t, we have

$$\frac{dy}{dt} = f'(x) \frac{dx}{dt}.$$

At any instant the time rate of change of y is equal to the time rate of change of x multiplied by the derivative of y with respect to x.

The last equation may also be written in the form

$$\frac{\frac{dy}{dt}}{\frac{dx}{dt}} = \frac{dy}{dx}.$$

The ratio of the time rate of change of y to that of x is equal to the derivative of y with respect to x. This ratio may be called the *rate of change of y with respect to x.*

From the relation $ds^2 = dx^2 + dy^2$ we have

$$\frac{ds}{dt} = \sqrt{\left(\frac{dx}{dt}\right)^2 + \left(\frac{dy}{dt}\right)^2}.$$

The time rate of change of length along the arc is equal to the square root of the sum of the squares of the time rate of x and the time rate of y.

In dealing with problems of rate it is often convenient to employ the following four steps: (1) *Draw a figure illustrating the problem and mark with suitable letters the quantities which vary with the time;* (2) *obtain one or more relations involving these quantities, one for each independent condition in the problem;* (3) *differentiate with respect to the time variable t;* (4) *substitute given quantities into one or more of the relations obtained and thus compute the required quantities.* The student should note these steps in the illustrative examples now to be given.

Example 1. Water is flowing into a vertical cylindrical tank of radius 7 ft. at the rate of 11 cu. ft. per second. How fast is the water rising? (Take $\pi = \frac{22}{7}$.)

Let h denote the depth of the water in the tank at the time t and let V denote its volume. Then

$$V = \pi \cdot 7^2 \cdot h = 154\, h.$$

Differentiating with respect to t, we have

$$\frac{dV}{dt} = 154\,\frac{dh}{dt}.$$

Now the rate of change of volume is 11 cu. ft. per second. Hence $dV/dt = 11$ if we suppose that V and t are measured in cubic feet and seconds respectively. Substituting in the last equation, we have

$$11 = 154\,\frac{dh}{dt}.$$

Hence
$$\frac{dh}{dt} = \frac{1}{14};$$

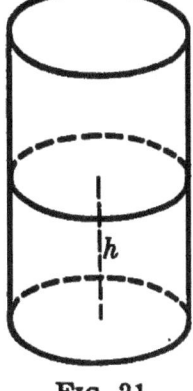

Fig. 31

or, the water is rising at the rate of $\frac{1}{14}$ of a foot per second.

Example 2. A ladder 25 ft. long rests against a vertical wall and on a horizontal floor. The bottom is moving outward at the rate of 3 ft. per second. How fast is the top moving downward when the bottom is 15 ft. from the wall?

Let OB and OA represent the wall and floor respectively and AB the ladder, B and A being at distances y and x from O at the general time t. Then we have

$$x^2 + y^2 = 25^2 = 625.$$

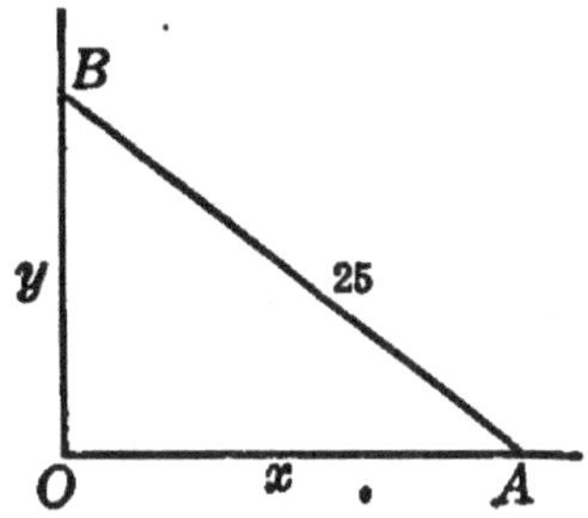

Hence $\qquad x\dfrac{dx}{dt} + y\dfrac{dy}{dt} = 0.$

Fig. 32

Now we are to take $x = 15$. Then $y = 20$. Moreover, $dx/dt = 3$. Substituting into the last equation, we have

$$15 \cdot 3 + 20 \cdot \frac{dy}{dt} = 0.$$

Hence $dy/dt = -2\frac{1}{4}$. Hence the top of the ladder is descending at the rate of $2\frac{1}{4}$ ft. per second.

Example 3. A point moves on the curve $y = 4\,x^2$ in such a way that the abscissa x is changing at the rate of 4 ft. per second. At what rates are the ordinates and arc lengths changing when $x = 2$?

From the equation $y = 4x^2$ we have

$$\frac{dy}{dt} = 8x \frac{dx}{dt}.$$

Taking $x = 2$ and $dx/dt = 4$, we have

$$\frac{dy}{dt} = 8 \cdot 2 \cdot 4 = 64.$$

Also, $\quad \dfrac{ds}{dt} = \sqrt{\left(\dfrac{dx}{dt}\right)^2 + \left(\dfrac{dy}{dt}\right)^2} = \sqrt{4^2 + 64^2} = 4\sqrt{257}.$

Hence y and s are changing at the respective rates of 64 and $4\sqrt{257}$ ft. per second.

54. Summary. In this chapter we have treated the following matters: definition of differential; differential of arc length both in rectangular coördinates and in polar coördinates; formulas for computing differentials; successive differentials; the interpretation of a derivative as a rate. The last topic in particular has important applications.

EXERCISES

1. A man 6 ft. tall walks at the rate of 4 mi. per hour on a horizontal pavement directly away from a light 12 ft. above the pavement. At what rate is the end of his shadow receding?
Ans. 8 mi. per hour.

2. A man is 40 ft. from a tower 30 ft. high and is approaching its base at the rate of 4 mi. per hour. How fast is he approaching its top? *Ans.* 3.2 mi. per hour.

3. A circular plate of metal is expanding under heat so that its radius is increasing at the rate of .007 in. per second. At what rate are its circumference and area increasing when the radius is 14 in.? (Take $\pi = 22/7$.) *Ans.* .044 in. per second; .616 sq. in. per second.

4. Water flows at the rate of 12 cu. ft. per minute into an inverted conical reservoir whose depth is 24 ft. and whose (upper) base is 8 ft. across. How fast is the water rising when it is 12 ft. deep? (Take $\pi = 22/7$.) *Ans.* $\frac{21}{22}$ ft. per minute.

5. In the previous problem, at what rate is the radius of the surface of the water increasing? *Ans.* $\frac{7}{44}$ ft. per minute.

6. In the same problem, at what rate is the area of the surface of the water increasing? (Take $\pi = 22/7$.)

Ans. 2 sq. ft. per minute.

7. If two trains start from the same point and one goes north at 40 mi. per hour and the other goes east at 30 mi. per hour, at what rate do they separate? *Ans.* 50 mi. per hour.

8. At what point on the parabola $y = 4x^2$ do the abscissa and ordinate increase at the same rate? *Ans.* At the point $(\frac{1}{8}, \frac{1}{16})$.

9. For what values of x is the rate of change of y with respect to x zero, when $y = \frac{1}{4}x^4 - 2x^3 + \frac{11}{2}x^2 - 6x + 25$?

Ans. 1, 2, 3.

10. How fast is the area of an equilateral triangle increasing when its side is 12 in. long and is increasing at the rate of 2 in. per hour?

Ans. $12\sqrt{3}$ sq. in. per hour.

11. How is the area of a triangle changing when its base is 10 in. and is increasing 2 in. per minute and its altitude is 12 in. and is decreasing 3 in. per minute? *Ans.* Decreasing 3 sq. in. per minute.

12. How fast is the arc length of the circle $x^2 + y^2 = 25$ changing at the point (3, 4) when the abscissa is increasing at the rate of 1 per second? *Ans.* $1\frac{1}{4}$ per second.

13. If the radius of a soap bubble is 3 in. and is increasing at the rate of 1 in. per minute, how fast is the volume increasing? (Take $\pi = 22/7$.) *Ans.* $113\frac{1}{7}$ cu. in. per minute.

14. An airplane is rising at an inclination of $30°$ to the horizontal. If its velocity along the path is 80 mi. per hour, at what rate is its vertical height increasing? *Ans.* 40 mi. per hour.

15. At what rate is the airplane in Ex. 14 moving horizontally?

Ans. $40\sqrt{3}$ mi. per hour.

16. At noon an airplane going north is directly above an express train going east and is $\frac{1}{4}$ mi. above it. If both airplane and train keep their course and move respectively 90 mi. per hour and 40 mi. per hour, the airplane remaining at the same height, at what rate are they separating at 1 P.M.? *Ans.* $\dfrac{38800}{\sqrt{155201}}$ mi. per hour.

17. The base of a triangle is 10 in. and is increasing 2 in. per minute. Its altitude is 12 in. and is decreasing 3 in. per minute. How will the area of the triangle be changing at the end of 2 min.?

Ans. Decreasing 15 sq. in. per minute.

18. If the edges of a cube are 20 in. and are increasing .01 in. per minute, at what rate are the area of its surface and its volume increasing?   *Ans.* 2.4 sq. in. per minute; 12 cu. in. per minute.

19. A train and a balloon start simultaneously from the same point, the former going 40 mi. per hour in a straight line and the latter rising vertically 8 mi. per hour. How fast are they separating?
Ans. 40.79 + mi. per hour.

20. A train and an airplane start simultaneously from the same point. The train moves in a horizontal line at the rate of 50 mi. per hour. The airplane moves upward over the same line at an inclination of 60° and with a speed of 80 mi. per hour. How fast are the train and airplane separating?   *Ans.* 70 mi. per hour.

CHAPTER V

MAXIMA AND MINIMA OF ALGEBRAIC FUNCTIONS

55. Illustrative example. A roofer wishes to make an open trapezoidal gutter of maximum capacity (that is, of maximum cross section), the sides and bottom being each 4 in. wide. What should be the width across the top?

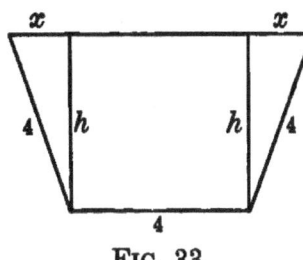

FIG. 33

Let the adjacent isosceles trapezoid of altitude h represent the cross section of the gutter. Let the distance across the top be $2x + 4$. We are to find the value of $2x + 4$ so that the area y of the trapezoid shall be the largest possible.

We have $$h = \sqrt{16 - x^2}.$$

Hence we have to determine the value of x for which the function y,
$$y = (x + 4)\sqrt{16 - x^2},$$
shall have the maximum value. By plotting points (using a scale on the y-axis whose unit is one fourth of that on the x-axis) we have the graph of the given function approximately in the form shown in the adjacent figure. We are to determine the value of x (between 0 and 4) for the highest point on the curve. When x is to the left of this point and is increasing, the value of y is increasing; to the right of the point the value of y is decreasing. Now when y is increasing as x increases, the rate of change of y with respect to x is positive; when y is decreasing as x increases, the rate of change of y with respect to x is negative. Hence the rate of change of y with respect to x is positive to the left of the point sought and is negative to the right of this point. But we

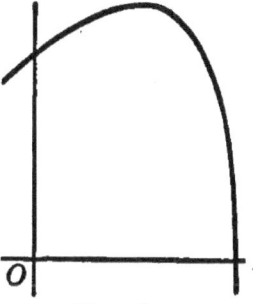

FIG. 34

have seen (§ 53) that the rate of change of y with respect to x is measured by the derivative dy/dx. Therefore dy/dx is positive to the left of the point sought and negative to its right.

Computing dy/dx from the given functional relation between x and y, we have

$$\frac{dy}{dx} = \frac{16 - 4x - 2x^2}{\sqrt{16-x^2}} = \frac{2(2-x)(4+x)}{\sqrt{16-x^2}}.$$

The value of x sought lies between 0 and 4, as we have seen from the graph. In the interior of the interval from 0 to 4 the derived function dy/dx is obviously continuous. Hence it can change from positive to negative only by passing through zero. Hence for the value of x which is sought we must have

$$\frac{2(2-x)(x+4)}{\sqrt{16-x^2}} = 0.$$

This requires that x shall have the value 2, since this is the only value of x in the interval 0 to 4 for which the last equation is satisfied, as one may verify by solving that equation.

Substituting 2 for x in $2x + 4$, the expression for the distance across the gutter, we find for that distance the required value 8.

In the solution of this problem we have used several principles which we shall now explain at greater length.

56. Increasing and decreasing of functions. We have seen (§ 53) that the derivative dy/dx of y with respect to x measures the rate of increase of y with respect to x. In the previous section we saw (by means of an example) that it may be an important matter to determine when y is increasing and when it is decreasing with respect to x. In studying such problems *it is convenient to suppose that x is increasing*; and we make this supposition. Then y is increasing when its derivative with respect to x is positive; it is decreasing when this derivative is negative. Hence we have the following theorem:

I. Let $y = f(x)$. Then y increases with increasing x when dy/dx is positive; it decreases with increasing x when dy/dx is negative.

Let us now suppose that both y and dy/dx are continuous in a given interval and let us consider the condition under which it is possible for y to change from an increasing function to a decreasing function or from a decreasing function to an increasing function. For the first of these changes dy/dx must pass from positive to negative values and for the second it must pass from negative to positive values. In either case it is clear that dy/dx must pass through the value 0. Hence we have the following theorem:

II. If both y and dy/dx are continuous in an interval and if y changes from an increasing to a decreasing function, or from a decreasing to an increasing function, as x passes a given value, then dy/dx must have the value 0 as x passes the given value.

From this theorem we have a ready means of finding those points for which y changes from an increasing to a decreasing function or from a decreasing to an increasing function, *provided that dy/dx is continuous at such points*. We form the derived function dy/dx, set it equal to 0, and solve for x. This gives all possible points. But it may also give additional points, as we shall presently see by means of an example, so that it is always necessary to test the points thus found.

In the case of the function $y = x^3$ we have $dy/dx = 3\,x^2$. This is equal to 0 when $x = 0$. But the function y increases as x increases through the point 0, since y then changes from negative to positive. Hence the derivative may be zero when the function does not change from increasing to decreasing or vice versa. This shows that the condition given in theorem II is not a *sufficient* condition; it is only a *necessary* condition.

57. Maximum and minimum of a function. The conception of maximum and minimum values of a function plays an important rôle in the applications of the differential calculus. It is necessary to have precise definitions of the terms *maximum* and *minimum*.

A function $f(x)$ is said to have a maximum for $x = a$ if $f(x)$ is continuous for $x = a$ and if $f(a)$ is greater than the values of $f(x)$ for x just preceding and just following $x = a$.

A function $f(x)$ is said to have a minimum for $x = a$ if $f(x)$ is continuous for $x = a$ and if $f(a)$ is less than the values of $f(x)$ for x just preceding and just following $x = a$.

What we are here calling *maximum* and *minimum* are sometimes called respectively *relative maximum* and *relative minimum*, the adjective *relative* having reference to the fact that the terms are defined with reference to the value of the function for a given x relative to its values for x near the given value of x.

If a function has a maximum for $x = a$ and if h is a sufficiently small positive quantity, then, from the definition, it follows that

$$f(a) > f(a \pm h).$$

Similarly, the condition for a minimum for $x = a$ is that

$$f(a) < f(a \pm h)$$

for h a sufficiently small positive quantity.

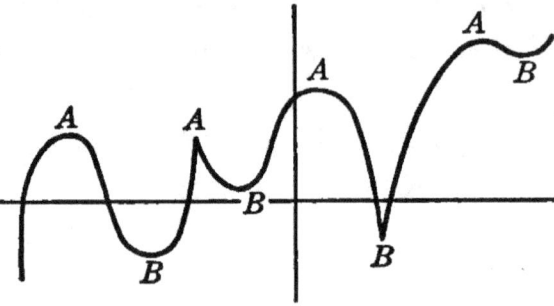

FIG. 35

From such a graph for $y = f(x)$ as that given in the foregoing figure it is obvious that $f(x)$ may have several maxima and several minima in a given interval. Here we have maxima at the points A and minima at the points B. Moreover, one minimum may be greater than another maximum.

If the function $y = f(x)$ has a maximum for $x = a$, then $f(a)$ is greater than $f(x)$ for x nearly equal to a. Hence $f(x)$ is increasing as x increases toward a and is decreasing as x increases away from a. Therefore $f(x)$ changes from an increasing to a decreasing function as x passes a. If, on the other hand, $f(x)$ has a minimum for $x = a$, then $f(a)$ is less than $f(x)$ for x nearly equal to a. Hence $f(x)$ is decreasing as x increases toward a and is increasing as x increases away from a. Remembering that y is increasing or decreasing according as $f'(x)$ is positive or negative, we may translate the results of this paragraph into analytical language as in the following theorem:

I. The function $f(x)$, continuous for $x = a$, has the maximum $f(a)$ for $x = a$ if $f'(x)$ changes from positive to negative as x increases and passes the value $x = a$. The function $f(x)$, continuous for $x = a$, has the minimum $f(a)$ for $x = a$ if $f'(x)$ changes from negative to positive as x increases and passes the value $x = a$. If

$f'(x)$ *does not change sign as x passes a, then $f(x)$ has neither a maximum nor a minimum for $x = a$.*

A value of x for which $f'(x)$ has the value 0 or at which $f'(x)$ is discontinuous is called a *critical value* of x for the function $f(x)$. From the foregoing theorem it follows that the critical values of x are the only ones which need to be examined in finding maxima and minima. If $f'(x)$ is continuous in an interval, then it can change sign only by passing through the value 0. The critical values are of the following two sorts: (1) values of x satisfying the equation $f'(x) = 0$; (2) values of x for which $f'(x)$ is discontinuous. In a particular example we may have critical values of only one kind, or both kinds may occur in the same example.

Theorem I is the fundamental theorem for maxima and minima. But when $f'(x)$ is continuous for a critical value $x = a$, we may sometimes conveniently employ an additional theorem. In this case we have $f'(a) = 0$. Let us consider the second derivative $f''(x)$. If $f''(x)$ is continuous for $x = a$ and is positive, then $f'(x)$ is increasing for $x = a$. Hence it passes from negative to positive values as x increases and passes a. Therefore in this case $f(x)$ has a minimum for $x = a$. If $f''(x)$ is continuous for $x = a$ and is negative, then $f'(x)$ is decreasing for $x = a$. Hence it passes from positive to negative values as x increases and passes a. In this case, therefore, $f(x)$ has a maximum for $x = a$. These considerations lead to the following theorem:

II. *If $f(x), f'(x), f''(x)$ are continuous for $x = a$, and if $f'(a) = 0$ and $f''(a) \neq 0$, then $f(x)$ has a maximum or a minimum for $x = a$ according as $f''(a)$ is negative or positive.*

We may extract from the foregoing principles the following working rule for finding the maxima and minima of a function $f(x)$:

Find the first derivative $f'(x)$ of the function. Find the critical values of one or both sorts by solving the equation $f'(x) = 0$ and by examining the function $f'(x)$ for points of discontinuity. Then treat each critical value separately. Let a be such a critical value. If the second derivative $f''(x)$ is conveniently computed, and if $f'(a) = 0$ and if $f''(a) \neq 0$, employ theorem II for testing the critical value. Otherwise, examine the first derivative $f'(x)$ and employ theorem I for testing the critical value. When a corresponds to a maximum or minimum, compute $f(a)$ for the required value of the function.

These general principles for finding maxima and minima apply to all classes of functions of one variable. In this chapter we shall give examples of their applications to problems involving algebraic functions, the only sorts of functions whose derivatives the student has so far learned to find.

The student should keep the working rule in mind while reading the solutions of the following illustrative examples.

Example 1. A carpenter wishes to build a box with a square base and open top and of such sort as to contain 108 sq. ft. of lumber. Neglecting the thickness of the lumber, find the dimensions of the box so that it shall be a maximum. What is the capacity of the box?

Let x be the length in feet of a side of the square base and let h be the height. Then the capacity y in cubic feet is x^2h. The area of the lumber in the box is $x^2 + 4hx$ sq. ft. Hence

$$x^2 + 4hx = 108, \quad \text{or} \quad h = \frac{108 - x^2}{4x}.$$

Therefore
$$y = f(x) = \tfrac{1}{4} x(108 - x^2).$$

This is the function whose maximum value is sought. We have

$$f'(x) = 27 - \tfrac{3}{4} x^2.$$

Since the derivative is continuous for all values of x, the totality of critical values is found by setting the derivative equal to zero and solving the equation. This gives

$$27 - \tfrac{3}{4} x^2 = 0, \quad \text{whence} \quad x = +6 \text{ or } -6.$$

The value -6 is obviously irrelevant to the present problem. Hence there is left only one critical value, namely, $x = 6$, to be examined.

Perhaps the student will be convinced, from physical considerations, that there is a box of maximum capacity. In that case he may take 6 as the value of x corresponding to the maximum sought, since the previous work leaves us with only one critical value. But the critical value may also be easily tested mathematically. We have $f''(x) = -\tfrac{3}{2} x$; therefore $f''(6) = -9$, a negative value. Hence $x = 6$ corresponds to a maximum value of $f(x)$. Substituting 6 for x in the expression for h, we have $h = 3$. Hence the dimensions of the box are 6 by 6 by 3 ft. Either from these dimensions or from the expression for $f(x)$ we see that the maximum capacity is 108 cu. ft.

Example 2. Investigate the function $f(x) = x^{\frac{2}{3}} + 2$ for maxima and minima.

We have $$f'(x) = \tfrac{2}{3} x^{-\frac{1}{3}} = \frac{2}{3 x^{\frac{1}{3}}}.$$

The function $f'(x)$ is not equal to zero for any value of x. It has a discontinuity for $x = 0$ since it becomes infinite as x approaches 0. Otherwise it is continuous. Therefore the only critical value to be examined is $x = 0$. If x is a small negative number, then $f'(x)$ is negative; if x is a small positive number, then $f'(x)$ is positive. Hence $f'(x)$ changes from negative to positive as x increases and passes the critical value $x = 0$. Therefore $f(x)$ has a minimum for $x = 0$. This minimum is 2.

The student should check the result by means of the adjoined graph of the function.

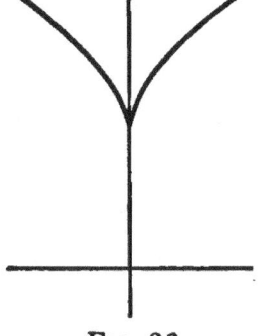

FIG. 36

Example 3. Investigate the function $f(x) = (x + 1)^2(x - 2)^3$ for maxima and minima.

We have $\qquad f'(x) = (x + 1)(x - 2)^2(5x - 1).$

The derivative is continuous for all values of x. Hence all the critical values are roots of the equation obtained by setting it equal to zero, namely, the equation

$$(x + 1)(x - 2)^2(5x - 1) = 0.$$

The roots of this equation, and hence the critical values of x for $f(x)$, are $-1, 2, \tfrac{1}{5}$.

When x is near -1 the second and third factors in the expression for $f'(x)$ are positive and negative respectively, while the factor $x + 1$ changes from negative to positive as x increases and passes $x = -1$. Hence, for such increase of x, $f'(x)$ changes from positive to negative. Therefore $f(x)$ has a maximum for $x = -1$.

When x is near 2 each of the three factors of $f'(x)$ is positive. Therefore $f'(x)$ does not change sign as x passes 2. Hence $f(x)$ has neither a maximum nor a minimum for $x = 2$.

When x is near $\tfrac{1}{5}$ the first and second factors of $f'(x)$ are both positive, while the third factor changes from negative to positive as x increases and passes $x = \tfrac{1}{5}$. Hence, for such increase of x, $f'(x)$ changes from negative to positive. Therefore $f(x)$ has a minimum for $x = \tfrac{1}{5}$.

58. Summary. In this chapter we have given the main principles by means of which to determine whether a function is increasing or decreasing with respect to the independent variable; and we have applied these in developing the main theorems concerning the maxima and minima of functions of one variable. As the illustrative examples show, these theorems may be made the basis of one of the most important and most interesting applications of the differential calculus.

EXERCISES

1. If the sum of two numbers is 10, what is each of them if their product is the largest possible? *Ans.* 5 and 5.

2. If the sum of two numbers is a, what is each of them if their product is a maximum? *Ans.* $\frac{1}{2}a$ and $\frac{1}{2}a$.

3. Show that a square is the maximum rectangle of given perimeter.

4. What number exceeds its square by the greatest possible amount? *Ans.* $\frac{1}{2}$.

5. What positive number gives the least possible sum when added to its reciprocal? *Ans.* 1.

6. Show that the sum of a positive number and its reciprocal is equal to or greater than 2.

Examine the following functions for maxima and minima:

7. $\dfrac{x}{1+x^2}$. *Ans.* Maximum when $x=1$; minimum when $x=-1$.

8. $\dfrac{1-x+x^2}{1+x-x^2}$. *Ans.* Minimum when $x=\frac{1}{2}$.

9. $3+(x-2)^{\frac{2}{3}}$. *Ans.* Minimum when $x=2$.

10. $5-(x-2)^{\frac{1}{3}}$. *Ans.* No maxima or minima.

11. What is the volume of the largest box that can be made by cutting equal squares out of the corners of a rectangular piece of cardboard 18 in. by 48 in. and turning up the edges for the sides of the box? *Ans.* 1600 cu. in.

12. A rectangular inclosure is to be fenced in along a straight river, no fencing being needed along the river. What shape is the most economical; that is, what is the ratio of length to breadth so

that the maximum area shall be inclosed by a given amount of fencing? *Ans.* Length along river = twice the width.

HINT. Denote the length of the fence by l, find the length and width in terms of l, and observe that the former is twice the latter.

13. The strength of a beam of rectangular cross section and of given length varies directly as the breadth and the square of the depth. What are the dimensions of the strongest beam that can be sawed from a log 24 in. in diameter? *Ans.* $8\sqrt{3}$ in. by $8\sqrt{6}$ in.

(Note that the diagonal of the cross section of the beam is equal to the diameter of the log.)

14. The deflection of a beam of rectangular cross section and of given length varies under a given load inversely as the product of the breadth and the cube of the depth. What are the dimensions of the beam of least deflection that can be sawed from a log 24 in. in diameter? *Ans.* 12 in. by $12\sqrt{3}$ in.

15. The stiffness of a beam of rectangular cross section and of given length varies directly as the product of the breadth and the cube of the depth. What are the dimensions of the beam of maximum stiffness that can be sawed from a log 24 in. in diameter?
Ans. 12 in. by $12\sqrt{3}$ in.

16. What is the shape of the rectangular cross section of given area of a channel of water when it is required that the wetted perimeter shall be a minimum? *Ans.* Width = twice the depth.

17. The work done in exploding a mixture of 1 cu. ft. of coal gas with x cubic feet of air is given by the formula $W = 83\,x - 3.2\,x^2$. For what value of x is this a maximum? What is the maximum work done? *Ans.* $x = 415/32$.

18. A tank is to be constructed with square base and open top and is to have a capacity of 343 cu. ft. If the bottom costs twice as much per square foot as the sides, what are the dimensions when the cost is a minimum? *Ans.* 7 ft. by 7 ft. by 7 ft.

19. What are the most economical proportions for a cylindrical tank closed at both ends? *Ans.* Diameter = height.

20. What is the shape of a closed rectangular box of given surface area and of maximum capacity if its base is a square? *Ans.* Cubical.

21. A Norman window has a horizontal base, a semicircular top, and vertical sides. What proportions yield the largest area for a given perimeter? *Ans.* Radius of semicircle = length of vertical sides.

22. Show that a square is the largest rectangle that can be inscribed in a given circle.

23. What are the dimensions of the right circular cone of maximum volume that can be inscribed in a sphere of radius a?

Ans. Altitude $= \frac{4}{3} a$; radius of base $= \frac{2}{3} a\sqrt{2}$.

24. What are the proportions of the most economical tent in the form of a right circular cone of given volume?

Ans. Height $= \sqrt{2}$ times radius of base.

25. What point on the parabola $y^2 = 4x$ is nearest to the point $(1, 0)$? *Ans.* $(0, 0)$.

26. A man wishes to row across a straight river 900 yd. wide and walk to a point on the other side 1500 yd. farther upstream. If he can walk 5 mi. an hour but can row only 4 mi. an hour, where should he land from the boat in order to make his destination in the shortest possible time, the motion of the stream being neglected?

Ans. 1200 yd. upstream.

27. If the cost of driving a steamboat through the water varies as the cube of her velocity relative to the water, what is the most economical rate per hour relative to the water when running against a stream which flows c miles per hour? *Ans.* $\frac{3}{2} c$ miles per hour.

28. A water tank with square base and vertical sides (open at the top) is to be constructed so as to have a capacity of 1000 cu. ft. If the bottom costs 60 cents per square foot and the sides cost 40 cents per square foot, what are the dimensions of the most economical tank? *Ans.* Side of base $= \frac{10}{3}\sqrt[3]{36}$; height $= \frac{5}{2}\sqrt[3]{36}$.

29. Two poles, with tops A and B, are standing in a vertical position on a horizontal plane. Where is the point P on the plane between them such that the sum of the squares of the distances PA and PB is a minimum? *Ans.* Midway between the poles.

30. Two poles, with tops A and B, are standing in a vertical position on a horizontal plane. The point Q on the plane between them is to be determined so that the sum of the distances QA and QB is a minimum. Show that the lines QA and QB will then make equal angles with the vertical line.

31. A trough of triangular cross section is to be made by placing together edge to edge two planks of equal width. At what angle should the planks meet in order to give maximum capacity?

Ans. At a right angle.

32. If the sum of the length and girth of a parcel-post package must not exceed 72 in., what are the dimensions of the largest rectangular box with square cross section that can be sent by parcel post? *Ans.* 12 in. by 12 in. by 24 in.

33. A picture $3\frac{1}{2}$ ft. in height is hung vertically on a wall. Its upper edge is 8 ft. higher than the eye of an observer. How far from the wall should he stand in order to get the most favorable view; that is, in order to make as large as possible the vertical angle subtended in his eye by the picture? *Ans.* 6 ft.

34. A rectangle is inscribed in the ellipse $b^2x^2 + a^2y^2 = a^2b^2$ with its sides parallel to the principal axes. What is its maximum area?
Ans. $2\,ab$.

35. A beam of length l, supported at both ends, carries a uniform load of w units per unit length. The bending moment M at distance x from one end is given by the formula
$$M = \tfrac{1}{2}\,wlx - \tfrac{1}{2}\,wx^2.$$
Show that the bending moment is a maximum at the center of the beam.

Examine for maxima and minima the functions in problems 36 to 45:

36. $(x-3)^3(x-2)^2$.
Ans. $x = 2$ gives maximum $= 0$; $x = \tfrac{12}{5}$ gives minimum.

37. $(x-1)^5(x+1)^4$.
Ans. Maximum for $x = -1$; minimum for $x = -\tfrac{1}{9}$.

38. $\dfrac{9}{x} + \dfrac{4}{3-x}$. *Ans.* Maximum for $x = \tfrac{9}{5}$; minimum for $x = 9$.

39. $x^3 - 6x^2 + 24x + 80$. *Ans.* No maxima or minima.

40. $x^3 - 18x^2 + 60x - 24$.
Ans. Maximum for $x = 2$; minimum for $x = 10$.

41. $x^{2n} - a^{2n}$, $n =$ positive integer. *Ans.* Minimum when $x = 0$.

42. $x^{2n+1} + c$, $n =$ positive integer or zero.
Ans. No maxima or minima.

43. $\dfrac{x^2 - 14x + 24}{x - 20}$. *Ans.* Maximum for $x = 8$; minimum for $x = 32$.

44. $x(x-a)^2(x+a)^3$, $a > 0$.
Ans. Maximum for $x = \tfrac{1}{2}a$; minima for $x = a$ and $-\tfrac{1}{3}a$.

45. $x^{\frac{2}{3}}(x-6)^2$.
Ans. Maximum for $x = 1\tfrac{1}{2}$; minima for $x = 0$ and 6.

CHAPTER VI

DIFFERENTIATION OF THE TRIGONOMETRIC FUNCTIONS

59. Introduction. In taking up the differentiation of a new class of functions, such as the trigonometric functions, one would expect to find it necessary to employ the Four-Step Rule (see § 23). In this chapter we shall find the derivative of the sine function and the cosine function by this rule; we shall also give an alternative method for finding the derivative of the cosine function. The other four trigonometric functions, namely, tangent, cotangent, secant, and cosecant, are so readily expressed in terms of the sine and cosine that it is desirable to use such expressions for them and to employ the general theorems in § 28 to effect their differentiation. In reading the chapter the student should carefully observe how the Four-Step Rule and these general theorems work together in finding the derivatives of the trigonometric functions. Before proceeding with the chapter he should review § 16 and the illustrative examples solved in that section.

60. Differentation of sin u.

Let $y = \sin u$.

Then $y + \Delta y = \sin(u + \Delta u)$

$\qquad = \sin u \cos \Delta u + \cos u \sin \Delta u,$

$\Delta y = \sin u \cos \Delta u + \cos u \sin \Delta u - \sin u$

$\qquad = \cos u \sin \Delta u - \sin u (1 - \cos \Delta u),$

$\dfrac{\Delta y}{\Delta u} = \cos u \dfrac{\sin \Delta u}{\Delta u} - \sin u \dfrac{1 - \cos \Delta u}{\Delta u},$

$\dfrac{dy}{du} = \cos u \lim\limits_{\Delta u = 0} \dfrac{\sin \Delta u}{\Delta u} - \sin u \lim\limits_{\Delta u = 0} \dfrac{1 - \cos \Delta u}{\Delta u}$

$\qquad = \cos u \cdot 1 - \sin u \cdot 0 = \cos u.$ [By § 16]

Therefore $$\frac{d(\sin u)}{du} = \cos u.$$

Considering u as a function of x and employing the principle in § 31, namely,
$$\frac{dy}{dx} = \frac{dy}{du} \cdot \frac{du}{dx},$$
we have $$\frac{d(\sin u)}{dx} = \cos u \frac{du}{dx}.$$

In the differential notation this may be written in the more compact form $$d(\sin u) = \cos u\, du.$$

61. Differentiation of cos u.

Let $\qquad y = \cos u.$

Then $\quad y + \Delta y = \cos(u + \Delta u)$
$$= \cos u \cos \Delta u - \sin u \sin \Delta u,$$
$$\Delta y = \cos u \cos \Delta u - \sin u \sin \Delta u - \cos u$$
$$= -\sin u \sin \Delta u - \cos u (1 - \cos \Delta u),$$
$$\frac{\Delta y}{\Delta u} = -\sin u \frac{\sin \Delta u}{\Delta u} - \cos u \frac{1 - \cos \Delta u}{\Delta u},$$
$$\frac{dy}{du} = -\sin u \cdot 1 - \cos u \cdot 0 = -\sin u.$$

Hence $$\frac{d(\cos u)}{dx} = -\sin u \frac{du}{dx},$$
$$d(\cos u) = -\sin u\, du.$$

We may also differentiate $\cos u$ in the following manner. We have
$$\cos u = \sin\left(\frac{\pi}{2} - u\right).$$

Differentiating both sides with respect to u, we have
$$\frac{d(\cos u)}{du} = \cos\left(\frac{\pi}{2} - u\right) \frac{d}{du}\left(\frac{\pi}{2} - u\right)$$
$$= -\cos\left(\frac{\pi}{2} - u\right)$$
$$= -\sin u.$$

Hence $\qquad d(\cos u) = -\sin u\, du.$

62. Differentiation of $\tan u$, $\cot u$, $\sec u$, $\csc u$. From the relation
$$\tan u = \frac{\sin u}{\cos u}$$
we have
$$\frac{d(\tan u)}{du} = \frac{\cos u \dfrac{d}{du}(\sin u) - \sin u \dfrac{d}{du}(\cos u)}{\cos^2 u}$$
$$= \frac{\cos^2 u + \sin^2 u}{\cos^2 u} = \frac{1}{\cos^2 u} = \sec^2 u.$$

Hence
$$\frac{d(\tan u)}{dx} = \sec^2 u \frac{du}{dx},$$
$$d(\tan u) = \sec^2 u \, du.$$

From the relation $\cot u = \dfrac{\cos u}{\sin u}$ we have, similarly,
$$\frac{d(\cot u)}{du} = \frac{-\cos^2 u - \sin^2 u}{\sin^2 u} = -\frac{1}{\sin^2 u} = -\csc^2 u.$$

Hence
$$\frac{d(\cot u)}{dx} = -\csc^2 u \frac{du}{dx},$$
$$d(\cot u) = -\csc^2 u \, du.$$

From the relation $\sec u = \dfrac{1}{\cos u}$
we have
$$\frac{d(\sec u)}{du} = \frac{-\dfrac{d}{du}(\cos u)}{\cos^2 u} = \frac{\sin u}{\cos^2 u}$$
$$= \frac{1}{\cos u} \cdot \frac{\sin u}{\cos u} = \sec u \tan u.$$

Hence
$$\frac{d(\sec u)}{dx} = \sec u \tan u \frac{du}{dx},$$
$$d(\sec u) = \sec u \tan u \, du.$$

From the relation $\csc u = \dfrac{1}{\sin u}$
we have, similarly,
$$\frac{d(\csc u)}{du} = -\frac{\cos u}{\sin^2 u} = -\frac{1}{\sin u} \cdot \frac{\cos u}{\sin u} = -\csc u \cot u.$$

Hence
$$\frac{d(\csc u)}{dx} = -\csc u \cot u \frac{du}{dx},$$
$$d(\csc u) = -\csc u \cot u\, du.$$

63. Illustrative examples. We shall now give some examples involving the differentiation of the trigonometric functions.

Example 1. Differentiate $\sec^2 \sqrt{a^2 + x^2}$.

We have
$$\frac{d}{dx}\sec^2\sqrt{a^2+x^2} = 2\sec\sqrt{a^2+x^2}\,\frac{d}{dx}\sec\sqrt{a^2+x^2}$$
$$= 2\sec\sqrt{a^2+x^2}\,\sec\sqrt{a^2+x^2}\,\tan\sqrt{a^2+x^2}\cdot\frac{d}{dx}\sqrt{a^2+x^2}$$
$$= 2\sec^2\sqrt{a^2+x^2}\,\tan\sqrt{a^2+x^2}\cdot\frac{x}{\sqrt{a^2+x^2}}$$
$$= \frac{2x}{\sqrt{a^2+x^2}}\sec^2\sqrt{a^2+x^2}\,\tan\sqrt{a^2+x^2}.$$

Example 2. Find dy/dx when
$$x = a(\theta - \sin\theta), \quad y = a(1 - \cos\theta).$$
We have $\quad \dfrac{dx}{d\theta} = a(1-\cos\theta), \quad \dfrac{dy}{d\theta} = a\sin\theta.$

Employing the formula in § 33, we have
$$\frac{dy}{dx} = \frac{\dfrac{dy}{d\theta}}{\dfrac{dx}{d\theta}} = \frac{a\sin\theta}{a(1-\cos\theta)} = \frac{\sin\theta}{1-\cos\theta}.$$

Example 3. Find the differential of arc length in the graph of $y = \sin x$.

We have $\quad \dfrac{dy}{dx} = \cos x.$

Employing the formula in § 49, we have
$$ds = \sqrt{1 + \left(\frac{dy}{dx}\right)^2}\,dx = \sqrt{1+\cos^2 x}\,dx.$$

Example 4. At what rate is $\tan x$ changing with respect to x when $x = \pi/6$?

If we write $y = \tan x$, then from § 53 it follows that the desired rate is the value of dy/dx when $x = \pi/6$. But $dy/dx = \sec^2 x$. Hence the required rate is $\sec^2 \pi/6 = \sec^2 30° = 4/3$.

Example 5. Investigate the function $y = \sin x(1 + \cos x)$ for maxima and minima.

We have
$$\frac{dy}{dx} = \cos x(1 + \cos x) + \sin x(-\sin x)$$
$$= \cos x + \cos^2 x - \sin^2 x$$
$$= \cos x + 2\cos^2 x - 1.$$

Since the derivative is continuous, it follows from § 57 that all the critical values are found by solving the equation obtained on setting the derivative equal to zero, namely, the equation
$$2\cos^2 x + \cos x - 1 = 0.$$

This is a quadratic equation in the unknown quantity $\cos x$. Writing it in factored form, we have
$$(2\cos x - 1)(\cos x + 1) = 0.$$

Hence $\quad 2\cos x - 1 = 0 \quad \text{or} \quad \cos x = -1.$

Therefore $\quad x = \pm\dfrac{\pi}{3} + 2n\pi \quad \text{or} \quad x = \pi + 2n\pi,$

where n is an integer.

Computing the second derivative, we have
$$\frac{d^2y}{dx^2} = -\sin x - 4\cos x \sin x.$$

This is negative when $x = \dfrac{\pi}{3} + 2n\pi$ and positive when $x = -\dfrac{\pi}{3} + 2n\pi$. Hence (theorem II of § 57) the function is a maximum when $x = \dfrac{\pi}{3} + 2n\pi$ and is a minimum when $x = -\dfrac{\pi}{3} + 2n\pi$.

For the other critical values $x = \pi + 2n\pi$ the second derivative vanishes. Hence we have to resort to theorem I of § 57. Writing the derivative in the form
$$\frac{dy}{dx} = (2\cos x - 1)(\cos x + 1),$$

we see that the first factor is negative and the second positive for x near $\pi + 2n\pi$; therefore it does not change sign as x passes this value. Hence there is neither a maximum nor a minimum for $x = \pi + 2n\pi$.

Example 6. Find the area between one arch of the sine curve and the x-axis.

The first arch rests on the interval from 0 to π of the x-axis, as the student may easily verify by constructing the graph of the function. Let x be any positive number less than π and let $A(x)$ denote the area bounded by the sine curve and the x-axis and the ordinate $x = x$. Then from § 25 it follows that

$$\frac{d}{dx} A(x) = \sin x.$$

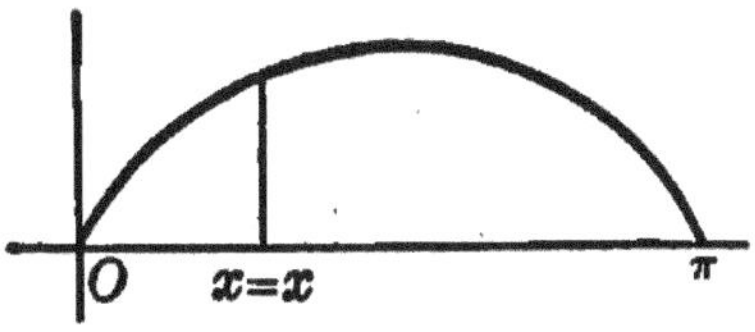

FIG. 37

The antiderivative of $\sin x$ is $-\cos x + c$, where c is an arbitrary constant. Hence

$$A(x) = -\cos x + c,$$

where c has such a value as to make $A(0) = 0$; that is, c has the value 1. Therefore

$$A(x) = 1 - \cos x.$$

From the definition of $A(x)$ it follows that $A(\pi)$ is the area to be found. Hence that area is 2.

64. Summary. In this chapter we have proved the following fundamental formulas for differentiation:

$$\frac{d(\sin u)}{dx} = \cos u \frac{du}{dx}.$$

$$\frac{d(\cos u)}{dx} = -\sin u \frac{du}{dx}.$$

$$\frac{d(\tan u)}{dx} = \sec^2 u \frac{du}{dx}.$$

$$\frac{d(\cot u)}{dx} = -\csc^2 u \frac{du}{dx}.$$

$$\frac{d(\sec u)}{dx} = \sec u \tan u \frac{du}{dx}.$$

$$\frac{d(\csc u)}{dx} = -\csc u \cot u \frac{du}{dx}.$$

EXERCISES

Differentiate the following functions with respect to x:

1. $\sin(ax+b)$.   Ans. $a\cos(ax+b)$.
2. $\cos(3x^2+7)$.   Ans. $-6x\sin(3x^2+7)$.
3. $\sin x \cos 2x$.   Ans. $\cos x \cos 2x - 2\sin x \sin 2x$.
4. $\tan x - x$.   Ans. $\tan^2 x$.
5. $\sin nx \sin^n x$.   Ans. $n\sin^{n-1} x \sin(n+1)x$.
6. $\sec^2 5x$.   Ans. $10 \sec^2 5x \tan 5x$.
7. $\sin^2 x$.   Ans. $\sin 2x$.
8. $\cot\sqrt{1-x}$.   Ans. $\dfrac{\csc^2\sqrt{1-x}}{2\sqrt{1-x}}$.
9. $\dfrac{\tan x - 1}{\sec x}$.   Ans. $\sin x + \cos x$.
10. $\sin^3 \dfrac{x}{3}$.   Ans. $\sin^2 \dfrac{x}{3} \cos \dfrac{x}{3}$.
11. $\dfrac{1+\cos x}{1-\cos x}$.   Ans. $-\dfrac{2\sin x}{(1-\cos x)^2}$.
12. $(x \csc x)^2$.   Ans. $2x\csc^2 x(1 - x\cot x)$.

Perform the following indicated differentiations:

13. $\dfrac{d}{dx}(x - \sin x \cos x) = 2\sin^2 x$.

14. $\dfrac{d}{dt}[\sin t(\cos^2 t + 2)] = 3\cos^3 t$.

15. $\dfrac{d}{dt}(a\cos \omega t - b\sin \omega t) = -\omega(a\sin \omega t + b\cos \omega t)$.

16. $\dfrac{d}{d\theta} a(\sin n\theta)^{\frac{1}{n}} = a\cos n\theta (\sin n\theta)^{\frac{1-n}{n}}$.

17. $\dfrac{d}{d\theta}\dfrac{1}{1-\cos\theta} = -\dfrac{\sin\theta}{(1-\cos\theta)^2}$.

18. $\dfrac{d}{d\theta}(\tfrac{1}{3}\tan^3\theta - \tan\theta + \theta) = \tan^4\theta$.

19. $\dfrac{d}{d\alpha}[\sin(\cos\alpha)] = -\sin\alpha \cos(\cos\alpha)$.

20. $\dfrac{d}{dt}[\sin(t+a)\cos(t-a)] = \cos 2t$.

Find the antiderivatives of the following functions:

21. $\cos x$. Ans. $\sin x + c$.

22. $\sec^2 x$. Ans. $\tan x + c$.

23. $\tan^2 x \ (= \sec^2 x - 1)$. Ans. $\tan x - x + c$.

24. $\cot^2 x$. Ans. $-\cot x - x + c$.

25. $\sin^2 x$. Ans. $\frac{1}{2}(x - \sin x \cos x) + c$.

26. $\cos^2 x$. Ans. $\frac{1}{2}(x + \sin x \cos x) + c$.

27. $\cos^3 x$. Ans. $\frac{1}{3} \sin x (\cos^2 x + 2) + c$.

28. $\cos ax$. Ans. $\frac{1}{a} \sin ax + c$.

29. $a \sin \alpha x + b \cos \beta x$. Ans. $-\frac{a}{\alpha} \cos \alpha x + \frac{b}{\beta} \sin \beta x + c$.

30. From the formula $\sin 2x = 2 \sin x \cos x$, derive by differentiation the formula $\cos 2x = \cos^2 x - \sin^2 x$.

31. From the formula $\sin(x+y) = \sin x \cos y + \cos x \sin y$, derive by differentiation with respect to x (y being treated as a constant) the formula $\cos(x+y) = \cos x \cos y - \sin x \sin y$.

32. Find dy/dx when
$$x = a \cos \theta + a\theta \sin \theta, \quad y = a \sin \theta - a\theta \cos \theta.$$
Ans. $\tan \theta$.

33. Find the expressions for the subnormal and the subtangent of the curve $y = \sin x$. Ans. $\sin x \cos x$, $\tan x$.

34. Find the points of intersection of the curves $y = \sin x$ and $y = \cos x$, the slopes of each at these points, and the angle with which these curves intersect.
Ans. $x = \frac{\pi}{4} + n\pi$; $\pm \frac{1}{2}\sqrt{2}$, $\mp \frac{1}{2}\sqrt{2}$; $\arctan(2\sqrt{2})$.

35. Find the angle at which the curves $y = \cos x$ and $y = \tan x$ intersect. Ans. $\frac{\pi}{2}$.

36. For what values of θ is $\tan \theta$ increasing twice as fast as θ?
Ans. $\frac{\pi}{4} + n\pi$.

37. Find the differential of arc length for the curve $y = \tan x$.
Ans. $ds = \sqrt{1 + \sec^4 x} \, dx$.

38. Find the differential of arc length for the curve $y = x \sin x + \cos x$.
Ans. $ds = \sqrt{1 + x^2 \cos^2 x} \, dx$.

Examine the following functions for maxima and minima:

39. $\cos x + \sin x$.   *Ans.* Maxima $\sqrt{2}$ when $x = \dfrac{\pi}{4} + 2n\pi$;

Minima $-\sqrt{2}$ when $x = \dfrac{5\pi}{4} + 2n\pi$.

40. $\sin 2x - x$.   Maxima when $x = \dfrac{\pi}{6} + 2n\pi$;

Minima when $x = -\dfrac{\pi}{6} + 2n\pi$.

41. $2x + \tan x$.   No maxima or minima.

42. $\tan x - \tfrac{1}{2} \tan^2 x$.   Maxima when $x = \dfrac{\pi}{4} + n\pi$.

43. Find the area between one arch of the cosine curve and the x-axis.   *Ans.* 2.

44. Find the area bounded by the curve $y = \sec^2 x$, the x-axis, the y-axis, and the line $x = a$ where a is positive and less than $\pi/2$.   *Ans.* $\tan a$.

45. Find the area bounded by the curve $y = \tan^2 x$, the x-axis, and the line $y = \dfrac{\pi}{4}$.   *Ans.* $1 - \dfrac{\pi}{4}$.

46. A fence 12 ft. high stands on a horizontal base 10 ft. from a vertical wall of stone. What is the length of the shortest ladder that will reach from the ground over the fence to the wall? (Use the angle between the ladder and the base as the independent variable.)

Ans. $(12^{\frac{2}{3}} + 10^{\frac{2}{3}})^{\frac{3}{2}}$.

47. Adjacent sides of a field meet in a right angle. A spring near the corner of the field is 48 ft. from one side and 64 ft. from the adjacent side. A straight road is to be run across the corner by the spring so as to cut off as little of the field as possible. How should the road lie? *Ans.* Its ends should be 96 and 128 ft. from the corner.

48. A passageway 128 in. wide meets at right angles a corridor which is 54 in. wide. What is the longest girder that can be moved horizontally around the corner if the width of the girder is negligible?   *Ans.* 250 in.

49. A steel girder 375 in. long is to be moved horizontally along a passageway 192 in. wide and into a corridor at right angles to it. If the width of the girder is negligible, what is the least width of the corridor?   *Ans.* 81 in.

50. Show that the function
$$y = a \cos mx + b \sin mx$$
satisfies the relation
$$\frac{d^2y}{dx^2} + m^2 y = 0.$$

51. Show that if y is any one of the functions
$$\sin x, \ x \sin x, \ \cos x, \ x \cos x,$$
then y satisfies the "differential equation"
$$\frac{d^4y}{dx^4} + 2\frac{d^2y}{dx^2} + y = 0.$$
Thence show that the same differential equation is satisfied by the function
$$y = c_1 \sin x + c_2 \, x \sin x + c_3 \cos x + c_4 \, x \cos x,$$
where c_1, c_2, c_3, c_4 are any constants whatever.

52. A point moves at uniform speed counterclockwise on the circle $x^2 + y^2 = a^2$, starting at time $t = 0$ from the point $(a, 0)$. Show that its distance from the x-axis is given by the equation $y = a \sin kt$, where k is a constant depending on the speed of rotation. Show that its distance x from the y-axis is $x = a \cos kt$. Show that the y-component of acceleration is $-k^2 y$ and that the x-component of acceleration is $-k^2 x$.

53. If an angle is increasing at a constant rate, show (1) that the sine and the tangent are increasing at the same rate when the angle is zero, and (2) that the sine is increasing one eighth as fast as the tangent when the angle is $\pi/3$.

54. For what angles is the sine increasing as fast as the cosine is decreasing if the angle itself is increasing at a constant rate?

$$Ans. \ \frac{\pi}{4} + n\pi, \text{ where } n \text{ is an integer.}$$

55. A revolving light sending out parallel rays is making one revolution per minute. If it is one mile from a straight shore, at what speed is the light traveling along the shore when it is $\frac{1}{2}$ mi. from the point on the shore nearest to the light? (Take $\pi = 22/7$.)

Ans. 55/7 mi. per minute.

CHAPTER VII

DIFFERENTIATION OF THE INVERSE TRIGONOMETRIC FUNCTIONS

65. Differentiation of arc sin u and arc cos u. In § 32 we have shown that the derivative of the inverse function and that of the direct function are reciprocals each of the other. We shall now employ this principle repeatedly in deriving the formulas for the differentiation of the inverse trigonometric functions. In connection with it we will also use the principle in § 31.

Let $\qquad y = \text{arc sin } u.$

Then $\qquad u = \sin y.$

Hence $\qquad \dfrac{du}{dy} = \cos y = \sqrt{1 - \sin^2 y} = \sqrt{1 - u^2}.$

Therefore $\qquad \dfrac{dy}{du} = \dfrac{1}{\sqrt{1-u^2}},$

by the principle for inverse functions. By the principle in § 31 we have

$$\frac{dy}{dx} = \frac{dy}{du}\frac{du}{dx}.$$

Hence $\qquad \dfrac{dy}{dx} = \dfrac{1}{\sqrt{1-u^2}}\dfrac{du}{dx}.$

Therefore $\qquad \dfrac{d(\text{arc sin } u)}{dx} = \dfrac{1}{\sqrt{1-u^2}}\dfrac{du}{dx},$

$$d(\text{arc sin } u) = \frac{du}{\sqrt{1-u^2}}.$$

Similarly, it may be shown that

$$\frac{d(\text{arc cos } u)}{dx} = -\frac{1}{\sqrt{1-u^2}}\frac{du}{dx},$$

$$d(\text{arc cos } u) = -\frac{du}{\sqrt{1-u^2}}.$$

66. Differentiation of arc tan u and arc cot u. Let
$$y = \text{arc tan } u.$$
Then
$$u = \tan y,$$
$$\frac{du}{dy} = \sec^2 y = 1 + \tan^2 y = 1 + u^2,$$
$$\frac{dy}{du} = \frac{1}{1+u^2}, \quad \frac{dy}{dx} = \frac{1}{1+u^2}\frac{du}{dx}.$$
Therefore
$$\frac{d(\text{arc tan } u)}{dx} = \frac{1}{1+u^2}\frac{du}{dx},$$
$$d(\text{arc tan } u) = \frac{du}{1+u^2}.$$

Similarly, it may be shown that
$$\frac{d(\text{arc cot } u)}{dx} = -\frac{1}{1+u^2}\frac{du}{dx},$$
$$d(\text{arc cot } u) = -\frac{du}{1+u^2}.$$

67. Differentiation of arc sec u and arc csc u. Let
$$y = \text{arc sec } u.$$
Then
$$u = \sec y,$$
$$\frac{du}{dy} = \sec y \tan y = \sec y \sqrt{\sec^2 y - 1} = u\sqrt{u^2-1},$$
$$\frac{dy}{du} = \frac{1}{u\sqrt{u^2-1}}, \quad \frac{dy}{dx} = \frac{1}{u\sqrt{u^2-1}}\frac{du}{dx}.$$
Therefore
$$\frac{d(\text{arc sec } u)}{dx} = \frac{1}{u\sqrt{u^2-1}}\frac{du}{dx},$$
$$d(\text{arc sec } u) = \frac{du}{u\sqrt{u^2-1}}.$$

Similarly, it may be shown that
$$\frac{d(\text{arc csc } u)}{dx} = -\frac{1}{u\sqrt{u^2-1}}\frac{du}{dx},$$
$$d(\text{arc csc } u) = -\frac{du}{u\sqrt{u^2-1}}.$$

68. Differentiation of arc vers u. The versed sine of v, written vers v, is defined by the relation

$$\text{vers } v = 1 - \cos v.$$

Then
$$\cos v = 1 - \text{vers } v.$$

If we let
$$y = \text{arc vers } u,$$

we have
$$u = \text{vers } y = 1 - \cos y.$$

Hence
$$\frac{du}{dy} = \sin y = \sqrt{1 - \cos^2 y} = \sqrt{1 - (1 - \text{vers } y)^2}$$
$$= \sqrt{1 - (1 - u)^2} = \sqrt{2u - u^2}.$$

Therefore
$$\frac{dy}{du} = \frac{1}{\sqrt{2u - u^2}}$$

Hence we have
$$\frac{d(\text{arc vers } u)}{dx} = \frac{1}{\sqrt{2u - u^2}} \frac{du}{dx},$$

$$d(\text{arc vers } u) = \frac{du}{\sqrt{2u - u^2}}.$$

Example 1. Differentiate $\text{arc sin } ax^2$.

We have
$$\frac{d(\text{arc sin } ax^2)}{dx} = \frac{1}{\sqrt{1 - a^2 x^4}} \frac{d}{dx}(ax^2) = \frac{2ax}{\sqrt{1 - a^2 x^4}}.$$

Example 2. Differentiate $\text{arc tan } \dfrac{x^2 + 1}{x^2 - 1}$.

We have
$$\frac{d}{dx}\left(\text{arc tan } \frac{x^2+1}{x^2-1}\right) = \frac{1}{1 + \left(\dfrac{x^2+1}{x^2-1}\right)^2} \frac{d}{dx}\left(\frac{x^2+1}{x^2-1}\right)$$
$$= \frac{(x^2-1)^2}{(x^2-1)^2 + (x^2+1)^2} \frac{(x^2-1)2x - (x^2+1)2x}{(x^2-1)^2}$$
$$= -\frac{2x}{x^4+1}.$$

69. Summary. In this chapter we have proved the following fundamental formulas for differentiation:

$$\frac{d(\text{arc sin } u)}{dx} = \frac{1}{\sqrt{1 - u^2}} \frac{du}{dx}.$$

$$\frac{d(\text{arc cos } u)}{dx} = -\frac{1}{\sqrt{1 - u^2}} \frac{du}{dx}.$$

$$\frac{d(\text{arc tan } u)}{dx} = \frac{1}{1+u^2}\frac{du}{dx}.$$

$$\frac{d(\text{arc cot } u)}{dx} = -\frac{1}{1+u^2}\frac{du}{dx}.$$

$$\frac{d(\text{arc sec } u)}{dx} = \frac{1}{u\sqrt{u^2-1}}\frac{du}{dx}.$$

$$\frac{d(\text{arc csc } u)}{dx} = -\frac{1}{u\sqrt{u^2-1}}\frac{du}{dx}.$$

$$\frac{d(\text{arc vers } u)}{dx} = \frac{1}{\sqrt{2u-u^2}}\frac{du}{dx}.$$

EXERCISES

The following are useful formulas which it is worth while to commit to memory. Prove them.

1. $\dfrac{d}{dx}\left(\text{arc sin }\dfrac{x}{a}\right) = \dfrac{1}{\sqrt{a^2-x^2}}.$

2. $\dfrac{d}{dx}\left(\text{arc cos }\dfrac{x}{a}\right) = -\dfrac{1}{\sqrt{a^2-x^2}}.$

3. $\dfrac{d}{dx}\left(\text{arc tan }\dfrac{x}{a}\right) = \dfrac{a}{a^2+x^2}.$

4. $\dfrac{d}{dx}\left(\text{arc cot }\dfrac{x}{a}\right) = -\dfrac{a}{a^2+x^2}.$

5. $\dfrac{d}{dx}\left(\text{arc sec }\dfrac{x}{a}\right) = \dfrac{a}{x\sqrt{x^2-a^2}}.$

6. $\dfrac{d}{dx}\left(\text{arc csc }\dfrac{x}{a}\right) = -\dfrac{a}{x\sqrt{x^2-a^2}}.$

7. $\dfrac{d}{dx}\left(\text{arc vers }\dfrac{x}{a}\right) = \dfrac{1}{\sqrt{2ax-x^2}}.$

Prove the following formulas:

8. $\dfrac{d}{dx}\left(\text{arc tan }\dfrac{2x}{1-x^2}\right) = \dfrac{2}{1+x^2}.$

9. $\dfrac{d}{dx}\left(\text{arc csc }\dfrac{1}{2x^2-1}\right) = \dfrac{2}{\sqrt{1-x^2}}.$

10. $\dfrac{d}{dx}\left(x\sqrt{a^2-x^2}+a^2\arcsin\dfrac{x}{a}\right)=2\sqrt{a^2-x^2}.$

11. $\dfrac{d}{dx}\left(\sqrt{a^2-x^2}+a\arcsin\dfrac{x}{a}\right)=\sqrt{\dfrac{a-x}{a+x}}.$

12. $\dfrac{d}{dx}\left(\arctan\dfrac{x+a}{1-ax}\right)=\dfrac{1}{1+x^2}.$

13. $\dfrac{d}{dx}\left(\operatorname{arc\,sec}\dfrac{1}{\sqrt{1-x^2}}\right)=\dfrac{1}{\sqrt{1-x^2}}.$

14. $\dfrac{d}{dx}(x\arcsin x)=\arcsin x+\dfrac{x}{\sqrt{1-x^2}}.$

15. $\dfrac{d}{dx}[\cos(\arccos x)]=1.$

Perform the following indicated differentiations:

16. $\dfrac{d}{dx}\operatorname{arc\,vers} 2x^2=\dfrac{2}{\sqrt{1-x^2}}.$

17. $\dfrac{d}{dx}\operatorname{arc\,vers}\dfrac{2x^2}{1+x^2}=\dfrac{2}{1+x^2}.$

18. $\dfrac{d}{dx}\left(\sqrt{2ax-x^2}+a\operatorname{arc\,vers}\dfrac{x}{a}\right)=\dfrac{1}{x}\sqrt{2ax-x^2}.$

19. $\dfrac{d}{dx}\left[(x-a)\sqrt{2ax-x^2}+a^2\operatorname{arc\,vers}\dfrac{x}{a}\right]=2\sqrt{2ax-x^2}.$

Find the antiderivatives of the following functions:

20. $\sqrt{a^2-x^2}.$ Ans. $\dfrac{x}{2}\sqrt{a^2-x^2}+\dfrac{a^2}{2}\arcsin\dfrac{x}{a}+c.$

21. $\sqrt{2ax-x^2}.$ Ans. $\dfrac{x-a}{2}\sqrt{2ax-x^2}+\dfrac{a^2}{2}\operatorname{arc\,vers}\dfrac{x}{a}+c.$

22. $\dfrac{1}{a^2+x^2}.$ Ans. $\dfrac{1}{a}\arctan\dfrac{x}{a}+c.$

23. $\dfrac{1}{\sqrt{a^2-x^2}}.$ Ans. $\arcsin\dfrac{x}{a}+c;$ or $-\arccos\dfrac{x}{a}+c.$

24. $\dfrac{1}{x\sqrt{x^2-a^2}}.$ Ans. $\dfrac{1}{a}\operatorname{arc\,sec}\dfrac{x}{a}+c.$

25. Find the antiderivative of $\dfrac{1}{x^2+2x+5}.$

We have $\dfrac{1}{x^2+2x+5}=\dfrac{1}{x^2+2x+1+4}=\dfrac{1}{2^2+(x+1)^2}.$

Since $d(x+1) = dx$ it follows from problem 22 that the required antiderivative is
$$\frac{1}{2} \arctan \frac{x+1}{2} + c.$$

The student may verify this by differentiation.

Find the antiderivatives of the following functions:

26. $\dfrac{1}{x^2 + 4x + 13}$. \hspace{2em} Ans. $\dfrac{1}{3} \arctan \dfrac{x+2}{3} + c.$

27. $\dfrac{1}{(x+2)\sqrt{x^2 + 4x + 3}}$. \hspace{2em} Ans. $\operatorname{arc\,sec}(x+2) + c.$

28. $\dfrac{1}{\sqrt{9-x^2}}$. \hspace{2em} Ans. $\arcsin \dfrac{x}{3} + c.$

29. $\dfrac{1}{\sqrt{5-4x-x^2}}$. \hspace{2em} Ans. $\arcsin \dfrac{x+2}{3} + c.$

30. By aid of the result in problem 20 find the area of that part of the circle $x^2 + y^2 = a^2$ which lies in the first quadrant, and thence show that the entire area of the circle is πa^2.

31. In a similar way find the area of the ellipse
$$\frac{x^2}{a^2} + \frac{y^2}{b^2} = 1. \hspace{3em} \text{Ans. } \pi ab.$$

32. Find the area bounded by the curve
$$y = \frac{1}{a^2 + x^2},$$
the x-axis, the y-axis, and the ordinate $x = t$ where t is a positive number. Find the limit of this area as t becomes infinite. (This is said to be the area bounded by the given curve and the y-axis and the positive part of the x-axis. Draw a figure showing this area extending off to infinity.) \hspace{1em} Ans. $\dfrac{1}{a} \arctan \dfrac{t}{a}, \dfrac{\pi}{2a}.$

33. Find the area bounded by the curve
$$y = \frac{1}{x^2 + 4x + 13},$$
the y-axis, and the positive part of the x-axis. \hspace{1em} Ans. $\dfrac{\pi}{6} - \dfrac{1}{3} \arctan \dfrac{2}{3}.$

34. What is the area bounded by the circle $x^2 + y^2 = a^2$ and the line $x + y = a$? \hspace{2em} Ans. $\tfrac{1}{4} a^2(\pi - 2).$

35. What is the area bounded by the ellipse
$$\frac{x^2}{a^2} + \frac{y^2}{b^2} = 1$$
and the line $\frac{x}{a} + \frac{y}{b} = 1$? *Ans.* $\tfrac{1}{4} ab (\pi - 2)$.

36. Show that the function arc tan x has no maximum or minimum.

37. Show that the function
$$\text{arc tan } \frac{1+x}{1-x}$$
has no maximum or minimum.

38. Investigate the function 2 arc tan $x - x$ for maxima and minima. *Ans.* Maximum for $x = 1$; minimum for $x = -1$.

CHAPTER VIII

DIFFERENTIATION OF THE LOGARITHMIC AND EXPONENTIAL FUNCTIONS

70. Differentiation of a logarithm. We shall now employ the Four-Step Rule for finding the derivative of a logarithm. In carrying out the work we need to employ the limit discussed in § 17. The student is therefore advised to review that section before proceeding with this one.

Let
$$y = \log_a u,$$
where a is a positive constant and u is a positive variable. Then we have
$$y + \Delta y = \log_a(u + \Delta u).$$
Hence
$$\Delta y = \log_a(u + \Delta u) - \log_a u$$
$$= \log_a \frac{u + \Delta u}{u} = \log_a\left(1 + \frac{\Delta u}{u}\right)$$
by the principle for the logarithm of a quotient. Therefore
$$\frac{\Delta y}{\Delta u} = \frac{1}{\Delta u} \log_a\left(1 + \frac{\Delta u}{u}\right)$$
$$= \frac{1}{u} \cdot \frac{u}{\Delta u} \log_a\left(1 + \frac{\Delta u}{u}\right)$$
$$= \frac{1}{u} \log_a\left(1 + \frac{\Delta u}{u}\right)^{\frac{u}{\Delta u}}$$
by the principle for the logarithm of a power, namely,
$$\log_a s^m = m \log_a s.$$
But
$$\lim_{\Delta u \to 0} \left(1 + \frac{\Delta u}{u}\right)^{\frac{u}{\Delta u}} = e,$$
as we see by taking $t = \Delta u/u$ in the limit
$$\lim_{t \to 0} (1 + t)^{\frac{1}{t}}$$

discussed in § 17. Therefore, from the equation giving the value of $\Delta y/\Delta u$ we have

$$\frac{dy}{du} = \frac{1}{u}\log_a e.$$

Hence $\quad \dfrac{d(\log_a u)}{dx} = \log_a e \cdot \dfrac{1}{u}\dfrac{du}{dx} = \dfrac{1}{\log a}\dfrac{1}{u}\dfrac{du}{dx},\quad$ [by § 17]

$$d(\log_a u) = \log_a e \cdot \frac{du}{u} = \frac{1}{\log a}\frac{du}{u}.$$

If a has the special value e, we have

$$\frac{d(\log u)}{dx} = \frac{1}{u}\frac{du}{dx},$$

$$d(\log u) = \frac{du}{u}.$$

Example 1. Differentiate $\log(a^2 + x^2)$. We have

$$\frac{d}{dx}\log(a^2 + x^2) = \frac{1}{a^2 + x^2}\frac{d}{dx}(a^2 + x^2) = \frac{2x}{a^2 + x^2}.$$

Example 2. Differentiate $\log_a(1 + x^2)$. We have

$$\frac{d}{dx}\log_a(1 + x^2) = \frac{1}{\log a} \cdot \frac{1}{1 + x^2}\frac{d}{dx}(1 + x^2)$$

$$= \frac{2x}{(1 + x^2)\log a}.$$

71. Differentiation of the simple exponential function. Let us write
$$y = a^u, \quad a > 0.$$

Taking the logarithms of both sides, using the base e, we have

$$\log y = u \log a.$$

Differentiating both sides with respect to x, we have

$$\frac{1}{y}\frac{dy}{dx} = \frac{du}{dx}\log a.$$

Multiplying both members by y, we have

$$\frac{dy}{dx} = y\frac{du}{dx}\log a = a^u \frac{du}{dx}\log a.$$

Therefore
$$\frac{d(a^u)}{dx} = \log a \cdot a^u \frac{du}{dx},$$
$$d(a^u) = \log a \cdot a^u \, du.$$

If a has the special value e, we have
$$\frac{d(e^u)}{dx} = e^u \frac{du}{dx},$$
$$d(e^u) = e^u \, du.$$

Example. Differentiate $e^{a^2+x^2}$.

We have $\dfrac{d}{dx} e^{a^2+x^2} = e^{a^2+x^2} \dfrac{d}{dx}(a^2+x^2) = 2x e^{a^2+x^2}$.

72. Differentiation of the general exponential function. Let
$$y = u^v,$$
where u and v are both functions of x, u being positive in value. Taking the logarithm of both sides to the base e, we have
$$\log y = v \log u.$$
We can differentiate both sides of this equation with respect to x, employing the principle for the derivative of a logarithm and that for the derivative of a product. Thus we have
$$\frac{1}{y}\frac{dy}{dx} = \log u \cdot \frac{dv}{dx} + v \cdot \frac{1}{u}\frac{du}{dx}.$$

Hence
$$\frac{dy}{dx} = y \log u \frac{dv}{dx} + v \cdot \frac{y}{u}\frac{du}{dx}$$
$$= u^v \log u \frac{dv}{dx} + v u^{v-1}\frac{du}{dx}.$$

Therefore
$$\frac{d(u^v)}{dx} = u^v \log u \frac{dv}{dx} + v u^{v-1}\frac{du}{dx}.$$

The derivative of a variable with a variable exponent is equal to the sum of the two terms, one of which is that which would be obtained if the base were a constant, and the other that which would be obtained if the exponent were a constant.

If we have the special case
$$y = u^n,$$

where n is a constant, we can carry out the same argument and derive the formula
$$\frac{d(u^n)}{dx} = nu^{n-1}\frac{du}{dx}.$$
This is a formula that we had already *proved* for the case when n is a positive integer. The present argument establishes this formula for all values of the constant exponent n.

Example. Differentiate x^{x^2}.

We have
$$\frac{d}{dx}x^{x^2} = x^2 x^{x^2-1}\frac{d}{dx}(x) + \log x \cdot x^{x^2}\frac{d}{dx}(x^2)$$
$$= x^2 x^{x^2-1} + \log x \cdot x^{x^2} \cdot 2x$$
$$= x^{x^2+1}(1 + 2\log x).$$

73. Logarithmic differentiation. The method of argument employed in the last two sections is often useful in differentiating certain classes of functions, especially products and quotients with fractional exponents. It is known as the method of logarithmic differentiation. We shall illustrate it by means of a few examples.

Example 1. Differentiate $y = \log(a^2 - x^2)^{\frac{1}{2}}$.

We have $y = \frac{1}{2}\log(a^2 - x^2)$.

Hence $\dfrac{dy}{dx} = \dfrac{1}{2}\cdot\dfrac{1}{a^2 - x^2}\dfrac{d}{dx}(a^2 - x^2) = -\dfrac{x}{a^2 - x^2}.$

Example 2. Differentiate $y = \log\sqrt{\dfrac{a^2 - x^2}{a^2 + x^2}}$.

We have
$$y = \tfrac{1}{2}\log(a^2 - x^2) - \tfrac{1}{2}\log(a^2 + x^2).$$

Hence $\dfrac{dy}{dx} = \dfrac{1}{2}\dfrac{1}{a^2 - x^2}\dfrac{d}{dx}(a^2 - x^2) - \dfrac{1}{2}\dfrac{1}{a^2 + x^2}\dfrac{d}{dx}(a^2 + x^2)$

$$= -\frac{x}{a^2 - x^2} - \frac{x}{a^2 + x^2}$$
$$= -\frac{2a^2 x}{a^4 - x^4}.$$

Example 3. Differentiate $y = \dfrac{(x-1)^{\frac{3}{2}}(x+1)^{\frac{1}{2}}}{(x-2)^{\frac{2}{3}}}$.

We have
$$\log y = \tfrac{3}{2}\log(x-1) + \tfrac{1}{2}\log(x+1) - \tfrac{2}{3}\log(x-2).$$

Differentiating with respect to x, we obtain

$$\frac{1}{y}\frac{dy}{dx} = \frac{3}{2(x-1)} + \frac{1}{2(x+1)} - \frac{2}{3(x-2)}$$

$$= \frac{4x^2 - 9x - 4}{3(x^2-1)(x-2)}.$$

Therefore
$$\frac{dy}{dx} = \frac{(x-1)^{\frac{3}{2}}(x+1)^{\frac{1}{2}}}{(x-2)^{\frac{2}{3}}} \cdot \frac{4x^2 - 9x - 4}{3(x^2-1)(x-2)}$$

$$= \frac{(x-1)^{\frac{1}{2}}(4x^2 - 9x - 4)}{3(x+1)^{\frac{1}{2}}(x-2)^{\frac{5}{3}}}.$$

Example 4. Differentiate $y = x^{\frac{1}{x}}$.

We have $\log y = \frac{1}{x} \log x.$

Therefore $\frac{1}{y}\frac{dy}{dx} = -\frac{1}{x^2} \log x + \frac{1}{x^2} = \frac{1}{x^2}(1 - \log x).$

Hence $\frac{dy}{dx} = x^{\frac{1}{x}} \cdot \frac{1}{x^2}(1 - \log x) = x^{\frac{1}{x} - 2}(1 - \log x).$

EXERCISES

Perform the following indicated differentiations:

1. $\dfrac{d}{dx} \log(x^2 + x + 1) = \dfrac{2x+1}{x^2+x+1}.$

2. $\dfrac{d}{dx} \log_a\left(\dfrac{x+k}{x-k}\right) = -\dfrac{2k}{x^2-k^2} \log_a e.$

3. $\dfrac{d}{dx} \log(\log x) = \dfrac{1}{x \log x}.$

4. $\dfrac{d}{dx} x \log x = 1 + \log x.$

5. $\dfrac{d}{dx} \log x^4 = \dfrac{4}{x}.$

6. $\dfrac{d}{dx} \log^4 x = \dfrac{4 \log^3 x}{x}.$

7. $\dfrac{d}{dx} \log(e^x + e^{-x}) = \dfrac{e^x - e^{-x}}{e^x + e^{-x}}.$

8. $\dfrac{d}{dx} \log \tan x = \dfrac{1}{\sin x \cos x}.$

9. $\dfrac{d}{dx} e^{ax} = ae^{ax}$.

10. $\dfrac{d}{dx} e^{ax^2+bx+c} = (2\,ax+b)e^{ax^2+bx+c}$.

11. $\dfrac{d}{dx} 8^{x^3+7} = 9 \log 2 \cdot x^2 8^{x^3+7}$.

12. $\dfrac{d}{dx} \dfrac{e^x + e^{-x}}{e^x - e^{-x}} = -\dfrac{4}{(e^x + e^{-x})^2}$.

13. $\dfrac{d}{dx} e^{\cos x} = -\sin x \cdot e^{\cos x}$.

14. $\dfrac{d}{dx} \log \dfrac{e^x}{e^x - 1} = -\dfrac{1}{e^x - 1}$.

15. $\dfrac{d}{dx} x^3 e^{x^2} = x^2 e^{x^2}(2\,x^2 + 3)$.

Find the derivatives of the functions in problems 16 to 26:

16. x^x.   Ans. $x^x(1 + \log x)$.

17. x^{x^x}.   Ans. $x^{x^x} x^x \left(\log x + \log^2 x + \dfrac{1}{x} \right)$.

18. x^{e^x}.   Ans. $x^{e^x} e^x \left(\dfrac{1}{x} + \log x \right)$.

19. e^{x^x}.   Ans. $e^{x^x} x^x (1 + \log x)$.

20. e^{e^x}.   Ans. $e^{e^x} e^x$.

21. $\dfrac{a}{2}\left(e^{\frac{x}{a}} - e^{-\frac{x}{a}}\right)$.   Ans. $\tfrac{1}{2}\left(e^{\frac{x}{a}} + e^{-\frac{x}{a}}\right)$.

22. $e^{-x} \log x$.   Ans. $e^{-x}\left(\dfrac{1}{x} - \log x\right)$.

23. $\log \dfrac{\sqrt{x^2+1}+x}{\sqrt{x^2+1}-x}$.   Ans. $\dfrac{2}{\sqrt{x^2+1}}$.

24. $\log(x + \sqrt{x^2 + a^2})$.   Ans. $(x^2 + a^2)^{-\frac{1}{2}}$.

25. $x(x+1)\sqrt{x-1}$.   Ans. $\tfrac{1}{2}(x-1)^{-\frac{1}{2}}(5x^2 - x - 2)$.

26. $\dfrac{x(x^2+1)}{\sqrt{x^2-1}}$.   Ans. $\dfrac{2x^4 - 3x^2 - 1}{(x^2-1)^{\frac{3}{2}}}$.

27. Show that the nth derivative of $x^{n-1} \log x$ is $(n-1)!/x$.

28. If $y = e^{-x} \cos x$, show that
$$\dfrac{d^4y}{dx^4} + 4y = 0.$$

29. If $y = a \cos \log x + b \sin \log x$, show that
$$x^2 \frac{d^2y}{dx^2} + x \frac{dy}{dx} + y = 0.$$

30. If $y = \sin(m \arcsin x)$, show that
$$(1 - x^2) \frac{d^2y}{dx^2} = x \frac{dy}{dx} - m^2 y.$$

31. If $y^2 = \sec 2x$, show that
$$\frac{d^2y}{dx^2} + y = 3 y^5.$$

32. If $y = ae^{mx} + be^{nx}$, show that
$$\frac{d^2y}{dx^2} - (m+n) \frac{dy}{dx} + mny = 0.$$

33. If $y = ae^x + be^{2x} + ce^{3x}$, show that
$$\frac{d^3y}{dx^3} - 6 \frac{d^2y}{dx^2} + 11 \frac{dy}{dx} - 6y = 0.$$

74. Applications involving exponential functions. The various sorts of applications of derivatives already treated occur in problems involving exponential functions. The latter have a special property which makes them useful also in applications to another sort of problem.

Let us consider the function
$$y = ce^{kx},$$
where c and k are constants. We have
$$\frac{dy}{dx} = cke^{kx} = ky.$$

Hence the rate of change of the given function y with respect to x is proportional to y, the factor of proportionality being k.

Let us consider the converse problem; that is, let y be any function of x whose rate of change with respect to x is k times the function itself, and let us find the general form of such a function. By hypothesis we have
$$\frac{dy}{dx} = ky.$$

Hence
$$\frac{1}{y}\frac{dy}{dx} = k.$$

A function having the first member for its derivative is $\log y$; a function having the second member for its derivative is kx. Since the derivative of $\log y$ and that of kx are equal (in accordance with the last equation), and since two functions having the same derivative differ by a constant (theorem IV of § 26), it follows that
$$\log y = kx + c_1.$$
Hence
$$y = e^{kx+c_1} = e^{c_1}e^{kx}.$$
Writing c for e^{c_1}, we have $y = ce^{kx}$.

The results in the last two paragraphs may be summarized into the following theorem:

The rate of change with respect to x of the function ce^{kx} is proportional to the function, the factor of proportionality being k. If y is any function of x whose rate of change with respect to x is k times the function itself, then y can be written in the form $y = ce^{kx}$.

When a variable y changes with respect to a variable x in such a way that the rate of change of y with respect to x is proportional to y, then y is said to vary with respect to x in accordance with the *compound-interest law*. (The law is also sometimes called the *law of organic growth*.)

Many natural phenomena follow the compound-interest law. We shall give a few examples.

Example 1. When the temperature of the atmosphere is the same at different heights, the rate of change of air pressure with respect to the height is proportional to the pressure. Find the formula for pressure in terms of the height.

Let p denote pressure and h denote height. Then we have
$$\frac{dp}{dh} = -kp,$$
where k is a positive constant. From the theorem of this section we have
$$p = p_0 e^{-kh},$$
where p_0 is a constant. If 0 is substituted for h it may be seen that p_0 is the value of p when $h = 0$; that is, p_0 is the pressure at the height $h = 0$.

Example 2. If a body has a temperature τ which is higher than the temperature τ_1 of the surrounding medium, then the body cools at a rate which is approximately proportional to the difference between its temperature and that of the surrounding medium. Find an approximate formula for its temperature in terms of the time t of cooling.

We have approximately
$$\frac{d\tau}{dt} = -k(\tau - \tau_1),$$
where k is a positive constant. Since $d\tau = d(\tau - \tau_1)$, this may be written in the form
$$\frac{d(\tau - \tau_1)}{dt} = -k(\tau - \tau_1).$$

Treating $\tau - \tau_1$ as the dependent variable and applying the theorem of this section, we have
$$\tau - \tau_1 = ce^{-kt},$$
where c is a constant. Let τ_0 denote the temperature of the body at time $t = 0$. Substituting τ_0 for τ and 0 for t in the last equation, we have
$$\tau_0 - \tau_1 = c.$$

Using this value of c, we have from the next preceding equation the required result
$$\tau = \tau_1 + (\tau_0 - \tau_1)e^{-kt}.$$

Example 3. When a body moves in a fluid (as air or water), it encounters a resistance which is approximately proportional to its velocity v, provided that the velocity is in a suitable range. Find a formula for the velocity v in terms of the time t. Also find a similar formula for the distance s.

Since the resistance is to be taken as proportional to v, we have an acceleration equal to $-kv$, where k is a positive constant. But the acceleration is also given by the derivative dv/dt (§ 41). Hence
$$\frac{dv}{dt} = -kv.$$

Applying the theorem, we have
$$v = v_0 e^{-kt},$$
where v_0 is the initial velocity (that is, the velocity at time $t = 0$). This is the first result required.

Now $v = ds/dt$. Hence we have
$$\frac{ds}{dt} = v_0 e^{-kt}.$$

From this it may be seen that
$$s = -\frac{v_0}{k} e^{-kt} + c,$$
where c is some constant. Let s_0 denote the value of s when $t = 0$. Then from the last equation we have
$$s_0 = -\frac{v_0}{k} + c, \quad \text{or} \quad c = s_0 + \frac{v_0}{k}.$$
Therefore
$$s = s_0 + \frac{v_0}{k}(1 - e^{-kt}).$$
This is the second expression required.

It is natural to measure s so that s_0 is zero. In this case we have
$$s = \frac{v_0}{k}(1 - e^{-kt}).$$

Example 4. If a disc is rotating rapidly in a fluid (as in the case of a gyroscope in air), it encounters a resistance which is approximately proportional to the square of the speed ω of rotation. Find an expression for ω in terms of the time t.

The angular acceleration is $-k\omega^2$, where k is a positive constant. But angular acceleration is given by the formula $d\omega/dt$ (§ 42). Hence we have
$$\frac{d\omega}{dt} = -k\omega^2.$$
From this equation we have
$$-\frac{d\omega}{\omega^2} = k\,dt.$$
Finding the antiderivative of each member and employing theorem IV of § 26, we have
$$\frac{1}{\omega} = kt + c.$$
If we let ω_0 be the value of ω when $t = 0$, we have
$$\frac{1}{\omega_0} = c.$$
Putting this value of c in the next preceding equation, we have
$$\frac{1}{\omega} = kt + \frac{1}{\omega_0} = \frac{k\omega_0 t + 1}{\omega_0}.$$
Therefore
$$\omega = \frac{\omega_0}{k\omega_0 t + 1}.$$
This is the expression sought.

EXERCISES

1. If a disk is rotating slowly in a fluid, it encounters a resistance which is approximately proportional to the angular speed ω, so that the acceleration is approximately $-k\omega$. Show that $\omega = \omega_0 e^{-kt}$, where ω_0 is the initial angular speed.

2. If a boat is moving in a straight line in water under its own inertia, it encounters a resistance which is approximately proportional to its velocity (until the velocity becomes very small). If the initial velocity is v_0, show that the expression for the velocity is of the form $v = v_0 e^{-kt}$.

3. Show that the distance which the boat of problem 2 moves in time t is approximately
$$\frac{v_0}{k}(1 - e^{-kt}).$$

4. If cane sugar is in solution in the presence of acids, it decomposes into other substances at a rate which is approximately proportional to the mass of the sugar still unchanged. If m is the original mass of the sugar and x is the mass that has been changed at time t, show that x is given by an expression of the form
$$x = m(1 - e^{-kt}).$$

5. Find the area bounded by the curve
$$y = \frac{a}{2}\left(e^{\frac{x}{a}} + e^{-\frac{x}{a}}\right),$$
the x-axis, and the ordinates $x = -a$ and $x = a$.   Ans. $a^2(e - e^{-1})$.

6. Find the area bounded by the curve
$$y = \frac{a}{2}\left(e^{\frac{x}{a}} + e^{-\frac{x}{a}}\right)$$
and the straight line $y = \frac{a}{2}(e + e^{-1})$.   Ans. $2a^2/e$.

7. Show that the differential ds of arc length for the curve
$$y = \frac{a}{2}\left(e^{\frac{x}{a}} + e^{-\frac{x}{a}}\right)$$
is given by the formula $ds = \frac{1}{2}\left(e^{\frac{x}{a}} + e^{-\frac{x}{a}}\right)dx$.

8. Find the length of the curve in the last problem from the point for which $x = 0$ to that for which $x = a$. (Use the value of ds/dx from the preceding problem.)   Ans. $\frac{a}{2}(e - e^{-1})$.

9. Find the length of the subtangent and the subnormal for the catenary
$$y = \frac{a}{2}\left(e^{\frac{x}{a}} + e^{-\frac{x}{a}}\right).$$

Ans. $\dfrac{a\left(e^{\frac{x}{a}} + e^{-\frac{x}{a}}\right)}{e^{\frac{x}{a}} - e^{-\frac{x}{a}}}$, $\dfrac{a}{4}\left(e^{\frac{2x}{a}} - e^{-\frac{2x}{a}}\right)$.

10. At what angle does the curve $y = \log x$ cut the x-axis? Ans. $\dfrac{\pi}{4}$.

11. If the distance described by a moving particle is given by the formula
$$s = ae^t + be^{-t},$$
show that the tangential acceleration is numerically equal to the distance passed over.

Investigate the following functions for maxima and minima:

12. $e^x + e^{-x} - x^2$.   Ans. Minimum for $x = 0$.

13. x^x.   Ans. Minimum for $x = 1/e$.

14. $\dfrac{x}{\log x}$.   Ans. Minimum for $x = e$.

15. $x^{\frac{1}{x}}$.   Ans. Maximum for $x = e$.

75. Summary. In this chapter we have proved the following fundamental formulas for differentiation:

$$\frac{d(\log_a u)}{dx} = \log_a e \cdot \frac{1}{u}\frac{du}{dx}.$$

$$\frac{d(\log u)}{dx} = \frac{1}{u}\frac{du}{dx}.$$

$$\frac{d(a^u)}{dx} = \log a \cdot a^u \frac{du}{dx}.$$

$$\frac{d(e^u)}{dx} = e^u \frac{du}{dx}.$$

$$\frac{d(u^v)}{dx} = vu^{v-1}\frac{du}{dx} + u^v \log u \cdot \frac{dv}{dx}.$$

We have also treated (in § 74) several important applications involving exponential functions.

CHAPTER IX

INTEGRATION

76. A fundamental process of summation. The differential calculus is based on the limit of a certain *quotient of differences*, namely

$$\lim_{\Delta x = 0} \frac{\Delta y}{\Delta x}.$$

Here Δx is the difference of two values of the independent variable x, and Δy is the difference of two values of the dependent variable y.

In a similar way the integral calculus is based on the limit of a *sum of products*; and this limit plays a fundamental rôle in the integral calculus similar in importance to that of the former limit in the differential calculus. The two types of limits are intimately related, as we shall see later. We shall now explain this limit of a sum of products by means of a geometric interpretation.

Let us consider the area A bounded by the curve $y = f(x)$, the x-axis, and the ordinates $x = a$ and $x = b$, as in the figure on page 133. We assume that $f(x)$ is a positive continuous function. We may divide this area into a large number of strips (of equal or unequal width) by means of ordinates drawn to the curve at a large number of intermediate points. The area A is equal to the sum of the areas of these strips.

Let us now replace the ith strip by the rectangle r_i whose height is equal to the least ordinate in the strip. We shall denote the sum of the areas of all these rectangles by

$$\sum_{a}^{b} r_i.$$

Then it is clear that $\sum_{a}^{b} r_i \leq A.$

In a similar way let us also replace the ith strip by the rectangle R_i whose height is equal to the greatest ordinate in the strip. We denote the sum of the areas of all these rectangles by

$$\sum_a^b R_i.$$

It is clear that

$$\sum_a^b R_i \geqq A.$$

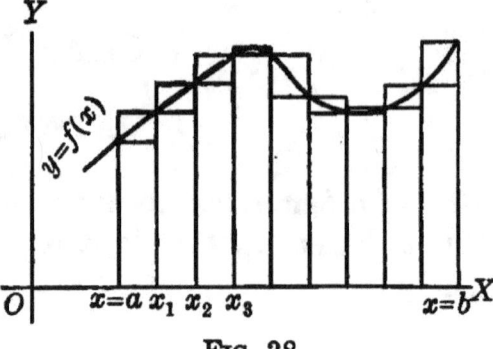

FIG. 38

Let us denote by ρ_i the small rectangle which gives the difference between R_i and r_i; and let us consider the sum

$$\sum_a^b \rho_i, \quad \text{or} \quad \sum_a^b (R_i - r_i),$$

of all these small rectangles. If every strip is very narrow, it may be seen from the figure that this sum is small. Let us now consider what value is approached by the sum

$$\sum_a^b \rho_i$$

as the number of strips is increased indefinitely and the width of each of them approaches zero. This sum represents the sum of the areas of a row of very small rectangles lying along the curve. It is apparent that the sum must approach zero as the number of strips increases indefinitely and the width of each approaches zero, since the sum of the bases of the small rectangles is $(b-a)$ and the greatest height of a small rectangle approaches zero. That is, we have

$$\lim \sum_a^b \rho_i = 0, \quad \text{or} \quad \lim \sum_a^b (R_i - r_i) = 0,$$

when the number of strips increases indefinitely and the width of each approaches zero.

But we have seen that

$$\sum_a^b R_i \geqq A \geqq \sum_a^b r_i.$$

Consider the three members of this inequality. The difference between the first and the last approaches zero, while A remains intermediate in value between the two. Hence

$$\lim \sum_a^b R_i = A, \quad \lim \sum_a^b r_i = A,$$

when the number of strips increases indefinitely and the width of each approaches zero. This result we shall employ presently in evaluating a fundamental limit.

Let us now consider another set of rectangles. The ith rectangle in the set (counted from the left) shall have the same base as the rectangles R_i and r_i. We shall denote the length of the base by Δx_i, since it is an increment of x; the subscript is used to distinguish the various increments corresponding to the various bases. Let ξ_i be the value of x corresponding to any point on the base of this ith rectangle. The corresponding ordinate is $f(\xi_i)$. This ith rectangle is then to have the base Δx_i and the height $f(\xi_i)$. Its area is $f(\xi_i)\Delta x_i$. Then it is clear that

$$R_i \geq f(\xi_i)\Delta x_i \geq r_i.$$

Taking the sum of the rectangles over the whole interval (a, b), we have

$$\sum_a^b R_i \geq \sum_a^b f(\xi_i)\Delta x_i \geq \sum_a^b r_i.$$

If the number of strips increases indefinitely and the width of each approaches zero, the first and third members of this inequality approach the value A, as we have already seen. Hence the second, being intermediate between these two, also approaches the value A. That is, we have

$$\lim \sum_a^b f(\xi_i)\Delta x_i = A$$

when the number of strips increases indefinitely and the width of each approaches zero.

77. The definite integral. The limiting process involved in the first member of the last equation is a fundamental one in the integral calculus. We have arrived at it from certain geometric considerations. We wish now to give an analytical formulation of it.

Let us consider the continuous function $y = f(x)$ when x ranges over the interval (a, b). Divide the interval (a, b) into n parts and denote the successive parts from the left to the right by
$$\Delta x_1, \Delta x_2, \Delta x_3, \cdots, \Delta x_n.$$
Let ξ_i be a value of x on the interval Δx_i. The corresponding value y_i of y is $f(\xi_i)$. Form the product $f(\xi_i)\Delta x_i$. Consider the sum of all such products for the intervals of (a, b), and denote the sum by
$$\sum_a^b f(\xi_i)\Delta x_i.$$
This sum has a limit when the number of intervals increases indefinitely and the length of each approaches zero (compare the previous section). That is, the limit
$$\lim_{\text{Every } \Delta x_i = 0} \sum_a^b f(\xi_i)\Delta x_i$$
exists.

This limit is called *the definite integral* of $f(x)$ from a to b. This definite integral is denoted by the symbol
$$\int_a^b f(x)dx.$$
The integral sign $\int$ is an elongated S; it stands for the word *sum*.

Summarizing our definition of the definite integral for the interval (a, b), we may write
$$\lim_{\text{Every } \Delta x_i = 0} \sum_a^b f(\xi_i)\Delta x_i = \int_a^b f(x)dx.$$
The integral from b to a we define by the formula
$$\int_b^a f(x)dx = -\int_a^b f(x)dx.$$

In §76 we saw that the area $ABCD$ bounded by the curve $y = f(x)$, the x-axis, and the ordinates $x = a$ and $x = b$ is represented by the same limit as that which we have just given as the definition of the definite integral
$$\int_a^b f(x)dx.$$
Hence
$$\int_a^b f(x)dx = \text{area of } ABCD.$$

In § 27 we saw that the antiderivative may also be used to compute the same area. Consequently there must be an intimate connection between the definite integral and the antiderivative. We shall now investigate that connection. (The student is advised to review §§ 26, 27, and 30.)

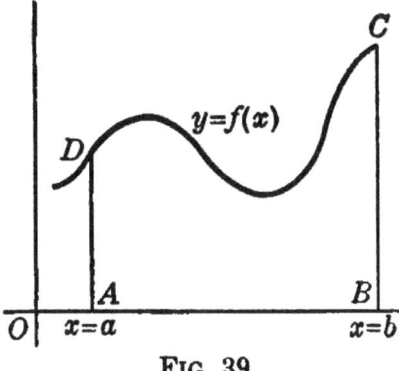

Fig. 39

Let $\phi(x)$ be any particular antiderivative of $f(x)$; and let $A(x)$ be that particular antiderivative (see § 25) which represents the area bounded by the curve $y = f(x)$, the x-axis, and the ordinates $x = a$ and $x = x$. Then (theorem IV of § 26) we have

$$A(x) = \phi(x) + c,$$

where c is a constant. Replacing x by a and remembering that $A(a) = 0$, we have

$$0 = \phi(a) + c, \quad \text{or} \quad c = -\phi(a).$$

Hence
$$A(x) = \phi(x) - \phi(a)$$
and
$$A(b) = \phi(b) - \phi(a).$$

But $A(b) =$ area of $ABCD$. Hence

$$\phi(b) - \phi(a) = \text{area of } ABCD.$$

But this area is also represented by the foregoing definite integral, as we have seen. Therefore

$$\int_a^b f(x)dx = \phi(b) - \phi(a).$$

That is, *the definite integral of $f(x)$ from a to b is numerically equal to the difference $\phi(b) - \phi(a)$ where $\phi(x)$ denotes any particular antiderivative of $f(x)$.*

Employing the definition of the integral from b to a, we see that
$$\int_b^a f(x)dx = \phi(a) - \phi(b).$$

From these results it follows that antiderivatives furnish a means for evaluating definite integrals. In many cases this is the most convenient means. For nearly all the definite integrals occurring in this text this means is the best one for the student to employ.

Certain important properties of definite integrals may now be readily established.

I. The sign of the integral is changed by interchanging the limits of integration; that is,

$$\int_a^b f(x)dx = -\int_b^a f(x)dx.$$

This is implied by the relation defining the integral from b to a. It is also implied in the equations

$$\int_a^b f(x)dx = \phi(b) - \phi(a), \quad \int_b^a f(x)dx = \phi(a) - \phi(b).$$

II. If c is a constant, we have

$$\int_a^b cf(x)dx = c\int_a^b f(x)dx.$$

Let $\phi(x)$ be an antiderivative of $f(x)$. Then $c\phi(x)$ is an antiderivative of $cf(x)$ (theorem V of § 26). Then

$$\int_a^b cf(x)dx = c\phi(b) - c\phi(a), \quad \int_a^b f(x)dx = \phi(b) - \phi(a).$$

From these relations the theorem follows at once.

III. If k lies between a and b, we have

$$\int_a^b f(x)dx = \int_a^k f(x)dx + \int_k^b f(x)dx.$$

Let $\phi(x)$ be an antiderivative of $f(x)$. Then

$$\int_a^k f(x)dx = \phi(k) - \phi(a), \quad \int_k^b f(x)dx = \phi(b) - \phi(k).$$

Hence

$$\int_a^k f(x)dx + \int_k^b f(x)dx = \phi(k) - \phi(a) + \phi(b) - \phi(k)$$
$$= \phi(b) - \phi(a)$$
$$= \int_a^b f(x)dx.$$

This is the result to be proved.

Theorems II and III may also be proved directly from the definition of the definite integral. It is a useful exercise to prove them in this way.

Example. Evaluate $\int_1^4 x^2\, dx$.

A particular antiderivative of x^2 is $\tfrac{1}{3} x^3$. Hence

$$\int_1^4 x^2\, dx = \tfrac{1}{3} x^3 \Big]_1^4 = \tfrac{1}{3}(64 - 1) = 21.$$

The symbol in the second member of this equation denotes the fact that the function $\tfrac{1}{3} x^3$ is to be *evaluated from 1 to 4*. In general we write

$$\phi(x)\Big]_a^b \quad \text{or} \quad \left[\phi(x)\right]_a^b \quad \text{or} \quad \phi(x)\Big]_{x=a}^{x=b} \quad \text{or} \quad \left[\phi(x)\right]_{x=a}^{x=b}$$

for the difference $\phi(b) - \phi(a)$, and this difference gives the result of evaluating $\phi(x)$ from a to b.

EXERCISES

Evaluate the following definite integrals : *

1. $\int_1^3 x\, dx = 4.$

2. $\int_0^{\frac{\pi}{2}} \sin x\, dx = 1.$

3. $\int_0^1 e^x\, dx = e - 1.$

4. $\int_0^{\alpha} \cos \theta\, d\theta = \sin \alpha.$

5. $\int_1^5 \frac{dx}{x} = \log 5.$

6. $\int_0^1 x^5\, dx = \tfrac{1}{6}.$

7. $\int_0^a \frac{dx}{a^2 + x^2} = \frac{\pi}{4a}.$

8. $\int_0^t \frac{dx}{\sqrt{a^2 - x^2}} = \arcsin \frac{t}{a},\ 0 < t < a.$

9. $\int_{-2}^1 \frac{dx}{x^2 + 4x + 13} = \frac{\pi}{12}.$

10. $\int_0^{\frac{\pi}{4}} \sec^2 x\, dx = 1.$

11. $\int_0^{\frac{\pi}{4}} \tan^2 x\, dx = 1 - \frac{\pi}{4}.$

12. $\int_0^{\frac{\pi}{2}} \sin^2 x\, dx = \frac{\pi}{4}.$

13. $\int_0^{\frac{\pi}{2}} \cos^2 x\, dx = \frac{\pi}{4}.$

14. $\int_0^1 (x + x^2)\, dx = \tfrac{5}{6}.$

15. Find the area bounded by the parabola $y^2 = 36\, x$ and the straight line $y = 6\, x$.

* It will be convenient to use antiderivatives already found, especially those in the formulas for differentiation and in the exercises at the ends of Chapters VI and VII.

Solution. The area sought is the difference of the areas $OABCO$ and $OABDO$. The point B is $(1, 6)$, as one sees by solving the two given equations simultaneously. But

area $OABDO = \int_0^1 6\, x\, dx = 3\, x^2 \Big]_0^1 = 3,$

area $OABCO = \int_0^1 \sqrt{36\, x}\, dx = 6 \int_0^1 x^{\frac{1}{2}} dx$
$= 4\, x^{\frac{3}{2}} \Big]_0^1 = 4.$

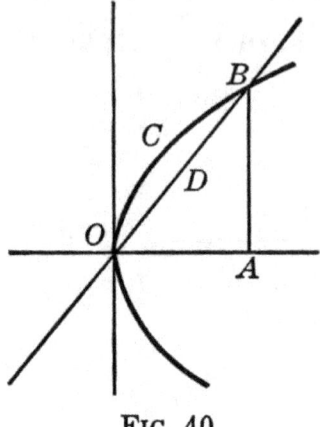

Fig. 40

Hence the area sought is 1.

16. Find the area bounded by the curve

$$y = \frac{1}{a^2 + x^2},$$

the x-axis, the y-axis, and the ordinate $x = a$. (Compare problem 7.)

Ans. $\pi/4\, a$.

17. (1) Find the area bounded by the curve

$$y = \frac{1}{\sqrt{a^2 - x^2}},$$

the x-axis, the y-axis, and the ordinate $x = t$, where $0 < t < a$. (2) Find the limit of this area as t approaches a. (Compare problem 8.)

Ans. (2) $\pi/2$.

18. Find the area of an arch of the curve $y = \sin^2 x$. (Compare problem 12.)

Ans. $\pi/2$.

19. Show that the area bounded by one arch of the curve $y = \cos^2 x$ is the same as that bounded by one arch of the curve $y = \sin^2 x$.

20. If c is a positive constant, show that the area bounded by the curve $y = cf(x)$, the x-axis, and the ordinates $x = a$ and $x = b$ is c times the area bounded by the curve $y = f(x)$, the x-axis, and the ordinates $x = a$ and $x = b$.

78. Integral functions; indefinite integrals. From the point of view of integration the antiderivative of a function $f(x)$ may be called the *integral function* of $f(x)$. From theorem IV of § 26 it follows that *every integral function of $f(x)$ is obtained by adding an arbitrary constant to a particular integral function.* Thus if $\phi(x)$ is an integral function of $f(x)$, the general integral function of $f(x)$ is $\phi(x) + C$. The integral function is denoted by the integral sign $\int$ without limits of integration. Thus we write

$$\int f(x)dx = \phi(x) + C,$$

where $\phi(x)$ is a particular integral function of $f(x)$ and C is an arbitrary constant. The function $f(x)$ between the integral sign $\int$ or $\int_a^b$ and the differential dx is called the *integrand*.

The process of finding integral functions is called *integration*. The process of evaluating a definite integral is also called integration.

The integral function of $f(x)$ is also called the *indefinite integral* of $f(x)$. The word *indefinite* refers to the fact that an arbitrary constant is involved in the indefinite integral. The integral between given limits is completely defined and hence is called the definite integral. Hereafter we shall generally use the term *indefinite integral of a function* rather than the term *antiderivative of a function*, since the former is the term which is usually employed.

For the problems which arise in this book the best method for finding the definite integral is to find the indefinite integral and then to evaluate it between the given limits. Consequently we shall need to develop means for finding indefinite integrals. There is no direct method for finding indefinite integrals. In seeking the indefinite integral of $f(x)$ one tries to recognize $f(x)$ as the derivative of some known function. This is accomplished by the aid of certain standard forms obtained from the formulas for differentiation.

The processes, methods, and theorems relating to integration and integrals constitute the *integral calculus*.

In mastering the integral calculus the student will find it useful to distinguish sharply between two things: (1) the general idea of summation underlying the definition of the definite integral, and (2) the practical processes of evaluating integrals by means of their connection with antiderivatives. The latter processes are the ones mainly in evidence in the computation of integrals by elementary means; the former idea is the fundamental one in seeing the uses and applications of definite integrals.

Many authors define the definite integral by means of the antiderivative or indefinite integral in accordance with the formula
$$\int_a^b f(x)dx = \phi(b) - \phi(a),$$
and pass thence to its property as the limit of a sum. But it seems to us best that the student shall meet the integral first in its most fundamental aspect, namely, as the limit of a sum in accordance with the discussion and definition in §§ 76 and 77. But before one can successfully use this notion in the applications of definite integrals it is necessary for him to learn to evaluate integrals; and this requires the finding of indefinite integrals. Consequently we shall now give the principal elementary means for effecting integration. This is nothing more than the systematic discussion of the methods of finding antiderivatives. It is more convenient to do this with the integral notation than without it. After these processes are learned we shall return to our conception of the integral as the limit of a sum. For the present, however, we must give attention to the practical processes of computation.

79. Standard elementary forms of integrals. In this section we shall give the standard elementary forms of integrals. They are all based on the previously derived formulas for differentiation. We list them here and will give the proofs in the following sections.

STANDARD ELEMENTARY FORMS

(1) $\int 0\,dv = C.$

(2) $\int dv = v + C.$

(3) $\int (du + dv - dw) = \int du + \int dv - \int dw.$

(4) $\int a\,dv = a \int dv, \quad a = \text{constant}.$

(5) $\int u\,dv = uv - \int v\,du.$

(6) $\int v^n\,dv = \dfrac{v^{n+1}}{n+1} + C$ if n is a constant and $\neq -1$.

(7) $\displaystyle\int \frac{dv}{v} = \log v + C$
$\phantom{(7) \displaystyle\int \frac{dv}{v}} = \log v + \log c \text{ if } \log c = C$
$\phantom{(7) \displaystyle\int \frac{dv}{v}} = \log cv.$

(8) $\displaystyle\int a^v \, dv = \frac{a^v}{\log a} + C, \quad a = \text{constant}.$

(9) $\displaystyle\int e^v \, dv = e^v + C.$

(10) $\displaystyle\int \sin v \, dv = -\cos v + C.$

(11) $\displaystyle\int \cos v \, dv = \sin v + C.$

(12) $\displaystyle\int \sec^2 v \, dv = \tan v + C.$

(13) $\displaystyle\int \csc^2 v \, dv = -\cot v + C.$

(14) $\displaystyle\int \sec v \tan v \, dv = \sec v + C.$

(15) $\displaystyle\int \csc v \cot v \, dv = -\csc v + C.$

(16) $\displaystyle\int \tan v \, dv = \log \sec v + C = -\log \cos v + C.$

(17) $\displaystyle\int \cot v \, dv = \log \sin v + C = -\log \csc v + C.$

(18) $\displaystyle\int \sec v \, dv = \log (\sec v + \tan v) + C.$

(19) $\displaystyle\int \csc v \, dv = \log (\csc v - \cot v) + C.$

(20) $\displaystyle\int \sin^2 v \, dv = \tfrac{1}{2} v - \tfrac{1}{2} \sin v \cos v + C.$

(21) $\displaystyle\int \cos^2 v \, dv = \tfrac{1}{2} v + \tfrac{1}{2} \sin v \cos v + C.$

(22) $\displaystyle\int \tan^2 v \, dv = \tan v - v + C.$

(23) $\displaystyle\int \cot^2 v \, dv = -\cot v - v + C.$

(24) $\displaystyle\int \frac{dv}{v^2+a^2} = \frac{1}{a} \arctan \frac{v}{a} + C.$

(25) $\displaystyle\int \frac{dv}{v^2-a^2} = \frac{1}{2a} \log \frac{v-a}{v+a} + C$ if $v^2 > a^2$,

$\qquad\qquad\quad = \frac{1}{2a} \log \frac{a-v}{a+v} + C$ if $v^2 < a^2$.

(26) $\displaystyle\int \frac{dv}{\sqrt{a^2-v^2}} = \arcsin \frac{v}{a} + C = -\arccos \frac{v}{a} + C'.$

(27) $\displaystyle\int \frac{dv}{\sqrt{v^2 \pm a^2}} = \log(v + \sqrt{v^2 \pm a^2}) + C.$

(28) $\displaystyle\int \frac{dv}{\sqrt{2av-v^2}} = \operatorname{arc\,vers} \frac{v}{a} + C.$

(29) $\displaystyle\int \frac{dv}{v\sqrt{v^2-a^2}} = \frac{1}{a} \operatorname{arc\,sec} \frac{v}{a} + C = -\frac{1}{a} \operatorname{arc\,csc} \frac{v}{a} + C'.$

(30) $\displaystyle\int \sqrt{a^2-v^2}\, dv = \tfrac{1}{2}\left(v\sqrt{a^2-v^2} + a^2 \arcsin \frac{v}{a}\right) + C.$

(31) $\displaystyle\int \sqrt{v^2 \pm a^2}\, dv = \tfrac{1}{2}[v\sqrt{v^2 \pm a^2} \pm a^2 \log(v + \sqrt{v^2 \pm a^2})] + C.$

80. Proof of formulas (1) to (9). These are based directly on corresponding formulas for differentiation. The constant C of integration is added in each case in accordance with theorem IV of § 26.

Formula (1) follows from the fact that the derivative of a constant is 0.

Formula (2) follows from the fact that the derivative of v with respect to v is 1. Hence

I. *The indefinite integral of the differential of a function is equal to that function plus a constant.*

Formulas (3) and (4) are the integral forms of theorems III and IV in § 28. Thus, from III we have

$$d(u+v-w) = du + dv - dw.$$

On integrating we have (3). From IV we have

$$d(av) = a\, dv.$$

On integrating we obtain (4). These two formulas yield the following two theorems:

II. The indefinite integral of the algebraic sum of several differential expressions is equal to the corresponding algebraic sum of the indefinite integrals of the several expressions taken separately

III. The indefinite integral of a constant times a differential is equal to that constant times the indefinite integral of the differential.

From formula V of § 28 we have
$$d(uv) = u\,dv + v\,du.$$
On integrating we have $uv = \int u\,dv + \int v\,du.$

On transposing and changing signs we have
$$\int u\,dv = uv - \int v\,du.$$

This is formula (5). It is the basis of the method of integration by parts, a method which we shall discuss in detail in § 83.

In formula VI of § 28 replace n by $n+1$. Then we have
$$d(v^{n+1}) = (n+1)v^n\,dv.$$
Integrating, we have
$$v^{n+1} + C' = (n+1)\int v^n\,dv.$$

Dividing by $n+1$ and interchanging the members of the resulting equation, we have formula (6).

In § 70 we saw that
$$d(\log v) = \frac{dv}{v}.$$

Writing this in integral form, we have (7). This formula (7) yields the following theorem:

IV. The indefinite integral of the differential of a function divided by that function is equal to the natural logarithm of that function ·s a constant.

71 we saw that $\quad d(a^v) = \log a \cdot a^v\,dv.$

(ıg, we have $\quad a^v + C' = \log a \cdot \int a^v\,dv.$

Dividing by log a and interchanging the members of the resulting equation, we have formula (8).

Formula (9) is the special case of formula (8) in which $a = e$.

In formulas (1) to (4) we have certain general principles of integration. Formula (5) affords the basis of the method of integration by parts which is to be discussed later. We shall now show by examples how to use formulas (6) to (9) in performing integrations.

Example 1. Show that $\int x^6 \, dx = \frac{1}{7} x^7 + C$.

We employ formula (6), taking $v = x$ and $n = 6$. Then $dv = dx$. The result to be proved is obtained by substituting these values in the second member of (6).

Example 2. Show that

$$\int (7x - 9)^4 \, dx = \tfrac{1}{35} (7x - 9)^5 + C.$$

Take $v = 7x - 9$ and $n = 4$. Then $dv = 7 \, dx$. We may therefore write

$$\int (7x - 9)^4 \, dx = \tfrac{1}{7} \int (7x - 9)^4 \, 7 \, dx$$

$$= \tfrac{1}{7} \int v^4 \, dv$$

$$= \frac{1}{7} \cdot \frac{v^5}{5} + C$$

$$= \tfrac{1}{35}(7x - 9)^5 + C.$$

Example 3. Show that

$$\int \frac{x \, dx}{\sqrt{a^2 + x^2}} = \sqrt{a^2 + x^2} + C.$$

The integral may be written in the form

$$\tfrac{1}{2} \int (a^2 + x^2)^{-\frac{1}{2}} \, 2x \, dx.$$

On taking $v = a^2 + x^2$ and $n = -\tfrac{1}{2}$ we have $dv = 2x \, dx$ and $n + 1 = \tfrac{1}{2}$. Hence

$$\tfrac{1}{2} \int (a^2 + x^2)^{-\frac{1}{2}} \, 2x \, dx = \tfrac{1}{2} \int v^{-\frac{1}{2}} \, dv$$

$$= \tfrac{1}{2} \frac{v^{\frac{1}{2}}}{\frac{1}{2}} + C$$

$$= (a^2 + x^2)^{\frac{1}{2}} + C.$$

Example 4. Show that $\int \frac{dx}{\sqrt{a-bx}} = -\frac{2}{b}\sqrt{a-bx} + C$.

We have
$$\int \frac{dx}{\sqrt{a-bx}} = -\frac{1}{b}\int (a-bx)^{-\frac{1}{2}}(-b\,dx)$$
$$= -\frac{1}{b}\frac{(a-bx)^{\frac{1}{2}}}{\frac{1}{2}} + C$$
$$= -\frac{2}{b}(a-bx)^{\frac{1}{2}} + C.$$

Here we have thought of v as being equal to $a - bx$, whence $dv = -b\,dx$, and have performed the integration accordingly.

The student is advised to learn to perform the integrations in this way without writing in the value of v.

Example 5. Show that $\int \frac{(2x+1)dx}{x^2+x+1} = \log(x^2+x+1) + C$.

Here the numerator is the differential of the denominator. Hence we think of the denominator as being v. Then $dv = (2x+1)dx$. Then we perform the integration at once by formula (7).

Example 6. Show that

$$\int \frac{(x^3+x)dx}{x^4+2x^2+7} = \frac{1}{4}\log(x^4+2x^2+7) + C.$$

We have $\int \frac{(x^3+x)dx}{x^4+2x^2+7} = \frac{1}{4}\int \frac{(4x^3+4x)dx}{x^4+2x^2+7}.$

The numerator of the last fraction is the differential of the denominator. Hence the desired result is obtained by applying formula (7).

Example 7. Show that

$$\int e^{x^2+x}(2x+1)dx = e^{x^2+x} + C.$$

We think of v as being $x^2 + x$. Then $dv = (2x+1)dx$. Then we apply formula (9) and write down the result at once.

Example 8. Show that

$$\int a^{\cos x} \sin x\, dx = -\frac{a^{\cos x}}{\log a} + C.$$

We take $v = \cos x$. Then $dv = -\sin x\,dx$. Hence
$$\int a^{\cos x} \sin x\,dx = -\int a^v dv$$
$$= -\frac{a^v}{\log a} + C.$$
$$= -\frac{a^{\cos x}}{\log a} + C.$$

EXERCISES

Verify the following integrations:

1. $\int \frac{dx}{x^3} = \int x^{-3} dx = -\frac{1}{2} x^{-2} + C.$

2. $\int \frac{dx}{x^5} = -\frac{1}{4} x^{-4} + C.$

3. $\int x^{\frac{3}{4}} dx = \frac{4}{7} x^{\frac{7}{4}} + C.$

4. $\int \frac{dx}{x^{\frac{2}{3}}} = 3 x^{\frac{1}{3}} + C.$

5. $\int \sqrt{2\,px}\,dx = \frac{2}{3} x \sqrt{2\,px} + C.$

6. $\int \sqrt[3]{x}\,dx = \frac{3}{4} x \sqrt[3]{x} + C.$

7. $\int (2\,x^3 + 7\,x^2 - 8\,x + 9)\,dx = \frac{1}{2} x^4 + \frac{7}{3} x^3 - 4\,x^2 + 9\,x + C.$
   (Integrate each term separately and employ formula (3)).

8. $\int \left(x + \frac{1}{x^2}\right) dx = \frac{1}{2} x^2 - \frac{1}{x} + C.$

9. $\int (a^2 + x^2)^2 dx = a^4 x + \frac{2}{3} a^2 x^3 + \frac{1}{5} x^5 + C.$
   (First expand the integrand.)

10. $\int (2\,ax + bx^2)^3 (a + bx) dx = \frac{1}{8} (2\,ax + bx^2)^4 + C.$

11. $\int \frac{dx}{\sqrt{1-x}} = -2\sqrt{1-x} + C.$

12. $\int \sin^2 x \cos x\,dx = \frac{1}{3} \sin^3 x + C.$
   (Apply formula (6), taking $v = \sin x$.)

13. $\int \cos^3 x \sin x\,dx = -\frac{1}{4} \cos^4 x + C.$

14. $\int (ax + b)^{\frac{2}{3}} dx = \frac{2}{5a} (ax + b)^{\frac{5}{3}} + C.$

15. $\int \sqrt[3]{1+x^2}\, x\,dx = \frac{3}{8} (1 + x^2)^{\frac{4}{3}} + C.$

16. $\int \frac{x\,dx}{x^2 + 1} = \frac{1}{2} \log(x^2 + 1) + C = \log(x^2 + 1)^{\frac{1}{2}} + C.$

17. $\int \frac{(x^2 - a^2) dx}{x^3 - 3\,a^2 x} = \frac{1}{3} \log(x^3 - 3\,a^2 x) + C.$

18. $\int \dfrac{x^3 + a^2}{x^4 + 4\,a^2 x}\, dx = \log\,(x^4 + 4\,a^2 x)^{\frac{1}{4}} + C.$

19. $\int \dfrac{x^3}{x - 1}\, dx = \dfrac{x^3}{3} + \dfrac{x^2}{2} + x + \log\,(x - 1) + C.$

(First divide the numerator by the denominator.)

20. $\int \dfrac{2\,x + 1}{2\,x - 1}\, dx = x + \log\,(2\,x - 1) + C.$

21. $\int \dfrac{t^3\,dt}{a + bt^4} = \dfrac{1}{4\,b}\log\,(a + bt^4) + C.$

22. $\int (1 + e^x)^3 e^x\, dx = \tfrac{1}{4}(1 + e^x)^4 + C.$

23. $\int 5\,e^x\, dx = 5\,e^x + C.$

24. $\int e^{ax}\, dx = \dfrac{1}{a}\,e^{ax} + C.$

25. $\int e^{\frac{x}{a}}\, dx = a e^{\frac{x}{a}} + C.$

26. $\int \left(e^{\frac{x}{a}} + e^{-\frac{x}{a}}\right) dx = a\left(e^{\frac{x}{a}} - e^{-\frac{x}{a}}\right) + C.$

27. $\int e^{-ax}\, dx = -\dfrac{1}{a}\,e^{-ax} + C.$

28. $\int a^x e^x\, dx = \dfrac{a^x e^x}{1 + \log a} + C.$

29. $\int \left(e^{\frac{x}{a}} + e^{-\frac{x}{a}}\right)^2 dx = \dfrac{a}{2}\left(e^{\frac{2x}{a}} - e^{-\frac{2x}{a}}\right) + 2\,x + C.$

30. $\int \dfrac{\sin x\, dx}{a + b\cos x} = -\dfrac{1}{b}\log\,(a + b\cos x) + C.$

31. $\int \dfrac{\sec^2 x}{1 + 4\tan x}\, dx = \dfrac{1}{4}\log\,(1 + 4\tan x) + C.$

32. $\int (\log x)^4\,\dfrac{dx}{x} = \dfrac{1}{5}(\log x)^5 + C.$

33. $\int \sin^4 ax \cos ax\, dx = \dfrac{1}{5\,a}\sin^5 ax + C.$

34. $\int e^{\tan x}\sec^2 x\, dx = e^{\tan x} + C.$

35. $\int e^{x^2 + 7} x\, dx = \tfrac{1}{2}\,e^{x^2 + 7} + C.$

36. $\int \dfrac{e^x - 1}{e^x + 1}\, dx = 2\log\,(e^x + 1) - x + C.$

37. Show that the area bounded by the curve $y = e^x$, the x-axis, and the ordinates $x = -t$ and $x = 0$ is $1 - e^{-t}$, where t is a positive number. Show that the limit of this area as t approaches infinity is 1. This is said to be the area bounded by the curve $y = e^x$ and the x-axis and the y-axis. Draw a graph to represent this area.

38. Find the area bounded by the curve
$$y = x^2 e^{x^3}$$
and the x-axis. *Ans.* $\frac{1}{3}$.

39. Find the area bounded by the catenary
$$y = \frac{a}{2}\left(e^{\frac{x}{a}} + e^{-\frac{x}{a}}\right),$$
the coördinate axes, and the line $x = a$. *Ans.* $\frac{a^2}{2e}(e^2 - 1)$.

40. What is the area bounded by the graphs of the four equations
$$y = \frac{1}{x}, \quad y = 0, \quad x = 1, \quad x = a. \qquad \textit{Ans. } \log a.$$

41. Show that $$\log t = \int_1^t \frac{dx}{x}.$$

(Note that this furnishes an analytical means by which it would be possible to define $\log t$ in terms of the simple function $1/x$. This affords one of the important methods of treating properties of logarithms in the higher mathematics.)

42. Show that $$\operatorname{arc tan} t = \int_0^t \frac{dx}{1 + x^2},$$
and hence observe that the function arc tan t can be defined analytically in terms of the simpler function
$$\frac{1}{1 + x^2}.$$

81. Proof of formulas (10) to (23). Formulas (10) to (15) are direct consequences of the formulas given in § 64 for the differentiation of the trigonometric functions. Formulas (22) and (23) are readily obtained from (12) and (13), respectively, in virtue of the relations
$$\tan^2 v = \sec^2 v - 1, \quad \cot^2 v = \csc^2 v - 1.$$

To prove (16) we write
$$\int \tan v \, dv = \int \frac{\sin v \, dv}{\cos v}$$
$$= -\int \frac{-\sin v \, dv}{\cos v}$$
$$= -\log \cos v + C \qquad \text{[by (7)]}$$
$$= -\log \frac{1}{\sec v} + C$$
$$= \log \sec v + C.$$

To prove (17) we write
$$\int \cot v \, dv = \int \frac{\cos v \, dv}{\sin v} = \log \sin v + C = -\log \csc v + C.$$

To prove (18) we write
$$\int \sec v \, dv = \int \sec v \, \frac{\sec v + \tan v}{\sec v + \tan v} \, dv$$
$$= \int \frac{(\sec v \tan v + \sec^2 v) \, dv}{\sec v + \tan v}$$
$$= \log(\sec v + \tan v) + C. \qquad \text{[By (7)]}$$

To prove (19) we write
$$\int \csc v \, dv = \int \csc v \, \frac{\csc v - \cot v}{\csc v - \cot v} \, dv$$
$$= \int \frac{(-\csc v \cot v + \csc^2 v) \, dv}{\csc v - \cot v}$$
$$= \log(\csc v - \cot v) + c. \qquad \text{[By (7)]}$$

To prove (20) we make use of the identity $\cos 2v = 1 - 2\sin^2 v$, or $\sin^2 v = \frac{1}{2}(1 - \cos 2v)$. Then
$$\int \sin^2 v \, dv = \frac{1}{2} \int (1 - \cos 2v) \, dv$$
$$= \frac{1}{2}(v - \frac{1}{2}\sin 2v) + C$$
$$= \frac{1}{2}v - \frac{1}{2}\sin v \cos v + C.$$

To prove (21) we write

$$\int \cos^2 v\, dv = \int (1 - \sin^2 v)\, dv$$
$$= v - (\tfrac{1}{2}v - \tfrac{1}{2}\sin v \cos v) + C$$
$$= \tfrac{1}{2} v + \tfrac{1}{2}\sin v \cos v + C.$$

Example. Evaluate the integral $\int \cos 3\, mx\, dx$.
We have

$$\int \cos 3\, mx\, dx = \frac{1}{3m}\int \cos 3\, mx\, 3\, m\, dx = \frac{1}{3m}\sin 3\, mx + C. \quad [\text{By (11)}]$$

EXERCISES

Perform the following indicated integrations:

1. $\int \sin mx\, dx = -\dfrac{1}{m}\cos mx + C.$

2. $\int \tan cx\, dx = \dfrac{1}{c}\log \sec cx + C.$

3. $\int \sec ax\, dx = \dfrac{1}{a}\log (\sec ax + \tan ax) + C.$

4. $\int \sec 4x \tan 4x\, dx = \tfrac{1}{4}\sec 4x + C.$

5. $\int \csc^2 3x\, dx = -\tfrac{1}{3}\cot 3x + C.$

6. $\int \dfrac{dx}{\cos^2 x} = \tan x + C.$

7. $\int \sec^2 x^4 \cdot x^3\, dx = \tfrac{1}{4}\tan x^4 + C.$

8. $\int \sin^2 3x\, dx = \tfrac{1}{2}x - \tfrac{1}{6}\sin 3x \cos 3x + C.$

9. $\int (\tan x + \cot x)\, dx = \log \tan x + C.$

10. $\int_{\frac{\pi}{4}}^{\frac{\pi}{2}} \dfrac{dx}{\sin^2 x} = 1.$

11. $\int_1^2 \cos(\log x)\, \dfrac{dx}{x} = \sin(\log 2).$

12. $\int \dfrac{dx}{1 + \sin x} = \int \dfrac{(1 - \sin x)\, dx}{\cos^2 x} = \tan x - \sec x + C.$

13. $\int \dfrac{dx}{1 + \cos x} = -\cot x + \csc x + C.$

14. $\int \cot \frac{x}{a} \, dx = a \log \sin \frac{x}{a} + C.$

15. $\int \sin^4 x \, dx = \frac{3x}{8} - \frac{\sin 2x}{4} + \frac{\sin 4x}{32} + C.$

(Change to multiple angles as in the proof of standard formula (20).)

16. $\int \cos^4 x \, dx = \frac{3x}{8} + \frac{\sin 2x}{4} + \frac{\sin 4x}{32} + C.$

82. Proof of formulas (24) to (29). Formulas (24), (26), (28), and (29) are readily proved by means of the formulas for differentiation given in § 69. In each case we replace u in the formula of § 69 by v/a. Thus

$$\frac{d(\text{arc vers } u)}{dx} = \frac{1}{\sqrt{2u - u^2}} \frac{du}{dx}.$$

Putting v/a for u, we have

$$\frac{d\left(\text{arc vers } \frac{v}{a}\right)}{dx} = \frac{1}{\sqrt{\frac{2v}{a} - \frac{v^2}{a^2}}} \cdot \frac{1}{a} \frac{dv}{dx}$$

$$= \frac{1}{\sqrt{2av - v^2}} \frac{dv}{dx}.$$

This establishes formula (28).

The student is advised to prove (24) and (26) and (29) in a similar way.

To prove (25) we observe that

$$\frac{1}{v^2 - a^2} = \frac{1}{2a}\left(\frac{1}{v-a} - \frac{1}{v+a}\right).$$

Hence if $v > a > 0$, we have

$$\int \frac{dv}{v^2 - a^2} = \frac{1}{2a} \int \left(\frac{1}{v-a} - \frac{1}{v+a}\right) dv$$

$$= \frac{1}{2a}\left(\int \frac{dv}{v-a} - \int \frac{dv}{v+a}\right)$$

$$= \frac{1}{2a}[\log(v-a) - \log(v+a)] + C$$

$$= \frac{1}{2a} \log \frac{v-a}{v+a} + C.$$

The student should establish the other part of formula (25).

We treat (27) under two cases. Writing $v = a \tan t$, we have $dv = a \sec^2 t\, dt$. Hence

$$\int \frac{dv}{\sqrt{v^2 + a^2}} = \int \frac{a \sec^2 t\, dt}{\sqrt{a^2 \tan^2 t + a^2}}$$

$$= \int \frac{\sec^2 t\, dt}{\sqrt{\tan^2 t + 1}}$$

$$= \int \sec t\, dt$$

$$= \log (\tan t + \sec t) + C'.$$

Now
$$\tan t = \frac{v}{a},\ \sec t = \sqrt{\tan^2 t + 1} = \sqrt{\frac{v^2}{a^2} + 1} = \frac{1}{a}\sqrt{v^2 + a^2}.$$

Then $\log (\tan t + \sec t) + C' = \log \left(\frac{v}{a} + \frac{1}{a}\sqrt{v^2 + a^2}\right) + C'$

$$= \log (v + \sqrt{v^2 + a^2}) - \log a + C'$$
$$= \log (v + \sqrt{v^2 + a^2}) + C$$

if $C = C' - \log a$. Hence

$$\int \frac{dv}{\sqrt{v^2 + a^2}} = \log (v + \sqrt{v^2 + a^2}) + C.$$

This is part of formula (27).

To prove the other part we set $v = a \sec t$. Then

$$\int \frac{dv}{\sqrt{v^2 - a^2}} = \int \frac{a \sec t \tan t\, dt}{\sqrt{a^2 \sec^2 t - a^2}}$$

$$= \int \sec t\, dt = \log (\sec t + \tan t) + C'$$

$$= \log \left(\frac{v}{a} + \frac{1}{a}\sqrt{v^2 - a^2}\right) + C'$$

$$= \log (v + \sqrt{v^2 - a^2}) + C.$$

EXERCISES

Perform the following indicated integrations:

1. $\int \dfrac{dx}{4x^2 - 9} = \dfrac{1}{2} \int \dfrac{d(2x)}{(2x)^2 - 3^2} = \dfrac{1}{12} \log \dfrac{2x - 3}{2x + 3} + C$, by (25).

2. $\int \dfrac{dx}{4x^2 + 9} = \dfrac{1}{6} \arctan \dfrac{2x}{3} + C.$

3. $\int \frac{dx}{x^2+4} = \frac{1}{2} \arctan \frac{x}{2} + C.$

4. $\int \frac{dx}{\sqrt{x^2+4}} = \log\left(x + \sqrt{x^2+4}\right) + C.$

5. $\int \frac{dx}{\sqrt{4-x^2}} = \arcsin \frac{x}{2} + C.$

6. $\int \frac{dx}{\sqrt{x^2-4}} = \log\left(x + \sqrt{x^2-4}\right) + C.$

7. $\int \frac{x\,dx}{x^4-4} = \frac{1}{8} \log \frac{x^2-2}{x^2+2} + C.$

8. $\int \frac{7x\,dx}{\sqrt{1-x^4}} = \frac{7}{2} \arcsin x^2 + C.$

9. $\int \frac{dx}{x^2+2x+5} = \int \frac{d(x+1)}{(x+1)^2+2^2} = \frac{1}{2} \arctan \frac{x+1}{2} + C.$

10. $\int \frac{dx}{x^2+4x+13} = \frac{1}{3} \arctan \frac{x+2}{3} + C.$

11. $\int \frac{dx}{\sqrt{2ax+x^2}} = \int \frac{dx}{\sqrt{(a+x)^2-a^2}} = \log\left(a+x+\sqrt{2ax+x^2}\right) + C.$

12. $\int \frac{dx}{\sqrt{x^2-2ax}} = \log\left(x - a + \sqrt{x^2-2ax}\right) + C.$

13. $\int \frac{dx}{x^2-6x+5} = \frac{1}{4} \log \frac{x-5}{x-1} + C.$

14. $\int \frac{(4x+1)dx}{x^2+4} = 4\int \frac{x\,dx}{x^2+4} + \int \frac{dx}{x^2+4}$

$= 2 \log (x^2+4) + \frac{1}{2} \arctan \frac{x}{2} + C.$

15. $\int \frac{(x+7)dx}{\sqrt{x^2+4}} = \sqrt{x^2+4} + 7 \log\left(x + \sqrt{x^2+4}\right) + C.$

16. $\int \frac{dx}{\sqrt{2-3x^2}} = \frac{1}{\sqrt{3}} \arcsin \sqrt{\frac{3}{2}}\, x + C.$

17. $\int \frac{dx}{x\sqrt{1-\log^2 x}} = \arcsin (\log x) + C.$

18. $\int \frac{\sin x\,dx}{a^2+\cos^2 x} = -\frac{1}{a} \arctan \left(\frac{\cos x}{a}\right) + C.$

19. $\int \frac{dx}{\sqrt{x^2+4x+29}} = \log\left(x + 2 + \sqrt{x^2+4x+29}\right) + C.$

20. $\displaystyle\int \frac{dx}{\sqrt{16 + 6x - x^2}} = \arcsin \frac{x-3}{5} + C.$

21. $\displaystyle\int \frac{\sqrt{1+x}}{\sqrt{1-x}}\, dx = \int \frac{1+x}{\sqrt{1-x^2}}\, dx = \arcsin x - \sqrt{1-x^2} + C.$

22. $\displaystyle\int \frac{\sqrt{1-x}}{\sqrt{1+x}}\, dx = \arcsin x + \sqrt{1-x^2} + C.$

23. $\displaystyle\int \frac{\arctan x\, dx}{1+x^2} = \frac{1}{2}(\arctan x)^2 + C.$

24. $\displaystyle\int_3^8 \frac{dx}{\sqrt{16 + 6x - x^2}} = \frac{\pi}{2}.$

25. $\displaystyle\int_0^1 \frac{\sqrt{1-x}}{\sqrt{1+x}} = \frac{\pi}{2} - 1.$

83. Integration by parts; proof of formulas (30) and (31). Formula (5) affords the basis of the method of integration by parts, one of the most useful methods of the integral calculus. By a convenient choice of u and v the integral which is to be evaluated is written in the form $\int u\, dv$. Then formula (5) makes its integration depend on the integration of $\int v\, du$. It sometimes happens that the latter is readily integrated while the former is not.

No complete general directions can be given for separating the given differential into the product of u by dv. But some useful suggestions may be made. Let x be the variable of integration. Since dv is itself a differential, the differential dx must always appear as a factor in dv. Moreover, dv must always be chosen so that it is possible to find v. One does this so that $v\, du$ can be integrated in as simple a manner as possible. It is often desirable to choose dv as the most complicated factor of the given differential that one can readily integrate. The following examples will show how the method is to be applied.

Example 1. Find $\int xe^x\, dx.$

Let $dv = e^x dx$. Then $u = x$. Hence
$$du = dx, \quad v = e^x.$$

Since v is an auxiliary function not yet fully defined, we complete its definition by taking for v the simplest value of $\int e^x\,dx$; thus we omit the constant of integration as a matter of convenience. With the given choice of u and v the integral becomes $\int u\,dv$. Substituting in the formula

$$\int u\,dv = uv - \int v\,du,$$

we have $$\int xe^x\,dx = xe^x - \int e^x\,dx.$$

A particular value of the last integral is e^x. Hence

$$\int xe^x\,dx = xe^x - e^x + C.$$

Example 2. Find $\int x^2 \cos x\,dx$.

We take $dv = \cos x\,dx$, $u = x^2$. Then $v = \sin x$ and $du = 2x\,dx$.

Hence $$\int x^2 \cos x\,dx = x^2 \sin x - 2\int x \sin x\,dx. \tag{1}$$

To evaluate the integral $\int x \sin x\,dx$ we apply again the method of integration by parts, taking $dv = \sin x\,dx$ and $u = x$. Then $v = -\cos x$ and $du = dx$. Hence

$$\int x \sin x\,dx = -x \cos x + \int \cos x\,dx = -x \cos x + \sin x + C'.$$

Substituting this value of $\int x \sin x\,dx$ in equation (1), we have

$$\int x^2 \cos x\,dx = x^2 \sin x + 2x \cos x - 2 \sin x + C.$$

Example 3. Find $\int e^{ax} \sin nx\,dx$.

Let $dv = e^{ax} dx$ and $u = \sin nx$. Then

$$v = \frac{1}{a} e^{ax}, \quad du = n \cos nx\,dx.$$

Hence $$\int e^{ax} \sin nx\,dx = \frac{1}{a} e^{ax} \sin nx - \frac{n}{a}\int e^{ax} \cos nx\,dx. \tag{1}$$

To evaluate the last integral we take $dv = e^{ax} dx$ and $u = \cos nx$. Then
$$v = \frac{1}{a} e^{ax}, \quad du = -n \sin nx\,dx.$$

Hence $$\int e^{ax} \cos nx\,dx = \frac{1}{a} e^{ax} \cos nx + \frac{n}{a}\int e^{ax} \sin nx\,dx.$$

Substituting this into (1), we have
$$\int e^{ax} \sin nx \, dx = \frac{1}{a} e^{ax} \sin nx - \frac{n}{a^2} e^{ax} \cos nx - \frac{n^2}{a^2} \int e^{ax} \sin nx \, dx.$$

Transposing the term containing the integral in the last member and inserting a constant of integration, we have
$$\left(1 + \frac{n^2}{a^2}\right) \int e^{ax} \sin nx \, dx = \frac{1}{a} e^{ax} \sin nx - \frac{n}{a^2} e^{ax} \cos nx + C'.$$

Multiplying through by a^2 and then dividing the resulting equation through by $a^2 + n^2$, we have
$$\int e^{ax} \sin nx \, dx = \frac{e^{ax}(a \sin nx - n \cos nx)}{a^2 + n^2} + C.$$

In a similar way it may be proved that
$$\int e^{ax} \cos nx \, dx = \frac{e^{ax}(n \sin nx + a \cos nx)}{a^2 + n^2} + C.$$

Example 4. Prove formula (30) of § 79.

Taking $dv = dx$ and $u = \sqrt{a^2 - x^2}$, we have
$$v = x \quad \text{and} \quad du = \frac{-x \, dx}{\sqrt{a^2 - x^2}}.$$

Hence
$$\int \sqrt{a^2 - x^2} \, dx = x\sqrt{a^2 - x^2} - \int \frac{-x^2 \, dx}{\sqrt{a^2 - x^2}}$$
$$= x\sqrt{a^2 - x^2} - \int \frac{a^2 - x^2 - a^2}{\sqrt{a^2 - x^2}} \, dx$$
$$= x\sqrt{a^2 - x^2} - \int \frac{a^2 - x^2}{\sqrt{a^2 - x^2}} \, dx + a^2 \int \frac{dx}{\sqrt{a^2 - x^2}}$$
$$= x\sqrt{a^2 - x^2} - \int \sqrt{a^2 - x^2} \, dx + a^2 \arcsin \frac{x}{a} + C'.$$

Transposing the integral in the second member and dividing through by 2, we have
$$\int \sqrt{a^2 - x^2} \, dx = \frac{1}{2}\left(x\sqrt{a^2 - x^2} + a^2 \arcsin \frac{x}{a}\right) + C.$$

Writing v in place of x, we have formula (30).

Example 5. Prove formula (31) of § 79.

The method is the same as that in the preceding example. The student is advised to carry out the work in detail.

Perform the following indicated integrations:

1. $\int \log x \, dx = x (\log x - 1) + C.$ (Take $dv = dx$.)
2. $\int x \log x \, dx = \dfrac{x^2}{2} \log x - \dfrac{x^2}{4} + C.$ (Take $dv = x \, dx$.)
3. $\int x^2 e^x \, dx = e^x (x^2 - 2x + 2) + C.$
4. $\int x a^x \, dx = a^x \left[\dfrac{x}{\log a} - \dfrac{1}{\log^2 a} \right] + C.$
5. $\int x^2 e^{-x} \, dx = -e^{-x}(x^2 + 2x + 2) + C.$
6. $\int_0^1 x e^x \, dx = 1.$
7. $\int_0^{\frac{\pi}{2}} x \sin x \, dx = 1.$
8. $\int \sin^2 x \, dx = \tfrac{1}{2}(x - \sin x \cos x) + C.$ (Take $dv = \sin x \, dx$.)
9. $\int \cos^2 x \, dx = \tfrac{1}{2}(x + \sin x \cos x) + C.$
10. $\int \text{arc sin } x \, dx = x \text{ arc sin } x + \sqrt{1-x^2} + C.$ (Take $dv = dx$.)
11. $\int \text{arc cos } x \, dx = x \text{ arc cos } x - \sqrt{1-x^2} + C.$
12. $\int \text{arc tan } x \, dx = x \text{ arc tan } x - \tfrac{1}{2} \log(1+x^2) + C.$
13. $\int \text{arc cot } x \, dx = x \text{ arc cot } x + \tfrac{1}{2} \log(1+x^2) + C.$
14. $\int \text{arc vers } x \, dx = (x-1) \text{ arc vers } x + \sqrt{2x-x^2} + C.$
15. $\int x \text{ arc tan } x \, dx = \dfrac{x^2+1}{2} \text{ arc tan } x - \dfrac{x}{2} + C.$
16. $\int_0^1 x^2 e^x \, dx = e - 2.$
17. $\int x^2 \sin x \, dx = 2 \cos x + 2x \sin x - x^2 \cos x + C.$
18. $\int x \tan^2 x \, dx = x \tan x - \dfrac{x^2}{2} + \log \cos x + C.$
19. $\int \dfrac{x^3 \, dx}{\sqrt{1-x^2}} = -\dfrac{1}{3}(x^2+2)(1-x^2)^{\frac{1}{2}} + C.$
20. $\int \dfrac{x^2 \, dx}{\sqrt{a^2-x^2}} = -\dfrac{x}{2} \sqrt{a^2-x^2} + \dfrac{a^2}{2} \text{ arc sin } \dfrac{x}{a} + C.$

21. $\int e^x \sin x \, dx = \frac{1}{2} e^x(\sin x - \cos x) + C.$

22. $\int e^x \cos x \, dx = \frac{1}{2} e^x(\sin x + \cos x) + C.$

23. $\int e^{-x} \sin x \, dx = -\frac{1}{2} e^{-x}(\sin x + \cos x) + C.$

24. $\int e^{3x} \cos 2x \, dx = \dfrac{e^{3x}}{13} (2 \sin 2x + 3 \cos 2x) + C.$

25. $\int e^{ax}(\sin ax + \cos ax) dx = \dfrac{e^{ax}}{a} \sin ax + C.$

84. Miscellaneous methods of integration. Many expressions can be integrated by two or more applications of standard formula (6). Thus if we wish to integrate

$$\int \sin^3 x \cos^2 x \, dx,$$

we may write

$$\int \sin^3 x \cos^2 x \, dx = \int \sin^2 x \cos^2 x \sin x \, dx$$
$$= \int (1 - \cos^2 x) \cos^2 x \sin x \, dx$$
$$= \int \cos^2 x \sin x \, dx - \int \cos^4 x \sin x \, dx$$
$$= -\tfrac{1}{3} \cos^3 x + \tfrac{1}{5} \cos^5 x + C.$$

Again, we have

$$\int \tan^4 x \, dx = \int \tan^2 x \tan^2 x \, dx$$
$$= \int \tan^2 x (\sec^2 x - 1) dx$$
$$= \int \tan^2 x \sec^2 x \, dx - \int \tan^2 x \, dx$$
$$= \int \tan^2 x \sec^2 x \, dx - \int (\sec^2 x - 1) dx$$
$$= \tfrac{1}{3} \tan^3 x - \tan x + x + C.$$

It often happens that a separation of the integrand into parts can be effected in such a way that the separate parts may be integrated by one or another of the standard forms. Thus we have

$$\int \frac{x^2\,dx}{\sqrt{a^2-x^2}} = \int \frac{a^2\,dx}{\sqrt{a^2-x^2}} - \int \frac{a^2-x^2}{\sqrt{a^2-x^2}}\,dx$$

$$= a^2 \int \frac{dx}{\sqrt{a^2-x^2}} - \int \sqrt{a^2-x^2}\,dx.$$

Now the two parts can be integrated by the standard formulas (26) and (30).

Expressions containing the radicals $\sqrt{a^2-x^2}$ and $\sqrt{x^2 \pm a^2}$ can often be integrated by a trigonometric substitution. The method has already been employed in the proof of the standard formula (27). In using this method,

when $\sqrt{a^2-x^2}$ occurs, put $x = a \sin t$;
when $\sqrt{x^2-a^2}$ occurs, put $x = a \sec t$;
when $\sqrt{x^2+a^2}$ occurs, put $x = a \tan t$.

One or the other of various substitutions may be suggested by the nature of the integrand. If we have the integral

$$\int \frac{dx}{e^{ax}+e^{-ax}},$$

we may put $e^{ax} = t$, whence $x = (1/a)\log t$. Then $dx = dt/(at)$, and we have

$$\int \frac{dx}{e^{ax}+e^{-ax}} = \int \frac{dt}{at(t+t^{-1})}$$

$$= \frac{1}{a}\int \frac{dt}{t^2+1}$$

$$= \frac{1}{a}\arctan t + C$$

$$= \frac{1}{a}\arctan e^{ax} + C.$$

In the integral
$$\int \frac{x\,dx}{a+bx}$$
we may write $a + bx = t$, whence

$$x = \frac{1}{b}(t-a), \quad dx = \frac{dt}{b}.$$

Then we have
$$\int \frac{x\,dx}{a+bx} = \int \frac{\frac{1}{b}(t-a)\frac{dt}{b}}{t}$$
$$= \frac{1}{b^2}\int\left(1-\frac{a}{t}\right)dt$$
$$= \frac{1}{b^2}(t - a\log t) + C$$
$$= \frac{1}{b^2}[(a+bx) - a\log(a+bx)] + C.$$

In the case of a definite integral it is sometimes desirable to avoid the change back to the original variable when the work is done by means of a substitution. This can be effected by changing the limits of integration to correspond with the change in the variable of integration. It is sufficient to illustrate the process by an example.

In the integral
$$\int_0^a \frac{dx}{(a^2+x^2)^{\frac{3}{2}}}$$
we may put $x = a\tan t$ or $t = \arctan \frac{x}{a}$. Then
$$(a^2+x^2)^{\frac{1}{2}} = a\sec t \quad \text{and} \quad dt = \frac{a}{a^2+x^2}\,dx.$$

Moreover, when $x = 0$ we have $t = 0$, and when $x = a$ we have $t = \pi/4$. Hence the limits of integration 0 and a with respect to x become 0 and $\pi/4$, respectively, when taken with respect to t. Therefore

$$\int_0^a \frac{dx}{(a^2+x^2)^{\frac{3}{2}}} = \frac{1}{a}\int_0^a \frac{1}{(a^2+x^2)^{\frac{1}{2}}} \cdot \frac{a\,dx}{a^2+x^2}$$
$$= \frac{1}{a}\int_0^{\frac{\pi}{4}} \frac{dt}{a\sec t}$$
$$= \frac{1}{a^2}\int_0^{\frac{\pi}{4}} \cos t\,dt$$
$$= \frac{1}{a^2}\Big[\sin t\Big]_0^{\frac{\pi}{4}} = \frac{1}{2\,a^2}\sqrt{2}.$$

No comprehensive rules can be given for the choice of the substitutions in such transformations of integrals. It is desirable to find them by inspection, especially in the simpler cases. In Chapter XVI we shall treat the matter more systematically. But in his earlier work the student should rely on direct inspection and the results of experience with both differentiation and integration.

EXERCISES

1. $\int \dfrac{\sin^2 x}{\cos^4 x}\, dx = \int \tan^2 x \sec^2 x\, dx = \tfrac{1}{3}\tan^3 x + C.$

2. $\int \sin^2 x \cos^3 x\, dx = \tfrac{1}{3}\sin^3 x - \tfrac{1}{5}\sin^5 x + C.$

3. $\int \sqrt{\sin x}\, \cos^3 x\, dx = \tfrac{2}{3}\sin^{\frac{3}{2}} x - \tfrac{2}{7}\sin^{\frac{7}{2}} x + C.$

4. $\int \sin 2x \cos 2x\, dx = \tfrac{1}{4}\sin^2 2x + C.$

5. $\int \tan^3 x\, dx = \tfrac{1}{2}\tan^2 x + \log \cos x + C.$

6. $\int \sec^2 x \tan^7 x\, dx = \tfrac{1}{8}\tan^8 x + C.$

7. $\int \dfrac{x^2\, dx}{\sqrt{x^2 - a^2}} = \dfrac{x}{2}\sqrt{x^2 - a^2} + \dfrac{a^2}{2}\log\left(x + \sqrt{x^2 - a^2}\right) + C.$

8. $\int \dfrac{x^2\, dx}{\sqrt{a^2 + x^2}} = \dfrac{x}{2}\sqrt{a^2 + x^2} - \dfrac{a^2}{2}\log\left(x + \sqrt{a^2 + x^2}\right) + C.$

9. $\int \dfrac{\sqrt{x^2 - a^2}}{x}\, dx = \sqrt{x^2 - a^2} - a\, \operatorname{arc\,sec} \dfrac{x}{a} + C.$

10. $\int \dfrac{dx}{x^2 \sqrt{x^2 + a^2}} = -\dfrac{\sqrt{x^2 + a^2}}{a^2 x} + C.$

11. $\int \dfrac{dx}{x \sqrt{x^2 + a^2}} = \dfrac{1}{a}\left[\log x - \log\left(a + \sqrt{a^2 + x^2}\right)\right] + C.$
    (Take $x = a/t$.)

12. $\int (1 + \log x)^{\frac{4}{3}} \dfrac{dx}{x} = \dfrac{3}{7}(1 + \log x)^{\frac{7}{3}} + C.$

13. $\int \dfrac{dx}{x\sqrt{x + 4}} = \dfrac{1}{2}\log \dfrac{\sqrt{x+4} - 2}{\sqrt{x+4} + 2} + C.$ (Take $x + 4 = t^2$.)

14. $\int_0^1 \dfrac{dx}{e^x + e^{-x}} = \operatorname{arc\,tan} e - \dfrac{\pi}{4}.$

15. $\int_0^{\log 2} \sqrt{e^x - 1}\, dx = \dfrac{4 - \pi}{2}.$ (Take $e^x - 1 = t^2$.)

16. $\int_0^a \sqrt{a^2 - x^2}\, dx = a^2 \int_0^{\frac{\pi}{2}} \cos^2 t\, dt = \dfrac{\pi}{4} a^2.$

85. Infinite limits of integration. Up to the present we have considered only those integrals which have finite limits of integration. It is convenient to define integrals with one or both limits infinite. Thus we *define* the following integrals by the indicated limits:

$$\int_a^{+\infty} f(x)dx = \lim_{t=+\infty} \int_a^t f(x)dx,$$

$$\int_{-\infty}^b f(x)dx = \lim_{t=-\infty} \int_t^b f(x)dx.$$

The integral
$$\int_{-\infty}^{+\infty} f(x)dx$$
is defined to be the sum of the two integrals

$$\int_{-\infty}^a f(x)dx + \int_a^\infty f(x)dx$$

where a is any convenient constant. (When ∞ (without the sign $+$ or $-$) is written as a limit of integration it is understood to stand for $+\infty$.)

If the limit involved in one of these definitions fails to exist in a particular case, we say that the corresponding integral does not exist.

Example 1. Find $\int_0^\infty \frac{dx}{1+x^2}$.

We have

$$\int_0^\infty \frac{dx}{1+x^2} = \lim_{t=+\infty} \int_0^t \frac{dx}{1+x^2} = \lim_{t=+\infty} \Big[\arctan x\Big]_0^t = \lim_{t=+\infty} \arctan t = \frac{\pi}{2}.$$

This integral defines the area bounded by the curve

$$y = \frac{1}{1+x^2}$$

and the y-axis and the positive part of the x-axis. (The student should draw the figure.)

Example. 2. Find $\int_1^\infty \frac{dx}{x^4}$.

We have

$$\int_1^\infty \frac{dx}{x^4} = \lim_{t=+\infty} \int_1^t \frac{dx}{x^4} = \lim_{t=+\infty} \Big[-\frac{1}{3x^3}\Big]_1^t = \lim_{t=+\infty} \Big(\frac{1}{3} - \frac{1}{3t^3}\Big) = \frac{1}{3}.$$

Example 3. Show that the integral $\int_1^\infty \frac{dx}{x}$ does not exist.

We have
$$\int_1^t \frac{dx}{x} = \log t \text{ if } t > 0.$$

As t becomes infinite so does $\log t$. Hence the limit
$$\lim_{t \to +\infty} \log t$$
fails to exist. The given integral therefore fails to exist.

86. Definite integrals of discontinuous functions. Such an integral as
$$\int_0^a \frac{dx}{\sqrt{a^2 - x^2}}$$
is not yet defined; for the integrand is discontinuous for $x = a$, becoming infinite for that value of x; and we have so far defined integrals only for continuous integrands. The foregoing integral is defined to be the value of the limit
$$\lim_{\epsilon \to 0} \int_0^{a-\epsilon} \frac{dx}{\sqrt{a^2 - x^2}},$$
where ϵ is understood to be a positive quantity. In general, if $f(x)$ is continuous in the interval (a, b) except for $x = b$, then we *define* the integral of $f(x)$ from a to b by the relation
$$\int_a^b f(x)dx = \lim_{\epsilon \to 0} \int_a^{b-\epsilon} f(x)dx,$$
provided that the latter limit exists. Similarly, if $f(x)$ is continuous in the interval (a, b) except for $x = a$, then we *define* the integral by the relation
$$\int_a^b f(x)dx = \lim_{\epsilon \to 0} \int_{a+\epsilon}^b f(x)dx,$$
ϵ again being a positive variable. In either case, if the limit fails to exist we say that the corresponding integral does not exist. If $f(x)$ is discontinuous for $x = c$, an interior point of the interval (a, b), and is otherwise continuous on (a, b) we *define* the integral by the relation
$$\int_a^b f(x)dx = \int_a^c f(x)dx + \int_c^b f(x)dx.$$
In case either (or both) of the last two integrals fails to exist, the integral in the first member is said not to exist.

Example 1. Find $\int_0^a \dfrac{x\,dx}{\sqrt{a^2-x^2}}$.

We have

$$\int_0^a \frac{x\,dx}{\sqrt{a^2-x^2}} = \lim_{\epsilon \to 0} \int_0^{a-\epsilon} \frac{x\,dx}{\sqrt{a^2-x^2}}$$
$$= \lim_{\epsilon \to 0}\left[-\sqrt{a^2-x^2}\right]_0^{a-\epsilon} = \lim_{\epsilon \to 0}\left[a - \sqrt{a^2 - (a-\epsilon)^2}\right] = a.$$

Example 2. Find $\int_0^2 \dfrac{x\,dx}{(x^2-1)^{\frac{2}{3}}}$.

Since the integrand is discontinuous for $x = 1$, we write

$$\int_0^2 \frac{x\,dx}{(x^2-1)^{\frac{2}{3}}} = \int_0^1 \frac{x\,dx}{(x^2-1)^{\frac{2}{3}}} + \int_1^2 \frac{x\,dx}{(x^2-1)^{\frac{2}{3}}}$$
$$= \lim_{\epsilon \to 0}\int_0^{1-\epsilon} \frac{x\,dx}{(x^2-1)^{\frac{2}{3}}} + \lim_{\epsilon \to 0}\int_{1+\epsilon}^{2} \frac{x\,dx}{(x^2-1)^{\frac{2}{3}}}$$
$$= \lim_{\epsilon \to 0}\left[\tfrac{3}{2}(x^2-1)^{\frac{1}{3}}\right]_0^{1-\epsilon} + \lim_{\epsilon \to 0}\left[\tfrac{3}{2}(x^2-1)^{\frac{1}{3}}\right]_{1+\epsilon}^{2}$$
$$= \lim_{\epsilon \to 0}\left[\tfrac{3}{2}\{(1-\epsilon)^2-1\}^{\frac{1}{3}} + \tfrac{3}{2}\right] + \lim_{\epsilon \to 0}\left[\tfrac{3}{2}\sqrt[3]{3} - \tfrac{3}{2}\{(1+\epsilon)^2-1\}^{\frac{1}{3}}\right]$$
$$= \tfrac{3}{2} + \tfrac{3}{2}\sqrt[3]{3} = \tfrac{3}{2}(1 + \sqrt[3]{3}).$$

Example 3. Show that the integral $\int_0^{2a} \dfrac{dx}{(x-a)^2}$ does not exist.

If this integral exists it is the sum of the following two integrals:

$$\int_0^a \frac{dx}{(x-a)^2}, \quad \int_a^{2a} \frac{dx}{(x-a)^2}. \tag{1}$$

If the first of these exists it is defined by the limit

$$\lim_{\epsilon \to 0}\int_0^{a-\epsilon} \frac{dx}{(x-a)^2} = \lim_{\epsilon \to 0}\left[-\frac{1}{x-a}\right]_0^{a-\epsilon} = \lim_{\epsilon \to 0}\left(\frac{1}{\epsilon} - \frac{1}{a}\right).$$

The last limit does not exist. Hence the first of the integrals (1) does not exist. (The student may show similarly that the second does not exist.) Hence the integral given in the problem does not exist.

If we should proceed to evaluate the given integral without regard to the discontinuity in the integrand, we should have the following incorrect result:

$$\int_0^{2a} \frac{dx}{(x-a)^2} = \left[-\frac{1}{x-a}\right]_0^{2a} = -\frac{1}{a} - \frac{1}{a} = -\frac{2}{a}.$$

This example shows the importance of adhering strictly to the definitions in treating integrals whose integrands are discontinuous.

EXERCISES

Evaluate the integrals in problems 1 to 12.

1. $\int_1^\infty \dfrac{dx}{x^2} = 1.$

2. $\int_0^\infty \dfrac{8 a^3 \, dx}{x^2 + 4 a^2} = 2\pi a^2.$

3. $\int_0^\infty \dfrac{dx}{a^2 + x^2} = \dfrac{\pi}{2a}.$

4. $\int_0^\infty e^{-ax} \, dx = \dfrac{1}{a}$ if $a > 0.$

5. $\int_{-\infty}^0 e^x \, dx = 1.$

6. $\int_1^\infty \dfrac{dx}{x^6} = \dfrac{1}{5}.$

7. $\int_0^a \dfrac{dx}{\sqrt{a-x}} = 2\sqrt{a}.$

8. $\int_{\sqrt{2}}^\infty \dfrac{dx}{x\sqrt{2x^2 - 1}} = \dfrac{\pi}{6}.$

9. $\int_{-\infty}^\infty \dfrac{dx}{x^2 + 4x + 5} = \pi.$

10. $\int_1^\infty \dfrac{dx}{(x+1)^n} = \dfrac{1}{(n-1)2^{n-1}}$ if $n > 1.$

11. $\int_0^\infty \dfrac{dx}{(a+x)^2} = \dfrac{1}{a}$ if $a > 0.$

12. $\int_0^8 \dfrac{x \, dx}{(x^2 - 1)^{\frac{2}{3}}} = \dfrac{9}{2}.$

13. Show that the following integrals do not exist:

$$\int_0^\infty e^x \, dx, \quad \int_0^1 \dfrac{dx}{x^2}, \quad \int_0^\infty \dfrac{dx}{x^3}.$$

14. Determine which of the following integrals exist:

$$\int_0^a \dfrac{dx}{a^2 - x^2}, \quad \int_0^\infty \dfrac{dx}{(x+1)^3}, \quad \int_0^\infty e^{ax} \, dx.$$

15. Find the area bounded by the curve

$$y = \dfrac{1}{a^2 + x^2}$$

and the x-axis.

Ans. $\dfrac{\pi}{a}.$

16. Find the area bounded by the curve

$$y(a^2 - x^2)^{\frac{1}{2}} = x,$$

the x-axis, and the ordinate $x = a.$

Ans. $a.$

CHAPTER X

THE SIMPLER APPLICATIONS OF INTEGRATION

87. Plane areas in rectangular coördinates. If a curve lies above the x-axis, as in the adjacent figure, then the area A bounded by the curve, the x-axis, and the ordinates $x = a$ and $x = b$ is given by the formula

$$A = \int_a^b f(x)dx,$$

as we saw in § 77.

FIG. 41

If the curve lies below the x-axis, as in the figure here given, the area A bounded by the curve, the x-axis, and the ordinates $x = a$ and $x = b$ is given by the formula

$$A = -\int_a^b f(x)dx,$$

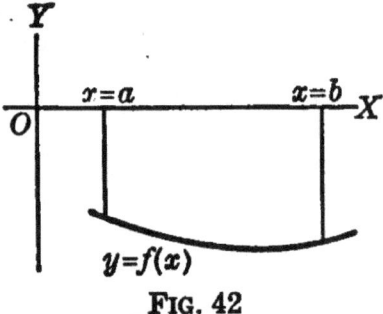

as one sees by observing that the curve $y = -f(x)$ lies above the x-axis and that the area bounded by it and the x-axis and the same ordinates is also equal to A.

FIG. 42

If the curve lies partly above the x-axis and partly below, as in the adjacent figure, then the integral

$$\int_a^b f(x)dx$$

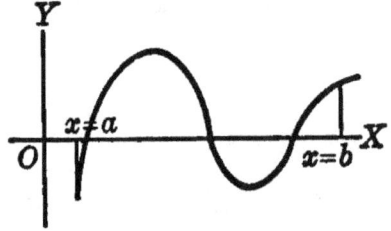

represents the sum of the areas above the x-axis minus the sum of the areas below the x-axis.

FIG. 43

In finding areas by integration it is often desirable to separate the required area into convenient parts and to find the

167

area of the parts separately and then take their sum. This is illustrated in the following problem.

Example. Find the area bounded by the parabola $y^2 = \tfrac{16}{3} x$ and the circle $x^2 + y^2 = 25$.

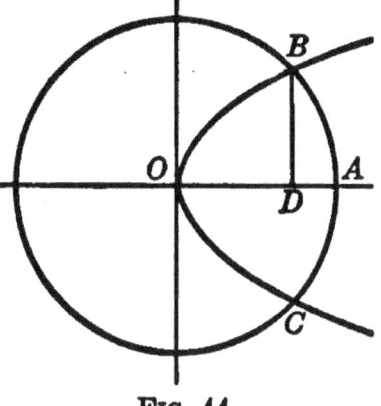

FIG. 44

These curves intersect in the points $(3, 4)$ and $(3, -4)$, as one sees by solving their equations simultaneously. Denote these points by B and C respectively; let O be the center of the circle and let A be the point $(5, 0)$ at which the circle cuts the x-axis. The required area is $OCAB$. It is evidently divided into two equal parts by the radius OA. We shall find the area of OAB, one of these parts. To find the latter area it is convenient to divide it into two parts by the line BD from B perpendicular to OA. Then

$$\text{area } ODB = \int_0^3 \frac{4}{\sqrt{3}} x^{\frac{1}{2}} \, dx = 8;$$

$$\text{area } DBA = \int_3^5 \sqrt{25 - x^2} \, dx$$

$$= \frac{1}{2}\left[x\sqrt{25 - x^2} + 25 \arcsin \frac{x}{5} \right]_3^5$$

$$= \frac{25\,\pi}{4} - 6 - \frac{25}{2} \arcsin \frac{3}{5}.$$

Hence area $OCAB = 2(\text{area } OAB) = 2(\text{area } ODB + \text{area } DBA)$

$$= \frac{25\,\pi}{2} + 4 - 25 \arcsin \frac{3}{5}.$$

88. Length in rectangular coördinates. In § 49 we saw that the differential ds of arc length is given by the formula

$$ds = \sqrt{1 + \left(\frac{dy}{dx}\right)^2} \, dx.$$

Let us suppose that we begin to measure arc length s from the point for which $x = a$, and let s_{ab} be the length of arc from the point for which $x = a$ to that for which $x = b$. On integrating the preceding formula for ds we have

$$\int_0^{s_{ab}} ds = \int_a^b \sqrt{1 + \left(\frac{dy}{dx}\right)^2}\, dx\, ;$$

or
$$s_{ab} = \int_a^b \sqrt{1 + \left(\frac{dy}{dx}\right)^2}\, dx. \tag{1}$$

In a similar way we may employ the formula

$$ds = \sqrt{\left(\frac{dx}{dy}\right)^2 + 1}\, dy$$

(from § 49) to prove that

$$s_{ab} = \int_c^d \sqrt{\left(\frac{dx}{dy}\right)^2 + 1}\, dy \tag{2}$$

if c and d are the values of y corresponding to the values a and b, respectively, for x.

Formulas (1) and (2) are the fundamental formulas for finding lengths of plane curves whose equations are given in rectangular coördinates.

Example 1. Find the circumference of the circle $x^2 + y^2 = a^2$. We have

$$\frac{dy}{dx} = -\frac{x}{y},\ 1 + \left(\frac{dy}{dx}\right)^2 = 1 + \frac{x^2}{y^2} = \frac{x^2 + y^2}{y^2} = \frac{a^2}{y^2} = \frac{a^2}{a^2 - x^2}.$$

One fourth of the circumference lies in the first quadrant and extends from the point for which $x = 0$ to that for which $x = a$. Hence

$$\text{circumference} = 4 \int_0^a \sqrt{1 + \left(\frac{dy}{dx}\right)^2}\, dx$$

$$= 4 \int_0^a \frac{a}{\sqrt{a^2 - x^2}}\, dx$$

$$= 4\, a \left[\arcsin \frac{x}{a}\right]_0^a$$

$$= 4\, a \cdot \frac{\pi}{2} = 2\, \pi a.$$

Example 2. Find the circumference of the circle defined by the parametric equations $x = a \cos \theta,\ y = a \sin \theta$.

This is the same circle as that given in the previous example. We have
$$\frac{dy}{dx} = -\cot\theta, \quad 1 + \left(\frac{dy}{dx}\right)^2 = 1 + \cot^2\theta = \csc^2\theta.$$

Since $dx = -a\sin\theta\, d\theta$ and the point (x, y) moves over a quadrant of the circle when θ ranges from $\frac{\pi}{2}$ to 0, we have
$$\text{circumference} = 4\int_{\frac{\pi}{2}}^{0} \csc\theta(-a\sin\theta)d\theta = -4a\int_{\frac{\pi}{2}}^{0} d\theta = 2\pi a.$$

89. Areas by the summation of thin strips. In finding areas it is sometimes convenient to go back to the definition of the integral as the limit of a sum. To the learner this method is useful in preparing the way for some of the most important applications of integration. The method is best illustrated by means of an example.

Example. Find the area A bounded by the two curves
$$y = \sqrt[4]{16 - x^2} \quad \text{and} \quad y = \sqrt[4]{16 - x^2} + x(4 - x).$$

The points of intersection are (4, 0) and (0, 4). A rough sketch of the curves between these points is given in the adjacent figure, the first equation being that of the lower curve. Let $g(x)$ be the length of that portion of the ordinate $x = x$ which is intercepted between these two curves. Then $g(x) = x(4 - x)$. Let the interval (0, 4) of the x-axis be divided into a large number of small intervals Δx_1, $\Delta x_2, \cdots, \Delta x_n$ and let ξ_i be a value of x on the interval Δx_i. Then
$$g(\xi_i)\Delta x_i$$

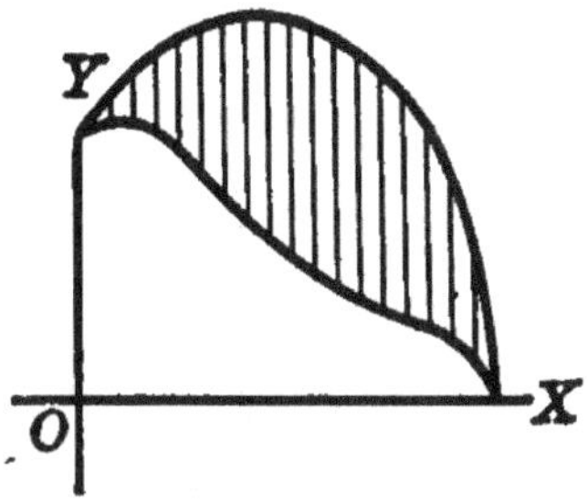

FIG. 45

is approximately the area of the corresponding vertical strip between the two curves. The area between the two curves is approximately equal to the sum
$$\sum_{0}^{4} g(\xi_i)\Delta x_i$$
of all the products $g(\xi_i)\Delta x_i$ formed for the n subintervals Δx_1, $\Delta x_2, \cdots, \Delta x_n$. The required area A is then given by the formula
$$A = \lim_{\text{Every } \Delta x_i \to 0} \sum_{0}^{4} g(\xi_i)\Delta x_i.$$

(Compare the similar argument in § 76, where the analysis is given in much greater detail.) But the last limit is equal to the integral
$$\int_0^4 g(x)dx, \quad \text{or} \quad \int_0^4 x(4-x)dx.$$
Hence
$$A = \int_0^4 x(4-x)dx = \left[2x^2 - \frac{x^3}{3}\right]_0^4 = 10\tfrac{2}{3}.$$

EXERCISES

1. Show that the area of the ellipse $a^2y^2 + b^2x^2 = a^2b^2$ is πab.

2. Show that the area of the ellipse $x = a\cos\theta$, $y = b\sin\theta$ is πab, using an integration with respect to θ.

3. Show that the area which is bounded by the hyperbola $a^2y^2 - b^2x^2 = -a^2b^2$ and the ordinate $x = m$ ($m > a$) is
$$\frac{b}{a}m\sqrt{m^2-a^2} - ab\log\left(\frac{m+\sqrt{m^2-a^2}}{a}\right).$$

4. Show that the area bounded by the curves $y = 2x^2$, $y = 2x$, and $y = 4x$ is $\tfrac{7}{3}$.

5. Show that the entire length of the curve $x^{\frac{2}{3}} + y^{\frac{2}{3}} = a^{\frac{2}{3}}$ is $6a$.

6. What is the length of the catenary
$$y = \frac{a}{2}\left(e^{\frac{x}{a}} + e^{-\frac{x}{a}}\right)$$
from $x = 0$ to $x = a$. *Ans.* $\dfrac{a}{2}(e - e^{-1})$.

7. Find the area between the two curves
$$y = \sqrt{a^2 - x^2} \quad \text{and} \quad y = \sqrt{a^2 - x^2} + x(a-x).$$
(Use summation of thin strips.) *Ans.* $\dfrac{a^3}{6}$.

8. Show that the area bounded by the curves $y = f(x)$ and $y = f(x) + c$ (c = positive constant) and the ordinates $x = a$ and $x = b$ ($b > a$) is $c(b-a)$.

9. If $g(x)$ is any function which is positive on the interval (a, b), show that the area bounded by the curves $y = f(x)$ and $y = f(x) + g(x)$ and the ordinates $x = a$ and $x = b$ ($b > a$) is
$$\int_a^b g(x)dx.$$

10. Find the area bounded by the curves
$$y = e^{x^2}, \quad y = (2x+1)e^{x^2}, \quad x = 5.$$
(Use summation of thin strips.) *Ans.* $e^{25} - 1$.

90. Volumes by the summation of thin slices. The method of finding areas by the summation of thin strips can be extended to space of three dimensions so as to afford a method of finding volumes by the summation of thin slices. This method has a wide range of usefulness in the applications of the integral calculus. (Review § 76 before proceeding.)

The solid whose volume is to be found we think of as being divided into a large number of thin slices by means of parallel planes. For the method to be effective we must be able to find the area of the cross section made by each of these planes. The planes themselves may lie in any direction, but it is more convenient to have them perpendicular to one of the axes. If this cannot be brought about with one of the original axes, a new set of axes may be chosen so as to have the cutting planes perpendicular to one of the axes. In developing the formula we shall suppose that the planes are perpendicular to the x-axis; but the student will remember that either the y-axis or the z-axis will serve equally well.

Let us consider a volume V bounded by some given surface and the planes $x = a$ and $x = b$. Let the interval (a, b) of the x-axis be divided into a large number of subintervals Δx_1, $\Delta x_2, \cdots, \Delta x_n$. Let planes be passed through the extremities of these intervals and perpendicular to the x-axis. They divide the given volume V into a large number of thin slices.

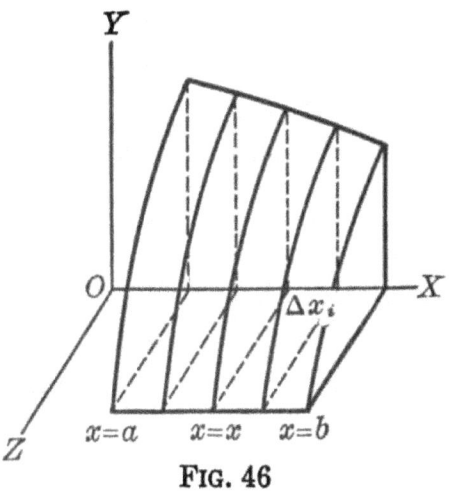

FIG. 46

Let us now consider a certain approximation to the volume of each of these thin slices. We shall find this by means of the function $F(x)$, where $F(x)$ is the area of the cross section of the given volume V by the plane $x = x$.

Now let ξ_i be any value of x on the interval Δx_i. Then the volume of the slice which has this interval for its altitude is approximately

$$F(\xi_i) \Delta x_i,$$

this being the volume of the solid of base $F(\xi_i)$ and altitude Δx_i. The sum of the volumes of all the slices is approximately

$$\sum_a^b F(\xi_i)\Delta x_i,$$

this symbol denoting the sum of the given approximate volumes of all the slices. Hence this sum denotes approximately the volume V itself.

Now suppose that the number of slices is increased indefinitely and that the thickness of each approaches zero. Then the foregoing sum approaches the volume V, as the learner will see by analyzing the situation geometrically and comparing it with the similar but simpler situation in § 76. Hence

$$V = \lim_{\text{Every } \Delta x_i \to 0} \sum_a^b F(\xi_i)\Delta x_i.$$

But in § 77 we saw that this limit is equal to

$$\int_a^b F(x)dx.$$

Hence this is the value of V. Therefore

If $F(x)$ is the area of the cross section of a given volume by the plane $x = x$, then the volume V of that portion of it which is cut off by the planes $x = a$ and $x = b$ ($b > a$) is given by the formula

$$V = \int_a^b F(x)dx.$$

Example 1. Find the volume of the ellipsoid

$$\frac{x^2}{a^2} + \frac{y^2}{b^2} + \frac{z^2}{c^2} = 1.$$

The section made by the plane parallel to the yz-plane and at a distance x from it, that is, the section made by the plane $x = x$, is an ellipse with the equation

$$\frac{y^2}{b^2} + \frac{z^2}{c^2} = 1 - \frac{x^2}{a^2}.$$

Here y and z are variables and x is a constant. Putting this equation in standard form, we have

$$\frac{y^2}{b^2\left(1-\dfrac{x^2}{a^2}\right)} + \frac{z^2}{c^2\left(1-\dfrac{x^2}{a^2}\right)} = 1.$$

Therefore the semi-axes of the ellipse are

$$b\sqrt{1-\frac{x^2}{a^2}} \quad \text{and} \quad c\sqrt{1-\frac{x^2}{a^2}}.$$

Its area is π times the product of these, or

$$\pi bc\left(1-\frac{x^2}{a^2}\right).$$

If x ranges from 0 to a, the thin slices cover half of the volume of the ellipsoid. Hence the required volume V is

$$V = 2\int_0^a \pi bc\left(1-\frac{x^2}{a^2}\right)dx = \frac{4}{3}\pi abc.$$

If $b=c=a$, the ellipsoid becomes a sphere of radius a and the volume is given by the well-known formula $\frac{4}{3}\pi a^3$.

Example 2. Prove that the volume V of a pyramid (or cone) of base B and height h is $\frac{1}{3}Bh$.

Let the cone be placed with its vertex at the origin and its base perpendicular to the x-axis. The area of a section made by a plane perpendicular to the x-axis at a distance x from the vertex is to the area of the base as x^2 is to h^2. Hence the area $F(x)$ of the cross section is Bx^2/h^2. Therefore

$$V = \int_0^h \frac{Bx^2}{h^2}\,dx = \frac{1}{3}\frac{B}{h^2}x^3\bigg]_0^h = \frac{1}{3}Bh.$$

Example 3. Find the volume V of the segment of a sphere of radius a cut off by a plane at a distance c from the center, c being less than a.

Place the sphere so that its center is at the origin of coördinates and so that the given cutting plane is perpendicular to the x-axis. The plane $x=x$ cuts from the sphere a circle of radius $\sqrt{a^2-x^2}$ and hence of area $\pi(a^2-x^2)$. Therefore

$$V = \int_c^a \pi(a^2-x^2)dx = \pi\left(\frac{2}{3}a^3 - a^2c + \frac{c^3}{3}\right).$$

If $c=0$ this gives $\frac{2}{3}\pi a^3$ as the volume of a hemisphere.

Example 4. The center of a variable square moves along the principal axis of the ellipse $b^2x^2 + a^2y^2 = a^2b^2$ from one end to the other. The plane of the square remains perpendicular to this axis and the extremities of one of its diagonals move along the ellipse. What is the volume V swept out by the square?

When the center of the square is at a distance x from the center of the ellipse the diagonal of the square is
$$\frac{2b}{a}\sqrt{a^2 - x^2}.$$

Hence its area is
$$\frac{2b^2}{a^2}(a^2 - x^2).$$

This is the function $F(x)$ of the theorem in this section. Hence
$$V = 2\int_0^a \frac{2b^2}{a^2}(a^2 - x^2)dx = \frac{8}{3}ab^2.$$

EXERCISES

1. What is the volume bounded by the surface
$$\frac{x^4}{a^4} + \frac{y^2}{b^2} + \frac{z^2}{c^2} = 1.$$
Ans. $\frac{8}{5}\pi abc$.

2. What is the volume of the segment of the ellipsoid
$$\frac{x^2}{a^2} + \frac{y^2}{b^2} + \frac{z^2}{c^2} = 1$$
cut off by the plane $x = k$, k being less than a?
Ans. $\pi bc\left(\frac{2}{3}a - k + \frac{k^3}{3\,a^2}\right).$

3. Find the volume bounded by the surface $y^2 + z^2 = 6x$ and the planes $x = a$ and $x = b$ ($b > a$). *Ans.* $3\pi(b^2 - a^2)$.

4. Find the volume bounded by the surface
$$\frac{x^2}{9} + \frac{y^2}{4} = 6z$$
and the planes $z = 0$ and $z = 3$. *Ans.* 162π.

5. The base of a variable isosceles triangle is a chord of a circle of radius a. It moves across the circle remaining perpendicular to a diameter. If the plane of the triangle is perpendicular to this diameter and if the triangle is of constant altitude h, what is the volume swept out by it when its base moves entirely across the circle?
Ans. $\frac{1}{2}\pi ha^2$.

6. If the altitude of the triangle in the foregoing problem should also vary and in such way as to keep the triangle equilateral, what is the volume swept out by it? *Ans.* $\frac{4}{3} a^3 \sqrt{3}$.

7. Find the volume which is generated by revolving the ellipse $b^2 x^2 + a^2 y^2 = a^2 b^2$ about its principal axis. *Ans.* $\frac{4}{3} \pi a b^2$.

8. Find the volume which is generated by revolving the ellipse $b^2 x^2 + a^2 y^2 = a^2 b^2$ about its minor axis. *Ans.* $\frac{4}{3} \pi a^2 b$.

9. A woodsman in felling a circular tree of radius a makes a cut whose base is a horizontal semicircle and whose other cut face is a plane making an angle β with the horizontal base. What is the volume which he cuts out? *Ans.* $\frac{2}{3} a^3 \tan \beta$.

10. Two great circles on a sphere of radius a cut each other at right angles. A variable square moves perpendicular to their common diameter and its vertices remain on the great circles. What volume is swept out by it? *Ans.* $\frac{8}{3} a^3$.

91. Mean value. The *arithmetic mean*, or *average value*, of a finite number of quantities is their sum divided by their number. Let us consider the arithmetic mean of n ordinates of the curve $y = f(x)$ spaced at equal intervals between $x = a$ and $x = b$, inclusive of $x = a$ but exclusive of $x = b$. Denote these ordinates from left to right by y_1, y_2, y_3, $\cdots$, y_n. Then their arithmetic mean is

$$\frac{y_1 + y_2 + \cdots + y_n}{n}.$$

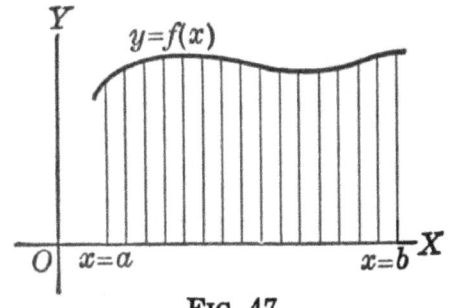

FIG. 47

Since the interval (a, b) of length $(b - a)$ is divided into n equal parts, the length Δx of each part is $(b - a)/n$. Hence

$$n = \frac{b - a}{\Delta x}.$$

Substituting this value of n in the foregoing arithmetic mean and simplifying, we find for it the value

$$\frac{y_1 \Delta x + y_2 \Delta x + \cdots + y_n \Delta x}{b - a}, \quad \text{or} \quad \frac{\sum_{a}^{b} y_i \Delta x}{b - a}.$$

Now let us suppose that the number of ordinates is indefinitely increased. Since they are evenly spaced on the fixed interval (a, b), the length Δx of each of them approaches zero. Hence the limit of the foregoing arithmetic mean is

$$\frac{\lim_{\Delta x \to 0} \sum_a^b y_i \Delta x}{b - a}.$$

From the definition of definite integral in § 77 it follows that this limit has the value

$$\frac{\int_a^b y\, dx}{b - a}.$$

This is called the *mean value* of the function y, or $f(x)$, on the interval (a, b).

The mean value of the function $f(x)$ on the interval (a, b) is

$$\frac{\int_a^b f(x)dx}{b - a}.$$

This mean value is the average height of the curve $y = f(x)$ on the interval (a, b); that is, it is equal to the altitude of the rectangle on the base $b - a$ and of area equal to the area bounded by the curve $y = f(x)$ and the x-axis and the ordinates $x = a$ and $x = b$.

The definition of mean value has been introduced by means of a geometrical illustration. But mean value, in the abstract sense of the formal definition, is independent of this geometric phrasing. The independent variable x may refer to any geometrical or physical magnitude, as time, distance, angle, area, volume, pressure, work done, velocity, acceleration, etc.

Example 1. If a body falls from rest it acquires in t seconds the velocity gt, where g denotes the acceleration due to gravity ($g =$ approximately 32.2 ft. per second per second). What is the average velocity of fall (1) for the first 6 seconds? (2) for the fourth second?

The mean value of the velocity v from time $t = 0$ to time $t = 6$ is

$$\frac{\int_0^6 gt\, dt}{6 - 0} = 3\, g.$$

$t = 3$ to time $t = 4$, is
$$\frac{\int_3^4 gt\, dt}{4-3} = \frac{7}{2}g.$$

Example 2. What is the mean value of the ordinates of a semicircle of radius a when they are uniformly spaced along the arc of the semicircle?

Let the diameter of the semicircle be the x-axis. Let θ be the angle made with the x-axis by the radius to the extremity of a given ordinate. Then this ordinate is of length $a \sin \theta$. We take θ to be the independent variable, since it varies proportionately to the arc length. Hence the mean value sought is equal to

$$\frac{\int_0^\pi a \sin \theta\, d\theta}{\pi - 0} = \frac{2a}{\pi}.$$

EXERCISES

1. What is the mean value of the ordinates of the ellipse
$$b^2 x^2 + a^2 y^2 = a^2 b^2$$
when they are uniformly spaced along the principal axis? *Ans.* $\frac{1}{4}\pi b$.

2. What is the mean value of the ordinates of the parabola $y^2 = 2px$ from $x = 0$ to $x = x_1$ when they are uniformly spaced on the x-axis? *Ans.* $\frac{2}{3}\sqrt{2\, px_1}$.

3. What is the mean value of the ordinates of one arch of the sine curve $y = \sin x$ when they are uniformly spaced on the x-axis?

Ans. $\dfrac{2}{\pi}$.

4. What is the mean value of a cross section of the ellipsoid
$$\frac{x^2}{a^2} + \frac{y^2}{b^2} + \frac{z^2}{c^2} = 1$$
made by uniformly spaced planes perpendicular to (1) the x-axis? (2) the y-axis? (3) the z-axis? *Ans.* (1) $\frac{2}{3}\pi bc$, (2) $\frac{2}{3}\pi ac$, (3) $\frac{2}{3}\pi ab$.

5. What is the mean value of the area of the square described in problem 10 of § 90? *Ans.* $\frac{4}{3}a^2$.

6. What is the mean value of the area of the isosceles triangle described in problem 5 of § 90? *Ans.* $\frac{1}{4}\pi ha$.

7. What is the mean value of the area of the equilateral triangle described in problem 6 of § 90? Ans. $\frac{2}{3} a^2 \sqrt{3}$.

8. What is the mean value of the function
$$y = \frac{a}{2}\left(e^{\frac{x}{a}} + e^{-\frac{x}{a}}\right)$$
from $x = 0$ to $x = a$? Ans. $\frac{a}{2}(e - e^{-1})$.

92. The constant of indefinite integration in geometrical problems. The use of this constant in geometrical problems is best illustrated by means of examples.

Example 1. Find the equations of those plane curves whose slope at every point (x, y) is equal to x.

Since $\frac{dy}{dx}$ is equal to the slope, we have
$$\frac{dy}{dx} = x, \quad \text{whence} \quad y = \frac{x^2}{2} + C.$$

For every value of C this is the equation of one of the curves described. The totality of these curves constitutes a *family* of parabolas, one member of the family corresponding to each value of the constant C.

Each curve of the family has the property that the slope at a point on one of them for which $x = a$ is equal to the slope at the point on any other for which $x = a$. This is obviously a general property of any family of curves whose slope at each point is defined by a given function of x.

Example 2. Find the equations of those plane curves whose slope at every point (x, y) is equal to x/y.

We have
$$\frac{dy}{dx} = \frac{x}{y}.$$

In order to integrate this equation we need to write it in such form that one of the variables appears alone on one side of the sign of equality and the other alone on the other side. In the present case this is accomplished by clearing of fractions. Thus we have
$$y\,dy = x\,dx.$$
Integrating both members and multiplying by 2, we have
$$y^2 = x^2 + C.$$
For varying values of C this gives a family of equilateral hyperbolas.

Example 3. Find the equations of those plane curves whose slope at every point (x, y) is equal to y/x.

We have
$$\frac{dy}{dx} = \frac{y}{x}.$$

Writing this equation in such form as to have the variables separated, we have
$$\frac{dy}{y} = \frac{dx}{x}.$$

Integrating, we have
$$\log y = \log x + C = \log cx$$

if $\log c = C$. Hence
$$y = cx$$

is the equation of the family of curves, one curve of the family corresponding to each value of the parameter c. Each of the "curves" is a straight line through the origin.

In these examples we have certain families, or systems, of curves, each family, or system, consisting of a set of curves obtained from a single equation involving a parameter by giving to that parameter arbitrarily chosen values. It is possible to have two systems of plane curves so related that each curve of one system cuts every curve of the other system at right angles. Each of two systems so related is said to constitute the *orthogonal trajectories* of the other system. At a point of intersection of a curve of one of these systems with a curve of the other system the slope of the one curve is the negative reciprocal of the slope of the other, since the two curves are perpendicular. This affords a means of finding the orthogonal trajectories of a given system of curves.

Example 4. Determine the orthogonal trajectories of the system of equilateral hyperbolas whose equations are of the form $y^2 - x^2 = C$.

On differentiating the given equation member by member and solving for dy/dx, we have
$$\frac{dy}{dx} = \frac{x}{y}.$$

At a given point (x, y) a curve of the given system has the slope x/y. The curve of the orthogonal system which passes through the same point (x, y) has its slope equal to the negative reciprocal of this,

21. $\int e^x \sin x \, dx = \tfrac{1}{2} e^x(\sin x - \cos x) + C.$

22. $\int e^x \cos x \, dx = \tfrac{1}{2} e^x(\sin x + \cos x) + C.$

23. $\int e^{-x} \sin x \, dx = -\tfrac{1}{2} e^{-x}(\sin x + \cos x) + C.$

24. $\int e^{3x} \cos 2x \, dx = \dfrac{e^{3x}}{13} (2 \sin 2x + 3 \cos 2x) + C.$

25. $\int e^{ax}(\sin ax + \cos ax) dx = \dfrac{e^{ax}}{a} \sin ax + C.$

96. Miscellaneous methods of integration. Many expressions can be integrated by two or more applications of standard formula (6). Thus if we wish to integrate

$$\int \sin^3 x \cos^2 x \, dx,$$

we may write

$$\int \sin^3 x \cos^2 x \, dx = \int \sin^2 x \cos^2 x \sin x \, dx$$

$$= \int (1 - \cos^2 x) \cos^2 x \sin x \, dx$$

$$= \int \cos^2 x \sin x \, dx - \int \cos^4 x \sin x \, dx$$

$$= -\tfrac{1}{3} \cos^3 x + \tfrac{1}{5} \cos^5 x + C.$$

Again, we have

$$\int \tan^4 x \, dx = \int \tan^2 x \tan^2 x \, dx$$

$$= \int \tan^2 x (\sec^2 x - 1) dx$$

$$= \int \tan^2 x \sec^2 x \, dx - \int \tan^2 x \, dx$$

$$= \int \tan^2 x \sec^2 x \, dx - \int (\sec^2 x - 1) dx$$

$$= \tfrac{1}{3} \tan^3 x - \tan x + x + C.$$

It often happens that a separation of the integrand into parts can be effected in such a way that the separate parts may

be integrated by one or another of the standard forms. Thus we have
$$\int \frac{x^2\, dx}{\sqrt{a^2-x^2}} = \int \frac{a^2\, dx}{\sqrt{a^2-x^2}} - \int \frac{a^2-x^2}{\sqrt{a^2-x^2}}\, dx$$
$$= a^2 \int \frac{dx}{\sqrt{a^2-x^2}} - \int \sqrt{a^2-x^2}\, dx.$$

Now the two parts can be integrated by the standard formulas (26) and (30).

Expressions containing the radicals $\sqrt{a^2-x^2}$ and $\sqrt{x^2 \pm a^2}$ can often be integrated by a trigonometric substitution. The method has already been employed in the proof of the standard formula (27). In using this method,

when $\sqrt{a^2-x^2}$ occurs, put $x = a \sin t$;
when $\sqrt{x^2-a^2}$ occurs, put $x = a \sec t$;
when $\sqrt{x^2+a^2}$ occurs, put $x = a \tan t$.

One or the other of various substitutions may be suggested by the nature of the integrand. If we have the integral
$$\int \frac{dx}{e^{ax} + e^{-ax}},$$
we may put $e^{ax} = t$, whence $x = (1/a) \log t$. Then $dx = dt/(at)$, and we have
$$\int \frac{dx}{e^{ax} + e^{-ax}} = \int \frac{dt}{at(t + t^{-1})}$$
$$= \frac{1}{a} \int \frac{dt}{t^2 + 1}$$
$$= \frac{1}{a} \arctan t + C$$
$$= \frac{1}{a} \arctan e^{ax} + C.$$

In the integral
$$\int \frac{x\, dx}{a + bx}$$
we may write $a + bx = t$, whence
$$x = \frac{1}{b}(t - a), \quad dx = \frac{dt}{b}.$$

Then we have
$$\int \frac{x\,dx}{a+bx} = \int \frac{\frac{1}{b}(t-a)\frac{dt}{b}}{t}$$
$$= \frac{1}{b^2}\int \left(1-\frac{a}{t}\right)dt$$
$$= \frac{1}{b^2}(t - a\log t) + C$$
$$= \frac{1}{b^2}[(a+bx) - a\log(a+bx)] + C.$$

In the case of a definite integral it is sometimes desirable to avoid the change back to the original variable when the work is done by means of a substitution. This can be effected by changing the limits of integration to correspond with the change in the variable of integration. It is sufficient to illustrate the process by an example.

In the integral
$$\int_0^a \frac{dx}{(a^2+x^2)^{\frac{3}{2}}}$$
we may put $x = a\tan t$ or $t = \arctan \frac{x}{a}$. Then
$$(a^2+x^2)^{\frac{1}{2}} = a\sec t \quad \text{and} \quad dt = \frac{a}{a^2+x^2}\,dx.$$

Moreover, when $x = 0$ we have $t = 0$, and when $x = a$ we have $t = \pi/4$. Hence the limits of integration 0 and a with respect to x become 0 and $\pi/4$, respectively, when taken with respect to t. Therefore
$$\int_0^a \frac{dx}{(a^2+x^2)^{\frac{3}{2}}} = \frac{1}{a}\int_0^a \frac{1}{(a^2+x^2)^{\frac{1}{2}}} \cdot \frac{a\,dx}{a^2+x^2}$$
$$= \frac{1}{a}\int_0^{\frac{\pi}{4}} \frac{dt}{a\sec t}$$
$$= \frac{1}{a^2}\int_0^{\frac{\pi}{4}} \cos t\,dt$$
$$= \frac{1}{a^2}\Big[\sin t\Big]_0^{\frac{\pi}{4}} = \frac{1}{2a^2}\sqrt{2}.$$

No comprehensive rules can be given for the choice of the substitutions in such transformations of integrals. It is desirable to find them by inspection, especially in the simpler cases. In Chapter XVIII we shall treat the matter more systematically. But in his earlier work the student should rely on direct inspection and the results of experience with both differentiation and integration.

EXERCISES

1. $\int \dfrac{\sin^2 x}{\cos^4 x}\, dx = \int \tan^2 x \sec^2 x\, dx = \tfrac{1}{3} \tan^3 x + C.$

2. $\int \sin^2 x \cos^3 x\, dx = \tfrac{1}{3} \sin^3 x - \tfrac{1}{5} \sin^5 x + C.$

3. $\int \sqrt{\sin x} \cos^3 x\, dx = \tfrac{2}{3} \sin^{\tfrac{3}{2}} x - \tfrac{2}{7} \sin^{\tfrac{7}{2}} x + C.$

4. $\int \sin 2x \cos 2x\, dx = \tfrac{1}{4} \sin^2 2x + C.$

5. $\int \tan^3 x\, dx = \tfrac{1}{2} \tan^2 x + \log \cos x + C.$

6. $\int \sec^2 x \tan^7 x\, dx = \tfrac{1}{8} \tan^8 x + C.$

7. $\int \dfrac{x^2\, dx}{\sqrt{x^2 - a^2}} = \dfrac{x}{2} \sqrt{x^2 - a^2} + \dfrac{a^2}{2} \log(x + \sqrt{x^2 - a^2}) + C.$

8. $\int \dfrac{x^2\, dx}{\sqrt{a^2 + x^2}} = \dfrac{x}{2} \sqrt{a^2 + x^2} - \dfrac{a^2}{2} \log(x + \sqrt{a^2 + x^2}) + C.$

9. $\int \dfrac{\sqrt{x^2 - a^2}}{x}\, dx = \sqrt{x^2 - a^2} - a \operatorname{arc\,sec} \dfrac{x}{a} + C.$

10. $\int \dfrac{dx}{x^2 \sqrt{x^2 + a^2}} = -\dfrac{\sqrt{x^2 + a^2}}{a^2 x} + C.$

11. $\int \dfrac{dx}{x \sqrt{x^2 + a^2}} = \dfrac{1}{a}[\log x - \log(a + \sqrt{a^2 + x^2})] + C.$
 (Take $x = a/t$.)

12. $\int (1 + \log x)^{\tfrac{4}{3}} \dfrac{dx}{x} = \dfrac{3}{7}(1 + \log x)^{\tfrac{7}{3}} + C.$

13. $\int \dfrac{dx}{x \sqrt{x + 4}} = \dfrac{1}{2} \log \dfrac{\sqrt{x + 4} - 2}{\sqrt{x + 4} + 2} + C.$ (Take $x + 4 = t^2$.)

14. $\int_0^1 \dfrac{dx}{e^x + e^{-x}} = \operatorname{arc\,tan} e - \dfrac{\pi}{4}.$

15. $\int_0^{\log 2} \sqrt{e^x - 1}\, dx = \dfrac{4 - \pi}{2}.$ (Take $e^x - 1 = t^2$.)

16. $\int_0^a \sqrt{a^2 - x^2}\, dx = a^2 \int_0^{\tfrac{\pi}{2}} \cos^2 t\, dt = \dfrac{\pi}{4} a^2.$

97. Infinite limits of integration. Up to the present we have considered only those integrals which have finite limits of integration. It is convenient to define integrals with one or both limits infinite. Thus we *define* the following integrals by the indicated limits:

$$\int_a^{+\infty} f(x)dx = \lim_{t=+\infty} \int_a^t f(x)dx,$$

$$\int_{-\infty}^b f(x)dx = \lim_{t=-\infty} \int_t^b f(x)dx.$$

The integral
$$\int_{-\infty}^{+\infty} f(x)dx$$
is defined to be the sum of the two integrals

$$\int_{-\infty}^a f(x)\,dx + \int_a^{\infty} f(x)\,dx$$

where a is any convenient constant. (When ∞ (without the sign $+$ or $-$) is written as a limit of integration it is understood to stand for $+\infty$.)

If the limit involved in one of these definitions fails to exist in a particular case, we say that the corresponding integral does not exist.

Example 1. Find $\int_0^{\infty} \dfrac{dx}{1+x^2}.$

We have

$$\int_0^{\infty} \frac{dx}{1+x^2} = \lim_{t=+\infty} \int_0^t \frac{dx}{1+x^2} = \lim_{t=+\infty} \Big[\arctan x\Big]_0^t = \lim_{t=+\infty} \arctan t = \frac{\pi}{2}.$$

This integral defines the area bounded by the curve

$$y = \frac{1}{1+x^2}$$

and the y-axis and the positive part of the x-axis. (The student should draw the figure.)

Example. 2. Find $\int_1^{\infty} \dfrac{dx}{x^4}.$

We have

$$\int_1^{\infty} \frac{dx}{x^4} = \lim_{t=+\infty} \int_1^t \frac{dx}{x^4} = \lim_{t=+\infty} \Big[-\frac{1}{3x^3}\Big]_1^t = \lim_{t=+\infty} \Big(\frac{1}{3} - \frac{1}{3t^3}\Big) = \frac{1}{3}.$$

Example 3. Show that the integral $\int_1^\infty \frac{dx}{x}$ does not exist.

We have
$$\int_1^t \frac{dx}{x} = \log t \text{ if } t > 0.$$

As t becomes infinite so does $\log t$. Hence the limit
$$\lim_{t = +\infty} \log t$$
fails to exist. The given integral therefore fails to exist.

98. Definite integrals of discontinuous functions. Such an integral as
$$\int_0^a \frac{dx}{\sqrt{a^2 - x^2}}$$
is not yet defined; for the integrand is discontinuous for $x = a$, becoming infinite for that value of x; and we have so far defined integrals only for continuous integrands. The foregoing integral is defined to be the value of the limit
$$\lim_{\epsilon = 0} \int_0^{a-\epsilon} \frac{dx}{\sqrt{a^2 - x^2}},$$
where ϵ is understood to be a positive quantity. In general, if $f(x)$ is continuous in the interval (a, b) except for $x = b$, then we *define* the integral of $f(x)$ from a to b by the relation
$$\int_a^b f(x)dx = \lim_{\epsilon = 0} \int_a^{b-\epsilon} f(x)dx,$$
provided that the latter limit exists. Similarly, if $f(x)$ is continuous in the interval (a, b) except for $x = a$, then we *define* the integral by the relation
$$\int_a^b f(x)dx = \lim_{\epsilon = 0} \int_{a+\epsilon}^b f(x)dx,$$
ϵ again being a positive variable. In either case, if the limit fails to exist we say that the corresponding integral does not exist. If $f(x)$ is discontinuous for $x = c$, an interior point of the interval (a, b), and is otherwise continuous on (a, b) we *define* the integral by the relation
$$\int_a^b f(x)dx = \int_a^c f(x)dx + \int_c^b f(x)dx.$$

In case either (or both) of the last two integrals fails to exist, the integral in the first member is said not to exist.

Example 1. Find $\int_0^a \dfrac{x\,dx}{\sqrt{a^2-x^2}}$.

We have

$$\int_0^a \frac{x\,dx}{\sqrt{a^2-x^2}} = \lim_{\epsilon \to 0}\int_0^{a-\epsilon} \frac{x\,dx}{\sqrt{a^2-x^2}}$$
$$= \lim_{\epsilon \to 0}\left[-\sqrt{a^2-x^2}\right]_0^{a-\epsilon} = \lim_{\epsilon \to 0}\left[a - \sqrt{a^2-(a-\epsilon)^2}\right] = a.$$

Example 2. Find $\int_0^2 \dfrac{x\,dx}{(x^2-1)^{\frac{2}{3}}}$.

Since the integrand is discontinuous for $x = 1$, we write

$$\int_0^2 \frac{x\,dx}{(x^2-1)^{\frac{2}{3}}} = \int_0^1 \frac{x\,dx}{(x^2-1)^{\frac{2}{3}}} + \int_1^2 \frac{x\,dx}{(x^2-1)^{\frac{2}{3}}}$$
$$= \lim_{\epsilon \to 0}\int_0^{1-\epsilon} \frac{x\,dx}{(x^2-1)^{\frac{2}{3}}} + \lim_{\epsilon \to 0}\int_{1+\epsilon}^2 \frac{x\,dx}{(x^2-1)^{\frac{2}{3}}}$$
$$= \lim_{\epsilon \to 0}\left[\tfrac{3}{2}(x^2-1)^{\frac{1}{3}}\right]_0^{1-\epsilon} + \lim_{\epsilon \to 0}\left[\tfrac{3}{2}(x^2-1)^{\frac{1}{3}}\right]_{1+\epsilon}^2$$
$$= \lim_{\epsilon \to 0}\left[\tfrac{3}{2}\{(1-\epsilon)^2-1\}^{\frac{1}{3}} + \tfrac{3}{2}\right] + \lim_{\epsilon \to 0}\left[\tfrac{3}{2}\sqrt[3]{3} - \tfrac{3}{2}\{(1+\epsilon)^2-1\}^{\frac{1}{3}}\right]$$
$$= \tfrac{3}{2} + \tfrac{3}{2}\sqrt[3]{3} = \tfrac{3}{2}(1+\sqrt[3]{3}).$$

Example 3. Show that the integral $\int_0^{2a} \dfrac{dx}{(x-a)^2}$ does not exist.

If this integral exists it is the sum of the following two integrals:

$$\int_0^a \frac{dx}{(x-a)^2}, \quad \int_a^{2a} \frac{dx}{(x-a)^2}. \qquad (1)$$

If the first of these exists it is defined by the limit

$$\lim_{\epsilon \to 0}\int_0^{a-\epsilon} \frac{dx}{(x-a)^2} = \lim_{\epsilon \to 0}\left[-\frac{1}{x-a}\right]_0^{a-\epsilon} = \lim_{\epsilon \to 0}\left(\frac{1}{\epsilon} - \frac{1}{a}\right).$$

The last limit does not exist. Hence the first of the integrals (1) does not exist. (The student may show similarly that the second does not exist.) Hence the integral given in the problem does not exist.

If we should proceed to evaluate the given integral without regard to the discontinuity in the integrand, we should have the following incorrect result:

$$\int_0^{2a} \frac{dx}{(x-a)^2} = \left[-\frac{1}{x-a}\right]_0^{2a} = -\frac{1}{a} - \frac{1}{a} = -\frac{2}{a}.$$

This example shows the importance of adhering strictly to the definitions in treating integrals whose integrands are discontinuous.

EXERCISES

Evaluate the integrals in problems 1 to 12.

1. $\int_1^\infty \frac{dx}{x^2} = 1.$

2. $\int_0^\infty \frac{8a^3\,dx}{x^2 + 4a^2} = 2\pi a^2.$

3. $\int_0^\infty \frac{dx}{a^2 + x^2} = \frac{\pi}{2a}.$

4. $\int_0^\infty e^{-ax}\,dx = \frac{1}{a}$ if $a > 0.$

5. $\int_{-\infty}^0 e^x\,dx = 1.$

6. $\int_1^\infty \frac{dx}{x^6} = \frac{1}{5}.$

7. $\int_0^a \frac{dx}{\sqrt{a-x}} = 2\sqrt{a}.$

8. $\int_{\sqrt{2}}^\infty \frac{dx}{x\sqrt{2x^2-1}} = \frac{\pi}{6}.$

9. $\int_{-\infty}^\infty \frac{dx}{x^2 + 4x + 5} = \pi.$

10. $\int_1^\infty \frac{dx}{(x+1)^n} = \frac{1}{(n-1)2^{n-1}}$ if $n > 1.$

11. $\int_0^\infty \frac{dx}{(a+x)^2} = \frac{1}{a}$ if $a > 0.$

12. $\int_0^3 \frac{x\,dx}{(x^2-1)^{\frac{2}{3}}} = \frac{9}{2}.$

13. Show that the following integrals do not exist:

$$\int_0^\infty e^x\,dx, \quad \int_0^1 \frac{dx}{x^2}, \quad \int_0^\infty \frac{dx}{x^3}.$$

14. Determine which of the following integrals exist:

$$\int_0^a \frac{dx}{a^2 - x^2}, \quad \int_0^\infty \frac{dx}{(x+1)^3}, \quad \int_0^\infty e^{ax}\,dx.$$

15. Find the area bounded by the curve

$$y = \frac{1}{a^2 + x^2}$$

and the x-axis. *Ans.* $\frac{\pi}{a}.$

16. Find the area bounded by the curve

$$y(a^2 - x^2)^{\frac{1}{2}} = x,$$

the x-axis, and the ordinate $x = a.$ *Ans.* $a.$

CHAPTER XI

THE SIMPLER APPLICATIONS OF INTEGRATION

99. Plane areas in rectangular coördinates. If a curve lies above the x-axis, as in the adjacent figure, then the area A bounded by the curve, the x-axis, and the ordinates $x = a$ and $x = b$ is given by the formula

$$A = \int_a^b f(x)dx,$$

as we saw in § 32.

If the curve lies below the x-axis, as in the figure here given, the area A bounded by the curve, the x-axis, and the ordinates $x = a$ and $x = b$ is given by the formula

$$A = -\int_a^b f(x)dx,$$

as one sees by observing that the curve $y = -f(x)$ lies above the x-axis and that the area bounded by it and the x-axis and the same ordinates is also equal to A.

If the curve lies partly above the x-axis and partly below, as in the adjacent figure, then the integral

$$\int_a^b f(x)dx$$

represents the sum of the areas above the x-axis minus the sum of the areas below the x-axis.

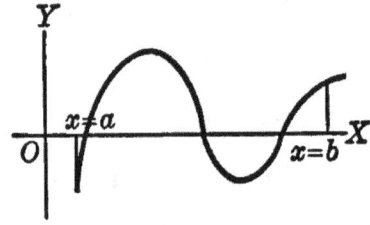

FIG. 45

In finding areas by integration it is often desirable to separate the required area into convenient parts and to find the area of the parts separately and then take their sum. This is illustrated in the following problem.

Example. Find the area bounded by the parabola $y^2 = \tfrac{16}{3} x$ and the circle $x^2 + y^2 = 25$.

These curves intersect in the points $(3, 4)$ and $(3, -4)$, as one sees by solving their equations simultaneously. Denote these points by B and C respectively; let O be the center of the circle and let A be the point $(5, 0)$ at which the circle cuts the x-axis. The required area is $OCAB$. It is evidently divided into two equal parts by the radius OA. We shall find the area of OAB, one of these parts. To find the latter area it is convenient to divide it into two parts by the line BD from B perpendicular to OA. Then

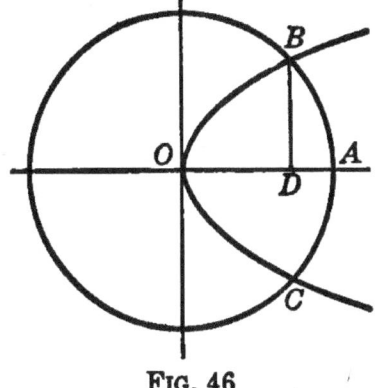

Fig. 46

$$\text{area } ODB = \int_0^3 \frac{4}{\sqrt{3}} x^{\tfrac{1}{2}} dx = 8;$$

$$\text{area } DBA = \int_3^5 \sqrt{25 - x^2}\, dx$$

$$= \frac{1}{2}\left[x\sqrt{25 - x^2} + 25 \arcsin \frac{x}{5} \right]_3^5$$

$$= \frac{25\,\pi}{4} - 6 - \frac{25}{2} \arcsin \frac{3}{5}.$$

Hence area $OCAB = 2(\text{area } OAB) = 2(\text{area } ODB + \text{area } DBA)$

$$= \frac{25\,\pi}{2} + 4 - 25 \arcsin \frac{3}{5}.$$

100. Length in rectangular coördinates. In § 62 we saw that the differential ds of arc length is given by the formula

$$ds = \sqrt{1 + \left(\frac{dy}{dx}\right)^2}\, dx.$$

Let us suppose that we begin to measure arc length s from the point for which $x = a$, and let s_{ab} be the length of arc from the point for which $x = a$ to that for which $x = b$. On integrating the preceding formula for ds we have

$$\int_0^{s_{ab}} ds = \int_a^b \sqrt{1 + \left(\frac{dy}{dx}\right)^2} dx;$$

or
$$s_{ab} = \int_a^b \sqrt{1 + \left(\frac{dy}{dx}\right)^2} dx. \tag{1}$$

In a similar way we may employ the formula

$$ds = \sqrt{\left(\frac{dx}{dy}\right)^2 + 1}\, dy$$

(from § 62) to prove that

$$s_{ab} = \int_c^d \sqrt{\left(\frac{dx}{dy}\right)^2 + 1}\, dy \tag{2}$$

if c and d are the values of y corresponding to the values a and b, respectively, for x.

Formulas (1) and (2) are the fundamental formulas for finding lengths of plane curves whose equations are given in rectangular coördinates.

Example 1. Find the circumference of the circle $x^2 + y^2 = a^2$.
We have

$$\frac{dy}{dx} = -\frac{x}{y}, \quad 1 + \left(\frac{dy}{dx}\right)^2 = 1 + \frac{x^2}{y^2} = \frac{x^2 + y^2}{y^2} = \frac{a^2}{y^2} = \frac{a^2}{a^2 - x^2}.$$

One fourth of the circumference lies in the first quadrant and extends from the point for which $x = 0$ to that for which $x = a$. Hence

$$\text{circumference} = 4 \int_0^a \sqrt{1 + \left(\frac{dy}{dx}\right)^2}\, dx$$

$$= 4 \int_0^a \frac{a}{\sqrt{a^2 - x^2}}\, dx$$

$$= 4\, a \left[\arcsin \frac{x}{a} \right]_0^a$$

$$= 4\, a \cdot \frac{\pi}{2} = 2\, \pi a.$$

Example 2. Find the circumference of the circle defined by the parametric equations $x = a \cos \theta$, $y = a \sin \theta$.

This is the same circle as that given in the previous example. We have
$$\frac{dy}{dx} = -\cot\theta, \quad 1 + \left(\frac{dy}{dx}\right)^2 = 1 + \cot^2\theta = \csc^2\theta.$$

Since $dx = -a\sin\theta\,d\theta$ and the point (x, y) moves over a quadrant of the circle when θ ranges from $\frac{\pi}{2}$ to 0, we have

$$\text{circumference} = 4\int_{\frac{\pi}{2}}^{0} \csc\theta(-a\sin\theta)d\theta = -4a\int_{\frac{\pi}{2}}^{0} d\theta = 2\pi a.$$

101. Areas by the summation of thin strips. In finding areas it is sometimes convenient to go back to the definition of the integral as the limit of a sum. To the learner this method is useful in preparing the way for some of the most important applications of integration. The method is best illustrated by means of an example.

Example. Find the area A bounded by the two curves
$$y = \sqrt[4]{16 - x^2} \quad \text{and} \quad y = \sqrt[4]{16 - x^2} + x(4 - x).$$

The points of intersection are $(4, 0)$ and $(0, 4)$. A rough sketch of the curves between these points is given in the adjacent figure, the first equation being that of the lower curve. Let $g(x)$ be the length of that portion of the ordinate $x = x$ which is intercepted between these two curves. Then $g(x) = x(4 - x)$. Let the interval $(0, 4)$ of the x-axis be divided into a large number of small intervals $\Delta x_1, \Delta x_2, \cdots, \Delta x_n$ and let ξ_i be a value of x on the interval Δx_i. Then
$$g(\xi_i)\Delta x_i$$

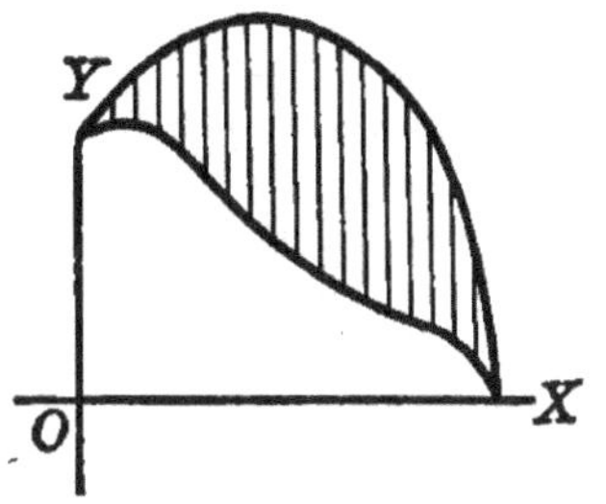

Fig. 47

is approximately the area of the corresponding vertical strip between the two curves. The area between the two curves is approximately equal to the sum
$$\sum_0^4 g(\xi_i)\Delta x_i$$

of all the products $g(\xi_i)\Delta x_i$ formed for the n subintervals $\Delta x_1, \Delta x_2, \cdots, \Delta x_n$. The required area A is then given by the formula

$$A = \lim_{\text{Every } \Delta x_i \to 0} \sum_0^4 g(\xi_i)\Delta x_i.$$

(Compare the similar argument in § 31, where the analysis is given in much greater detail.) But the last limit is equal to the integral

$$\int_0^4 g(x)dx, \quad \text{or} \quad \int_0^4 x(4-x)dx.$$

Hence
$$A = \int_0^4 x(4-x)dx = \left[2x^2 - \frac{x^3}{3}\right]_0^4 = 10\tfrac{2}{3}.$$

EXERCISES

1. Show that the area of the ellipse $a^2y^2 + b^2x^2 = a^2b^2$ is πab.

2. Show that the area of the ellipse $x = a\cos\theta$, $y = b\sin\theta$ is πab, using an integration with respect to θ.

3. Show that the area which is bounded by the hyperbola $a^2y^2 - b^2x^2 = -a^2b^2$ and the ordinate $x = m$ ($m > a$) is

$$\frac{b}{a}m\sqrt{m^2 - a^2} - ab\log\left(\frac{m + \sqrt{m^2 - a^2}}{a}\right).$$

4. Show that the area bounded by the curves $y = 2x^2$, $y = 2x$, and $y = 4x$ is $\tfrac{7}{3}$.

5. Show that the entire length of the curve $x^{\frac{2}{3}} + y^{\frac{2}{3}} = a^{\frac{2}{3}}$ is $6a$.

6. What is the length of the catenary

$$y = \frac{a}{2}\left(e^{\frac{x}{a}} + e^{-\frac{x}{a}}\right)$$

from $x = 0$ to $x = a$.   *Ans.* $\dfrac{a}{2}(e - e^{-1})$.

7. Find the area between the two curves

$$y = \sqrt{a^2 - x^2} \quad \text{and} \quad y = \sqrt{a^2 - x^2} + x(a - x).$$

(Use summation of thin strips.)   *Ans.* $\dfrac{a^3}{6}$.

8. Show that the area bounded by the curves $y = f(x)$ and $y = f(x) + c$ (c = positive constant) and the ordinates $x = a$ and $x = b$ ($b > a$) is $c(b-a)$.

9. If $g(x)$ is any function which is positive on the interval (a, b), show that the area bounded by the curves $y = f(x)$ and $y = f(x) + g(x)$ and the ordinates $x = a$ and $x = b$ ($b > a$) is

$$\int_a^b g(x)dx.$$

10. Find the area bounded by the curves

$$y = e^{x^2}, \quad y = (2x + 1)e^{x^2}, \quad x = 5.$$

(Use summation of thin strips.)   *Ans.* $e^{25} - 1$.

102. Volumes by the summation of thin slices. The method of finding areas by the summation of thin strips can be extended to space of three dimensions so as to afford a method of finding volumes by the summation of thin slices. This method has a wide range of usefulness in the applications of the integral calculus. (Review § 31 before proceeding.)

The solid whose volume is to be found we think of as being divided into a large number of thin slices by means of parallel planes. For the method to be effective we must be able to find the area of the cross section made by each of these planes. The planes themselves may lie in any direction, but it is more convenient to have them perpendicular to one of the axes. If this cannot be brought about with one of the original axes, a new set of axes may be chosen so as to have the cutting planes perpendicular to one of the axes. In developing the formula we shall suppose that the planes are perpendicular to the x-axis; but the student will remember that either the y-axis or the z-axis will serve equally well.

Let us consider a volume V bounded by some given surface and the planes $x = a$ and $x = b$. Let the interval (a, b) of the x-axis be divided into a large number of subintervals Δx_1, Δx_2, $\cdots$, Δx_n. Let planes be passed through the extremities of these intervals and perpendicular to the x-axis. They divide the given volume V into a large number of thin slices.

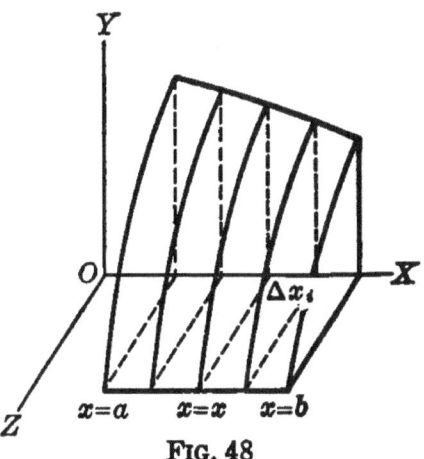

FIG. 48

Let us now consider a certain approximation to the volume of each of these thin slices. We shall find this by means of the function $F(x)$, where $F(x)$ is the area of the cross section of the given volume V by the plane $x = x$. Now let ξ_i be any value of x on the interval Δx_i. Then the volume of the slice which has this interval for its altitude is approximately

$$F(\xi_i)\Delta x_i,$$

this being the volume of the solid of base $F(\xi_i)$ and altitude Δx_i. The sum of the volumes of all the slices is approximately

$$\sum_a^b F(\xi_i)\Delta x_i,$$

this symbol denoting the sum of the given approximate volumes of all the slices. Hence this sum denotes approximately the volume V itself.

Now suppose that the number of slices is increased indefinitely and that the thickness of each approaches zero. Then the foregoing sum approaches the volume V, as the learner will see by analyzing the situation geometrically and comparing it with the similar but simpler situation in § 31. Hence

$$V = \lim_{\text{Every } \Delta x_i \to 0} \sum_a^b F(\xi_i)\Delta x_i.$$

But in § 32 we saw that this limit is equal to

$$\int_a^b F(x)dx.$$

Hence this is the value of V. Therefore

If $F(x)$ is the area of the cross section of a given volume by the plane $x = x$, then the volume V of that portion of it which is cut off by the planes $x = a$ and $x = b$ ($b > a$) is given by the formula

$$V = \int_a^b F(x)dx.$$

Example 1. Find the volume of the ellipsoid

$$\frac{x^2}{a^2} + \frac{y^2}{b^2} + \frac{z^2}{c^2} = 1.$$

The section made by the plane parallel to the yz-plane and at a distance x from it, that is, the section made by the plane $x = x$, is an ellipse with the equation

$$\frac{y^2}{b^2} + \frac{z^2}{c^2} = 1 - \frac{x^2}{a^2}.$$

Here y and z are variables and x is a constant. Putting this equation in standard form, we have

$$\frac{y^2}{b^2\left(1-\dfrac{x^2}{a^2}\right)} + \frac{z^2}{c^2\left(1-\dfrac{x^2}{a^2}\right)} = 1.$$

Therefore the semi-axes of the ellipse are

$$b\sqrt{1-\frac{x^2}{a^2}} \quad \text{and} \quad c\sqrt{1-\frac{x^2}{a^2}}.$$

Its area is π times the product of these,

$$\pi bc\left(1-\frac{x^2}{a^2}\right).$$

If x ranges from 0 to a, the thin slices cover half of the volume of the ellipsoid. Hence the required volume V is

$$V = 2\int_0^a \pi bc\left(1-\frac{x^2}{a^2}\right)dx = \frac{4}{3}\pi abc.$$

If $b=c=a$, the ellipsoid becomes a sphere of radius a and the volume is given by the well-known formula $\frac{4}{3}\pi a^3$.

Example 2. Prove that the volume V of a pyramid (or cone) of base B and height h is $\frac{1}{3}Bh$.

Let the cone be placed with its vertex at the origin and its base perpendicular to the x-axis. The area of a section made by a plane perpendicular to the x-axis at a distance x from the vertex is to the area of the base as x^2 is to h^2. Hence the area $F(x)$ of the cross section is Bx^2/h^2. Therefore

$$V = \int_0^h \frac{Bx^2}{h^2}dx = \frac{1}{3}\frac{B}{h^2}x^3\bigg]_0^h = \frac{1}{3}Bh.$$

Example 3. Find the volume V of the segment of a sphere of radius a cut off by a plane at a distance c from the center, c being less than a.

Place the sphere so that its center is at the origin of coördinates and so that the given cutting plane is perpendicular to the x-axis. The plane $x=x$ cuts from the sphere a circle of radius $\sqrt{a^2-x^2}$ and hence of area $\pi(a^2-x^2)$. Therefore

$$V = \int_c^a \pi(a^2-x^2)dx = \pi\left(\frac{2}{3}a^3 - a^2c + \frac{c^3}{3}\right).$$

If $c=0$ this gives $\frac{2}{3}\pi a^3$ as the volume of a hemisphere.

Example 4. The center of a variable square moves along the principal axis of the ellipse $b^2x^2 + a^2y^2 = a^2b^2$ from one end to the other. The plane of the square remains perpendicular to this axis and the extremities of one of its diagonals move along the ellipse. What is the volume V swept out by the square?

When the center of the square is at a distance x from the center of the ellipse the diagonal of the square is

$$\frac{2\,b}{a}\sqrt{a^2 - x^2}.$$

Hence its area is

$$\frac{2\,b^2}{a^2}(a^2 - x^2).$$

This is the function $F(x)$ of the theorem in this section. Hence

$$V = 2\int_0^a \frac{2\,b^2}{a^2}(a^2 - x^2)dx = \frac{8}{3}ab^2.$$

EXERCISES

1. What is the volume bounded by the surface

$$\frac{x^4}{a^4} + \frac{y^2}{b^2} + \frac{z^2}{c^2} = 1. \qquad \text{Ans. } \tfrac{8}{5}\pi abc.$$

2. What is the volume of the segment of the ellipsoid

$$\frac{x^2}{a^2} + \frac{y^2}{b^2} + \frac{z^2}{c^2} = 1$$

cut off by the plane $x = k$, k being less than a?

$$\text{Ans. } \pi bc\left(\frac{2}{3}a - k + \frac{k^3}{3\,a^2}\right).$$

3. Find the volume bounded by the surface $y^2 + z^2 = 6\,x$ and the planes $x = a$ and $x = b$ ($b > a$). Ans. $3\,\pi(b^2 - a^2)$.

4. Find the volume bounded by the surface

$$\frac{x^2}{9} + \frac{y^2}{4} = 6\,z$$

and the planes $z = 0$ and $z = 3$. Ans. $162\,\pi$.

5. The base of a variable isosceles triangle is a chord of a circle of radius a. It moves across the circle remaining perpendicular to a diameter. If the plane of the triangle is perpendicular to this diameter and if the triangle is of constant altitude h, what is the volume swept out by it when its base moves entirely across the circle?

Ans. $\tfrac{1}{2}\pi ha^2$.

6. If the altitude of the triangle in the foregoing problem should also vary and in such way as to keep the triangle equilateral, what is the volume swept out by it? *Ans.* $\frac{4}{3} a^3 \sqrt{3}$.

7. Find the volume which is generated by revolving the ellipse $b^2 x^2 + a^2 y^2 = a^2 b^2$ about its principal axis. *Ans.* $\frac{4}{3} \pi a b^2$.

8. Find the volume which is generated by revolving the ellipse $b^2 x^2 + a^2 y^2 = a^2 b^2$ about its minor axis. *Ans.* $\frac{4}{3} \pi a^2 b$.

9. A woodsman in felling a circular tree of radius a makes a cut whose base is a horizontal semicircle and whose other cut face is a plane making an angle β with the horizontal base. What is the volume which he cuts out? *Ans.* $\frac{2}{3} a^3 \tan \beta$.

10. Two great circles on a sphere of radius a cut each other at right angles. A variable square moves perpendicular to their common diameter and its vertices remain on the great circles. What volume is swept out by it? *Ans.* $\frac{8}{3} a^3$.

103. Mean value. The *arithmetic mean*, or *average value*, of a finite number of quantities is their sum divided by their number. Let us consider the arithmetic mean of n ordinates of the curve $y = f(x)$ spaced at equal intervals between $x = a$ and $x = b$, inclusive of $x = a$ but exclusive of $x = b$. Denote these ordinates from left to right by y_1, $y_2, y_3, \cdots, y_n$. Then their arithmetic mean is

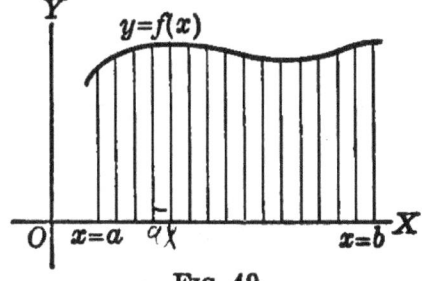

FIG. 49

$$\frac{y_1 + y_2 + \cdots + y_n}{n}.$$

Since the interval (a, b) of length $(b - a)$ is divided into n equal parts, the length Δx of each part is $(b - a)/n$. Hence

$$n = \frac{b - a}{\Delta x}.$$

Substituting this value of n in the foregoing arithmetic mean and simplifying, we find for it the value

$$\frac{y_1 \Delta x + y_2 \Delta x + \cdots + y_n \Delta x}{b - a}, \quad \text{or} \quad \frac{\sum_{a}^{b} y_i \Delta x}{b - a}.$$

Now let us suppose that the number of ordinates is indefinitely increased. Since they are evenly spaced on the fixed interval (a, b), the length Δx of each of them approaches zero. Hence the limit of the foregoing arithmetic mean is

$$\frac{\lim\limits_{\Delta x \to 0} \sum\limits_{a}^{b} y_i \Delta x}{b - a}.$$

From the definition of definite integral in § 32 it follows that this limit has the value

$$\frac{\int_a^b y\, dx}{b - a}.$$

This is called the *mean value* of the function y, or $f(x)$, on the interval (a, b).

The mean value of the function $f(x)$ on the interval (a, b) is

$$\frac{\int_a^b f(x)\, dx}{b - a}.$$

This mean value is the average height of the curve $y = f(x)$ on the interval (a, b); that is, it is equal to the altitude of the rectangle on the base $b - a$ and of area equal to the area bounded by the curve $y = f(x)$ and the x-axis and the ordinates $x = a$ and $x = b$.

The definition of mean value has been introduced by means of a geometrical illustration. But mean value, in the abstract sense of the formal definition, is independent of this geometric phrasing. The independent variable x may refer to any geometrical or physical magnitude, as time, distance, angle, area, volume, pressure, work done, velocity, acceleration, etc.

Example 1. If a body falls from rest it acquires in t seconds the velocity gt, where g denotes the acceleration due to gravity ($g =$ approximately 32.2 ft. per second per second). What is the average velocity of fall (1) for the first 6 seconds? (2) for the fourth second?

The mean value of the velocity v from time $t = 0$ to time $t = 6$ is

$$\frac{\int_0^6 gt\, dt}{6 - 0} = 3\, g.$$

The mean value of v for the fourth second, that is, from time $t = 3$ to time $t = 4$, is

$$\frac{\int_3^4 gt\, dt}{4 - 3} = \frac{7}{2} g.$$

Example 2. What is the mean value of the ordinates of a semicircle of radius a when they are uniformly spaced along the arc of the semicircle?

Let the diameter of the semicircle be the x-axis. Let θ be the angle made with the x-axis by the radius to the extremity of a given ordinate. Then this ordinate is of length $a \sin \theta$. We take θ to be the independent variable, since it varies proportionately to the arc length. Hence the mean value sought is equal to

$$\frac{\int_0^\pi a \sin \theta\, d\theta}{\pi - 0} = \frac{2a}{\pi}.$$

EXERCISES

1. What is the mean value of the ordinates of the ellipse

$$b^2 x^2 + a^2 y^2 = a^2 b^2$$

when they are uniformly spaced along the principal axis? *Ans.* $\frac{1}{4}\pi b$.

2. What is the mean value of the ordinates of the parabola $y^2 = 2px$ from $x = 0$ to $x = x_1$ when they are uniformly spaced on the x-axis? *Ans.* $\frac{2}{3}\sqrt{2px_1}$.

3. What is the mean value of the ordinates of one arch of the sine curve $y = \sin x$ when they are uniformly spaced on the x-axis?

Ans. $\dfrac{2}{\pi}$.

4. What is the mean value of a cross section of the ellipsoid

$$\frac{x^2}{a^2} + \frac{y^2}{b^2} + \frac{z^2}{c^2} = 1$$

made by uniformly spaced planes perpendicular to (1) the x-axis? (2) the y-axis? (3) the z-axis? *Ans.* (1) $\frac{2}{3}\pi bc$, (2) $\frac{2}{3}\pi ac$, (3) $\frac{2}{3}\pi ab$.

5. What is the mean value of the area of the square described in problem 10 of §102? *Ans.* $\frac{4}{3} a^2$.

6. What is the mean value of the area of the isosceles triangle described in problem 5 of §102? *Ans.* $\frac{1}{4}\pi ha$.

7. What is the mean value of the area of the equilateral triangle described in problem 6 of § 102? Ans. $\frac{2}{3} a^2 \sqrt{3}$.

8. What is the mean value of the function
$$y = \frac{a}{2}\left(e^{\frac{x}{a}} + e^{-\frac{x}{a}}\right)$$
from $x = 0$ to $x = a$? Ans. $\frac{a}{2}(e - e^{-1})$.

104. The constant of indefinite integration in geometrical problems. The use of this constant in geometrical problems is best illustrated by means of examples.

Example 1. Find the equations of those plane curves whose slope at every point (x, y) is equal to x.

Since $\frac{dy}{dx}$ is equal to the slope, we have
$$\frac{dy}{dx} = x, \quad \text{whence} \quad y = \frac{x^2}{2} + C.$$

For every value of C this is the equation of one of the curves described. The totality of these curves constitutes a *family* of parabolas, one member of the family corresponding to each value of the constant C.

Each curve of the family has the property that the slope at a point on one of them for which $x = a$ is equal to the slope at the point on any other for which $x = a$. This is obviously a general property of any family of curves whose slope at each point is defined by a given function of x.

Example 2. Find the equations of those plane curves whose slope at every point (x, y) is equal to x/y.

We have
$$\frac{dy}{dx} = \frac{x}{y}.$$

In order to integrate this equation we need to write it in such form that one of the variables appears alone on one side of the sign of equality and the other alone on the other side. In the present case this is accomplished by clearing of fractions. Thus we have
$$y \, dy = x \, dx.$$
Integrating both members and multiplying by 2, we have
$$y^2 = x^2 + C.$$
For varying values of C this gives a family of equilateral hyperbolas.

Example 3. Find the equations of those plane curves whose slope at every point (x, y) is equal to y/x.

We have
$$\frac{dy}{dx} = \frac{y}{x}.$$

Writing this equation in such form as to have the variables separated, we have
$$\frac{dy}{y} = \frac{dx}{x}.$$

Integrating, we have
$$\log y = \log x + C = \log cx$$

if $\log c = C$. Hence
$$y = cx$$

is the equation of the family of curves, one curve of the family corresponding to each value of the parameter c. Each of the "curves" is a straight line through the origin.

In these examples we have certain families, or systems, of curves, each family, or system, consisting of a set of curves obtained from a single equation involving a parameter by giving to that parameter arbitrarily chosen values. It is possible to have two systems of plane curves so related that each curve of one system cuts every curve of the other system at right angles. Each of two systems so related is said to constitute the *orthogonal trajectories* of the other system. At a point of intersection of a curve of one of these systems with a curve of the other system the slope of the one curve is the negative reciprocal of the slope of the other, since the two curves are perpendicular. This affords a means of finding the orthogonal trajectories of a given system of curves.

Example 4. Determine the orthogonal trajectories of the system of equilateral hyperbolas whose equations are of the form $y^2 - x^2 = C$.

On differentiating the given equation member by member and solving for dy/dx, we have
$$\frac{dy}{dx} = \frac{x}{y}.$$

At a given point (x, y) a curve of the given system has the slope x/y. The curve of the orthogonal system which passes through the same point (x, y) has its slope equal to the negative reciprocal of this,

and hence equal to $-y/x$. For the orthogonal system we therefore have
$$\frac{dy}{dx} = -\frac{y}{x}.$$

From this we have
$$\frac{dy}{y} = -\frac{dx}{x}.$$

Integrating, we have $\quad \log y = -\log x + c'$;

hence $\quad\quad\quad \log y + \log x = \log c$

if $c = e^{c'}$. Therefore
$$\log xy = \log c, \quad \text{whence} \quad xy = c.$$

Hence the curves of the orthogonal system are the equilateral hyperbolas obtained from the equation $xy = c$ by varying the parameter c.

EXERCISES

1. Find the equations of those plane curves whose slope at every point (x, y) is equal to $3\,x^2$. $\quad$ *Ans.* $y = x^3 + C$.

2. Find the equations of those plane curves whose slope at every point (x, y) is equal to $\tfrac{1}{2}\!\left(e^{\frac{x}{a}} - e^{-\frac{x}{a}}\right)$. $\quad$ *Ans.* $y = \dfrac{a}{2}\!\left(e^{\frac{x}{a}} + e^{-\frac{x}{a}}\right) + C$.

3. What is the equation of the system of plane curves whose slope at every point (x, y) is equal to $-x^2/y^2$? $\quad$ *Ans.* $x^3 + y^3 = C$.

4. What is the equation of the system of plane curves whose slope at every point (x, y) is equal to $2\,xy$? $\quad$ *Ans.* $y = ce^{x^2}$.

5. What is the equation of the system of plane curves whose slope at every point (x, y) is equal to $e^x e^y$? $\quad$ *Ans.* $e^x + e^{-y} = C$.

6. What are the orthogonal trajectories of the system of curves defined in each of the foregoing problems?
$$\text{Ans. } y = \frac{1}{3\,x} + C, \quad y = a\log\frac{e^{\frac{x}{a}} + 1}{e^{\frac{x}{a}} - 1} + C,$$
$$\frac{1}{y} - \frac{1}{x} = C, \quad x = Ce^{-y^2}, \quad e^y - e^{-x} = C.$$

7. What are the orthogonal trajectories of the circles $x^2 + y^2 = a^2$? $\quad$ *Ans.* $y = cx$.

8. Find the equations of those curves whose subnormal is the constant a. $\quad$ *Ans.* $y^2 = 2\,ax + C$.

9. Find the equations of the curves whose subtangent is equal to the constant a.

Ans. $y = ce^{\frac{x}{a}}$.

10. What is the equation of the curve whose subtangent is the constant a, when the curve is subject to the further condition that it shall pass through the point $(0, 1)$?

Ans. $y = e^{\frac{x}{a}}$.

105. The constant of indefinite integration in physical problems. The applications of antiderivatives are of course applications of indefinite integrals, since these are two names for the same thing. In treating antiderivatives we gave a number of geometrical and physical applications, and especially of physical applications. We shall now add further applications to physical problems. These are best given by means of examples. (In this connection the student will do well to review § 88.)

Example 1. A projectile is thrown with an initial velocity of magnitude v_0 and inclined at an angle α with the horizontal. Discuss its motion, the resistance of air being neglected.

Let O be the point of projection. Take as the y-axis the vertical line through O, the positive direction being upward. Take as the x-axis the horizontal line through O with which the velocity makes the angle α. It is evident that the projectile remains in the plane of these axes. If (x, y) is a point of the path of the projectile, then its motion is completely determined when we find x and y as functions of the time t measured from the instant of projection.

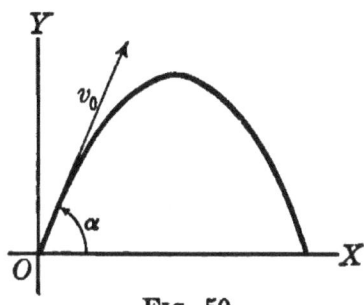

FIG. 50.

Since there is no force acting in a horizontal direction on the projectile, the x-component of acceleration is zero. The vertical acceleration is that due to gravity. Hence the y-component of acceleration is $-g$. These two facts give the equations

$$\frac{d^2x}{dt^2} = 0, \quad \frac{d^2y}{dt^2} = -g. \tag{1}$$

Integrating, we have

$$\frac{dx}{dt} = c_1, \quad \frac{dy}{dt} = -gt + c_2, \tag{2}$$

where c_1 and c_2 are constants of integration.

The initial x-component of velocity is $v_0 \cos \alpha$, and the initial y-component of velocity is $v_0 \sin \alpha$. These are the values of dx/dt and dy/dt, respectively, at time $t = 0$. Hence
$$c_1 = v_0 \cos \alpha, \quad c_2 = v_0 \sin \alpha.$$
Putting these values in equations (2) and integrating, we have
$$x = v_0 \cos \alpha \cdot t + c_3, \quad y = v_0 \sin \alpha \cdot t - \tfrac{1}{2} gt^2 + c_4, \qquad (3)$$
where c_3 and c_4 are constants of integration. But x and y have the value 0 when $t = 0$. Hence $c_3 = 0 = c_4$. Putting these values into equations (3), we have, finally,
$$x = v_0 \cos \alpha \cdot t, \quad y = v_0 \sin \alpha \cdot t - \tfrac{1}{2} gt^2. \qquad (4)$$
These equations describe fully the position of the projectile at time t.

The components of velocity may be obtained by differentiating with respect to t. They are
$$v_x = v_0 \cos \alpha, \quad v_y = v_0 \sin \alpha - gt. \qquad (5)$$

To determine the path of integration we eliminate t from equations (4) and obtain the equation of the path in the form
$$y = x \tan \alpha - \frac{gx^2}{2 v_0^2 \cos^2 \alpha}.$$

This is the equation of a parabola whose principal axis is parallel to the y-axis.

Example 2. When a particle moves in a straight line so that its acceleration is proportional to the distance of the particle from a fixed point O on the line of its motion and is directed toward that point, it is said to be in *simple harmonic motion*. Derive formulas for the velocity v and the distance s from O in terms of the time t, the velocity when $s = 0$ being v_0 ($v_0 \neq 0$) and time t being measured from an instant when $s = 0$.

The acceleration a is $-k^2 s$, where k^2 denotes a positive constant. From the relations
$$v = \frac{ds}{dt} \quad \text{and} \quad a = \frac{dv}{dt}$$
we have the (frequently useful) relation
$$v \, dv = a \, ds.$$
Hence in this problem we have
$$v \, dv = - k^2 s \, ds.$$

Integrating and using the fact that $v = v_0$ when $s = 0$, we have

$$v^2 = v_0^2 - k^2 s^2.$$

Substituting for v the value ds/dt, we have

$$\frac{ds}{dt} = \sqrt{v_0^2 - k^2 s^2},$$

or

$$dt = \frac{ds}{\sqrt{v_0^2 - k^2 s^2}}.$$

Hence

$$t = \frac{1}{k} \arcsin \frac{ks}{v_0} + C.$$

But $s = 0$ when $t = 0$. Hence $C = 0$. Therefore

$$t = \frac{1}{k} \arcsin \frac{ks}{v_0}.$$

Solving this equation for s, we have

$$s = \frac{v_0}{k} \sin kt.$$

Differentiating with respect to t, we have

$$v = \frac{ds}{dt} = v_0 \cos kt.$$

From the formula for s it follows that the particle moves back and forth over a distance $2 v_0/k$, the fixed point O being at the center of this interval.

EXERCISES

1. A vector is rotating about the fixed point O with a constant angular acceleration α. If at time $t = 0$ the vector makes an angle θ_0 with a fixed line through O and if it has then the angular velocity ω_0, show that the angle θ which it makes with the given fixed line at time t is defined by the formula

$$\theta = \tfrac{1}{2} \alpha t^2 + \omega_0 t + \theta_0.$$

2. A point moves in a straight line so that the velocity v is given by the formula $v = 100\, t - 16\, t^2$. Find the acceleration. Find s if $s = 0$ when $t = 0$. *Ans.* $100 - 32\, t,\ 50\, t^2 - \tfrac{16}{3} t^3$.

3. A body moves so that the x-component of acceleration is given by the formula $6 - t$ while the y-component of acceleration is given by the formula $6 + t$. If the initial x-component and y-component of velocity are both 2, what is the speed of the body at the end of 2 sec.? *Ans.* 20.

4. If a body is projected vertically upward with an initial velocity v_0, for how long does it continue to rise, and what is the greatest height attained by it?

Ans. $\dfrac{v_0}{g}$, $\dfrac{v_0^2}{2g}$.

5. The *range* of a projectile is the distance from the point of projection to the point at which it again reaches the horizontal plane containing that point. Show that the range of the projectile described in illustrative example 3 is

$$\dfrac{v_0^2}{g} \sin 2\alpha.$$

6. What is the greatest height attained by the same projectile?

Ans. $\dfrac{v_0^2 \sin^2 \alpha}{2g}$.

7. What time is required by the same projectile to attain its greatest height? to reach again the horizontal plane from which it was projected?

Ans. $\dfrac{v_0 \sin \alpha}{g}$, $\dfrac{2 v_0 \sin \alpha}{g}$.

8. If a stone is thrown horizontally from the top of a vertical cliff 161 ft. high, what velocity is required to carry it across a stream 300 ft. wide at the foot of the cliff? (Take $g = 32.2$ ft. per second per second.)

Ans. $30\sqrt{10}$ ft. per second.

9. A particle moves along the circumference of a circle with constant speed. Show that its projection on a fixed diameter moves with simple harmonic motion. (The projection of a point on a line is the foot of the perpendicular from the point to the line.)

10. How long does it require the particle in illustrative example 4 to move entirely across the interval on which it oscillates?

Ans. $\dfrac{\pi}{k}$.

11. A point with simple harmonic motion is oscillating back and forth over an interval with center at O. When it is 3 ft. from O it has a velocity of 4 ft. per second. When it is 4 ft. from O it has a velocity of 3 ft. per second. What is the length of the interval on which it oscillates, and what is the time required for it to pass from one end of the interval to the other end and return to the starting-point?

Ans. 10 ft., 2π sec.

CHAPTER XII

DIFFERENTIAL EQUATIONS

106. Definitions. An *ordinary differential* equation is an equation of the form

$$F(x, y, y', y'', \cdots, y^{(n)}) = 0, \qquad (1)$$

where $y^{(i)} = d^i y/dx^i$.

The *order* of the differential equation (1) is that of the derivative of highest order occurring in the function F. The *degree* of (1) is the degree of the derivative of highest order, when the left member of (1) has been simplified until neither derivatives raised to fractional powers nor quotients involving derivatives occur.

For example,
$$\frac{dy}{dx} = x^2$$

is a differential equation of the first order and the first degree, while

$$\frac{\left[1 + \left(\frac{dy}{dx}\right)^2\right]^{\frac{3}{2}}}{\frac{d^2y}{dx^2}} - r(x) = 0$$

is of the second order and second degree.

A function $y(x)$ is said to be a *solution* of a differential equation of the nth order if $y(x)$ can be differentiated n times and if, when y, dy/dx, d^2y/dx^2, $\cdots$, $d^n y/dx^n$ are substituted in (1), this equation becomes an identity in x.

A differential equation (1) is said to be *integrable* when it is possible to exhibit a formula representing all functions $y(x)$ which are solutions of (1). At present it is possible to deduce formulas for only certain types of differential equations. In the sections below several of the most important classes of integrable differential equations are treated.

107. Differential equations in which the variables are separable. The simplest type of integrable differential equation is of the form
$$\frac{dy}{dx} = f(x) \cdot g(y), \tag{1}$$
where $f(x)$ and $g(y)$ are continuous functions of x and y respectively. If either $f(x)$ or $g(y)$ is identically zero, the general solution for y is
$$y = \text{constant}. \tag{2}$$

If neither $f(x)$ nor $g(y)$ vanishes identically, we can divide by $g(y) \neq 0$ and obtain
$$\frac{dy}{g(y)} = f(x)dx \tag{3}$$
equivalent to (1) but in which the variables are separated. The general solution of (3) is found by equating any indefinite integral of the left member to the general indefinite integral of the right member. Symbolically this solution may be written in the form
$$\int_{y_0}^{y} \frac{dy}{g(y)} = \int_{x_0}^{x} f(x)dx + c, \tag{4}$$
where x_0 and y_0 are fixed values lying in the intervals of definition of $f(x)$ and $g(y)$ respectively, and c is an arbitrary constant.

We have already solved a number of examples of this type when $g(y) = 1$. For example, we solved the equation $dA/dx = x^2$ (in § 29), and the equation $dy/dx = x$ (in § 104). We have also solved a number of equations of the general type. For example, in § 88 we incidentally solved the following differential equations:
$$\frac{dy}{dx} = ky, \quad \frac{dp}{dh} = -kp, \quad \frac{d\tau}{dt} = -k(\tau - \tau_1).$$

We shall illustrate the method by some examples.

Example 1. Solve the equation
$$\sqrt{1 - x^2}\, dy = \sqrt{1 - y^2}\, dx.$$
This may be written in the form
$$\frac{dy}{\sqrt{1 - y^2}} = \frac{dx}{\sqrt{1 - x^2}}.$$
Integrating, we have $\arcsin y = \arcsin x + C,$
or $y = \sin(\arcsin x + C).$

Example 2. Solve the equation
$$(1-x)y\,dx + (1+y)x\,dy = 0.$$
The equation may be written in the form
$$\frac{1+y}{y}dy = \frac{x-1}{x}dx \quad \text{or} \quad \left(1+\frac{1}{y}\right)dy = \left(1-\frac{1}{x}\right)dx.$$
Integrating, we have $y + \log y = x - \log x + C$,
whence
$$\log xy - x + y = C.$$

Example 3. Solve the equation
$$\frac{dy}{dx} = \frac{1+y^2}{1+x^2}.$$
This may be written in the form
$$\frac{dy}{1+y^2} = \frac{dx}{1+x^2}.$$
Integrating, we have
$$\text{arc tan } y = \text{arc tan } x + \text{arc tan } c,$$
the constant of integration being written as the arc tan function of the arbitrary constant c. Then
$$y = \tan(\text{arc tan } x + \text{arc tan } c).$$
Hence
$$y = \frac{x+c}{1-cx}.$$

EXERCISES

Solve each of the following differential equations, and in each case verify the result by differentiation:

1. $y\,dx + x\,dy = 0.$ Ans. $xy = c.$
2. $y\,dx - x\,dy = 0.$ Ans. $y = cx.$
3. $x\,dx + y\,dy = 0.$ Ans. $x^2 + y^2 = c.$
4. $(y+1)dx - (x+1)dy = 0.$ Ans. $y+1 = c(x+1).$
5. $y^2\,dx + (x+1)dy = 0.$ Ans. $y = \dfrac{1}{\log[c(x+1)]}.$
6. $(1+x^2)y\,dy - (1+y^2)x\,dx = 0.$ Ans. $y^2 + 1 = c(x^2+1).$
7. $\sin x \cos y\,dx - \cos x \sin y\,dy = 0.$ Ans. $\cos y = c \cos x.$
8. $x\dfrac{dy}{dx} + 2y = xy\dfrac{dy}{dx}.$ Ans. $x^2y = ce^y.$
9. $dx = 2y(1+x^2)dy.$ Ans. $y^2 = \text{arc tan } x + c.$

108. Homogeneous differential equations. Because of the extreme simplicity of the differential equations treated in the preceding section it is desirable to be able to reduce as many types of differential equations as possible to the standard form (1) of that section. One important class of differential equations which can be reduced to this form is known as homogeneous. A homogeneous differential equation of the *first order* is one which can be written in the form

$$\frac{dy}{dx} = f\left(\frac{y}{x}\right), \tag{1}$$

where the right-hand member is a homogeneous function of degree zero.

In order to reduce such a differential equation to the standard form (1) of § 107 we have only to set $y = ux$. This substitution leads to a new differential equation in the new variables x and u which is of the form

$$x \frac{du}{dx} + u = f(u),$$

which reduces at once to

$$\frac{dx}{x} = \frac{du}{f(u) - u};$$

whence we have by integration

$$\log x - \log c = \int_{u_0}^{u} \frac{du}{f(u) - u},$$

or

$$\log\left(\frac{x}{c}\right) = \int_{u_0}^{u} \frac{du}{f(u) - u},$$

or

$$x = c e^{\int_{u_0}^{u} \frac{du}{f(u) - u}}.$$

Example. Solve the equation

$$(x^2 + y^2)dx = xy\, dy.$$

This equation is homogeneous, since it may be written in the form

$$\frac{dy}{dx} = \frac{x^2 + y^2}{xy} = \frac{1 + \frac{y^2}{x^2}}{\frac{y}{x}}.$$

We may therefore make the substitution $y = ux$. When this is done, the equation reduces to
$$\frac{dx}{x} = u\, du.$$

Then we have by integration with $u_0 = 0$,
$$\log x - \log c = \frac{u^2}{2} = \frac{y^2}{2x^2}.$$

Hence
$$x = c \cdot e^{\frac{y^2}{2x^2}}.$$

EXERCISES

Solve the following differential equations:

1. $(x + y)dx = (x - y)dy.$ Ans. $\arctan \frac{y}{x} - \log \sqrt{x^2 + y^2} = c.$
2. $x\frac{dy}{dx} - y = \sqrt{x^2 - y^2}.$ Ans. $\arcsin \frac{y}{x} = \log x + c.$
3. $(x^2 - y^2)dy = 2xy\, dx.$ Ans. $x^2 + y^2 = cy.$
4. $x\, dy = (y + \sqrt{x^2 + y^2})dx.$ Ans. $x^2 = c^2 - 2cy.$
5. $y^2\, dx + (xy + x^2)dy = 0.$ Ans. $xy^2 = c(x + 2y).$
6. $(x + y)dx + x\, dy = 0.$
7. $(mx + y)dx = (x - my)dy.$ Ans. $m \log \sqrt{x^2 + y^2} - \tan^{-1}\frac{y}{x} = c.$

109. Linear differential equations of the first order. A differential equation of the special form
$$\frac{dy}{dx} + P(x)y - Q(x) = 0 \tag{1}$$

is called a *linear differential equation* of the first order. If $y(x)$ is any solution of (1), its value can be obtained as follows: Transpose $Q(x)$ to the right member of (1), multiply both members of the resulting equation by $e^{\int P dx}$ and note that the result has the form
$$\frac{d}{dx}\left[y \cdot e^{\int P dx}\right] = Q \cdot e^{\int P dx},$$

from which it follows that
$$y \cdot e^{\int P dx} = \int Q \cdot e^{\int P dx} dx + c. \tag{2}$$

Hence
$$y = e^{-\int P dx}\left[\int Q \cdot e^{\int P dx} dx + c\right]. \tag{3}$$

Example 1. Solve the equation $dy/dx + 2y = x$.

In this equation $P = 2$, $Q = x$.

Hence $\quad e^{\int P dx} = e^{\int 2 dx} = e^{2x}$,

and $\quad y = e^{-2x} \int x e^{2x} dx + c e^{-2x} = \dfrac{x}{2} - \dfrac{1}{4} + c e^{-2x}$.

Example 2. A boat is being towed through the water at a velocity v_0, when the rope breaks. If the resistance of the water is proportional to the velocity of the boat, find the equation of motion after the rope breaks.

Let $\quad M$ = mass of the boat,

$\quad\quad\quad v$ = velocity of the boat at time t.

Then the force acting on the boat is kv, which must be negative, since the boat is being retarded. But force is equal to mass times acceleration. Therefore
$$\text{Force} = M \left(\frac{dv}{dt}\right),$$

and $\quad\quad\quad M\left(\dfrac{dv}{dt}\right) = -kv$.

This may be written in the form
$$\frac{dv}{dt} + \frac{kv}{M} = 0.$$

This is of the form (1). $P = k/M$, $Q = 0$.

Therefore $\quad\quad\quad v = c e^{-kt/M}$.

But $v = v_0$ when $t = 0$.

Therefore $\quad\quad\quad c = v_0$ and $v = v_0 e^{-kt/M}$.

EXERCISES

Solve the following differential equations:

1. $\dfrac{dy}{dx} - y = e^x$. $\quad\quad$ Ans. $y = xe^x + ce^x$.

2. $\dfrac{dy}{dx} - y = e^{-x}$. $\quad\quad$ Ans. $y = xe^{-x} + ce^{-x}$.

3. $\dfrac{dy}{dx} - \dfrac{y}{x} = x^2$. $\quad\quad$ Ans. $y = \dfrac{x^3}{2} + cx$.

4. $\dfrac{dy}{dx} + \dfrac{y}{x^2 + 1} = 0$. $\quad\quad$ Ans. $y = c e^{\text{arc tan } x}$.

5. $\dfrac{dy}{dx} - y = \sin x$. $\quad\quad$ Ans. $y = \dfrac{\cos x + \sin x}{2} + ce^x$.

6. A boat starting from rest is acted on by a constant force. If the resistance of the water is proportional to the velocity of the boat, find the equation of motion of the boat.

7. In Ex. 6 the proportionality factor is 5, the mass of the boat 10 tons, and the force applied is 500 lb. Find the velocity of the boat at the end of 4 sec.

8. In electricity we have the following facts:

(1) The sum of all the potential differences in a closed circuit is zero.

(2) The potential difference due to a current i flowing through a resistance R is iR.

(3) The potential difference due to an inductance L in a circuit in which a current is changing is $L(di/dt)$.

Using these facts, set up and solve the differential equation of an alternating circuit which has an electromotive force E, resistance R, and inductance L.

110. Linear differential equations of the second order. A differential equation of the form

$$\frac{d^2y}{dx^2} + P\left(\frac{dy}{dx}\right) + Qy - R = 0 \tag{1}$$

where P, Q, and R are continuous functions of x is called an *ordinary linear differential equation* of the second order. If R is equal to zero, the equation is called *homogeneous*; otherwise it is called *nonhomogeneous*.

Let us first consider the homogeneous differential equation

$$\frac{d^2y}{dx^2} + P\left(\frac{dy}{dx}\right) + Qy = 0. \tag{2}$$

I. *If $y_1(x)$ and $y_2(x)$ are any two solutions of the homogeneous equation (2), and if c_1 and c_2 are any two constants, then*

$$y(x) = c_1 y_1(x) + c_2 y_2(x) \tag{3}$$

is also a solution.

PROOF. From (3) we have, by differentiation,

$$\frac{dy}{dx} = c_1 \frac{dy_1}{dx} + c_2 \frac{dy_2}{dx}, \tag{4}$$

$$\frac{d^2y}{dx^2} = c_1 \frac{d^2y_1}{dx^2} + c_2 \frac{d^2y_2}{dx^2}. \tag{5}$$

A substitution from (3), (4), (5) in (2) gives

$$\frac{d^2y}{dx^2} + P\frac{dy}{dx} + Qy =$$
$$c_1\left(\frac{d^2y_1}{dx^2} + P\frac{dy_1}{dx} + Qy_1\right) + c_2\left(\frac{d^2y_2}{dx^2} + P\frac{dy_2}{dx} + Qy_2\right). \quad (6)$$

The right member of (6) vanishes because of the assumption that y_1 and y_2 are solutions of (2). Hence y is also a solution of (2).

The right member of (3) is called a linear combination of y_1 and y_2 with constant coefficients. Thus theorem I shows that every linear combination of two solutions of (2) is again a solution of (2). The converse, that every solution of (2) may be expressed as a linear combination of two given solutions, is not always true, as the following example will show:

Consider the differential equation

$$\frac{d^2y}{dx^2} + y = 0.$$

$y_1 = \sin x$ and $y_2 = 3\sin x$ are solutions of this equation. Also $y = \sin x - \cos x$ is a solution, but in this case it is not possible to choose constants c_1 and c_2 such that $y = c_1 y_1 + c_2 y_2$. However we can prove the following theorem:

II. *Every solution y of (2) can be written in the form (3) where y_1 and y_2 are any solutions of (2) which satisfy the condition*

$$y_1\frac{dy_2}{dx} - y_2\frac{dy_1}{dx} \neq 0, \quad (7)$$

and where c_1 and c_2 are suitably chosen constants.

PROOF. If y_1 and y_2 are any two solutions of (2), then the function

$$D = y_1\frac{dy_2}{dx} - y_2\frac{dy_1}{dx}$$

satisfies the relation

$$D = D_0 \cdot e^{-\int_{x_0}^{x} P\,dx}, \quad (8)$$

where D_0 is the value of D for $x = x_0$. For from the definition of D and (2) we have

$$\frac{dD}{dx} = y_1 \frac{d^2 y_2}{dx^2} - y_2 \frac{d^2 y_1}{dx^2} \tag{9}$$

$$= y_1 \left(-Qy_2 - P \frac{dy_2}{dx} \right) - y_2 \left(-Qy_1 - P \frac{dy_1}{dx} \right)$$

$$= -P \cdot D.$$

Hence
$$D = K e^{-\int_{x_0}^{x} P \, dx}, \tag{10}$$

where K is a constant whose value, found by setting $x = x_0$ in (10), is precisely D_0.

From (8) it is evident that D is always zero or never zero. We restrict attention to the case $D \neq 0$.

Suppose y is any solution of (2). We may define

$$D_1 = y_1 \frac{dy}{dx} - y \frac{dy_1}{dx},$$
$$D_2 = y_2 \frac{dy}{dx} - y \frac{dy_2}{dx}, \tag{11}$$

from which we have

$$D_1 = D_{10} e^{-\int_{x_0}^{x} P \, dx}, \quad D_2 = D_{20} e^{-\int_{x_0}^{x} P \, dx}, \tag{12}$$

where D_{10} and D_{20} are the values of D_1 and D_2 when $x = x_0$.

Regard the equations (11) as two simultaneous, non-homogeneous, linear equations in two unknowns $(dy/dx, -y)$. Since the determinant of the coefficients of these unknowns is $D \neq 0$ (by hypothesis), we can solve the equations and obtain

$$y = -\frac{D_2}{D} y_1 + \frac{D_1}{D} y_2. \tag{13}$$

Because of (8) and (12) it is evident that (13) reduces to

$$y = -\frac{D_{20}}{D_0} y_1 + \frac{D_{10}}{D_0} y_2. \tag{14}$$

Hence the theorem is proved.

If y_1 and y_2 are two solutions of (2) satisfying the condition $D \neq 0$, then the solution $y = c_1 y_1 + c_2 y_2$ is called the general solution of the homogeneous equation (2). Methods for obtaining y_1 and y_2 will be given later.

We now return to the nonhomogeneous equation (1). By a *particular integral* of (1) we shall mean a function of x which contains no constants of integration but which satisfies (1). We can now prove the following theorem:

III. If Y is a particular integral of (1), then the most general solution of (1) is given by

$$y = Y + c_1 y_1 + c_2 y_2, \tag{15}$$

where y_1 and y_2 are two integrals of (2) such that $D \neq 0$ and c_1 and c_2 are arbitrary constants.

PROOF. In view of our hypothesis we have

$$\frac{d^2 Y}{dx^2} + P \frac{dY}{dx} + QY = R. \tag{16}$$

Suppose now that y is any solution of (1) so that

$$\frac{d^2 y}{dx^2} + P \frac{dy}{dx} + Qy = R. \tag{17}$$

If we subtract (16) from (17) and write the difference in the form

$$\frac{d^2(y - Y)}{dx^2} + P \frac{d(y - Y)}{dx} + Q(y - Y) = 0, \tag{18}$$

we see that $y - Y$ is a solution of the homogeneous equation (2). Hence by theorem II this difference can be written in the form

$$y - Y = c_1 y_1 + c_2 y_2,$$

where y_1 and y_2 are solutions of (2) satisfying the condition $D \neq 0$. Hence

$$y = Y + c_1 y_1 + c_2 y_2 \tag{19}$$

and the theorem is proved.

The sum $c_1 y_1 + c_2 y_2$ in (19) is called the *complementary function*. Because of theorem II the problem of integrating the homogeneous equation (2) will be completely solved when we have given a method for finding the integrals y_1 and y_2 which appear in the complementary function. If, in addition, a method is given for finding a particular integral Y of (1), then theorem III shows that the problem of integrating the nonhomogeneous equation will be solved. Such methods are given below for the special types of linear differential equations which are particularly important in the applications.

111. Method for finding the solution of a homogeneous linear differential equation of the second order with constant coefficients. Let us now assume that $y = e^{mx}$ is a solution of

$$\frac{d^2y}{dx^2} + P\frac{dy}{dx} + Qy = 0. \tag{1}$$

Then $dy/dx = me^{mx}$, $d^2y/dx^2 = m^2 e^{mx}$, and (1) becomes

$$e^{mx}(m^2 + Pm + Q) = 0. \tag{2}$$

Since e^{mx} cannot be zero for any value of m or x, we must have

$$m^2 + Pm + Q = 0 \tag{3}$$

if (1) is to be satisfied.

Now equation (3) is a quadratic in m, and hence there are two values, m_1 and m_2, which will determine solutions for equation (1). These values may be

(I) real and distinct;
(II) real and equal;
(III) imaginary.

If the values for m in (3) are real and distinct, we shall have the two solutions $y_1 = e^{m_1 x}$ and $y_2 = e^{m_2 x}$, and, in view of theorem II, § 110, the general solution of (1) will be $y = c_1 e^{m_1 x} + c_2 e^{m_2 x}$ since it is easy to show that $D \neq 0$ for the solutions y_1 and y_2.

Example. Solve the differential equation

$$\frac{d^2y}{dx^2} + 3\frac{dy}{dx} + 2y = 0.$$

For this example equation (3) becomes

$$m^2 + 3m + 2 = 0,$$

from which we have $m = -2$ and $m = -1$. Hence the two solutions for the above equation are $y_1 = e^{-2x}$ and $y_2 = e^{-x}$. These then give the solution $y = c_1 e^{-2x} + c_2 e^{-x}$, which is the general solution, since $D = e^{-2x} e^{-x} \neq 0$.

If the values for m in (3) are real and equal, we obtain only one solution, namely $y = e^{mx}$. However, a simple substitution shows that $y = xe^{mx}$ is also a solution. Therefore $y = (c_1 - c_2 x)e^{mx}$ is also a solution.

The student should show that this is the general solution.

Example. Solve the differential equation

$$\frac{d^2y}{dx^2} + 4\frac{dy}{dx} + 4 = 0.$$

For this example equation (3) becomes

$$m^2 + 4m + 4 = 0.$$

This equation gives the value $m = -2$ counted twice. Hence $y = e^{-2x}$ is a solution. But a substitution from $y = xe^{-2x}$ in the original equation shows that this is also a solution. Hence the general solution is $y = (c_1 - c_2 x)e^{-2x}$.

If the values of m in (3) are imaginary, they will have the form $m_1 = a + ib$, $m_2 = a - ib$. Then $y_1 = e^{(a+ib)x}$, $y_2 = e^{(a-ib)x}$, from which we obtain the general solution

$$y = c_1 e^{(a+ib)x} + c_2 e^{(a-ib)x}$$
$$= e^{ax}(c_1 e^{ibx} + c_2 e^{-ibx}). \qquad (4)$$

Equation (4) is usually written in another form obtained by the use of the following identities, which we set down without proof:

$$e^{ibx} = \cos bx + i \sin bx,$$
$$e^{-ibx} = \cos bx - i \sin bx.$$

When a substitution of these values is made in (4) it becomes

$$y = e^{ax}[(c_1 + c_2)\cos bx + i(c_1 - c_2) \sin bx]$$
$$= e^{ax}(A \cos bx + B \sin bx) \qquad (5)$$

when we have set $c_1 + c_2 = A$, $i(c_1 - c_2) = B$.

Example. Solve the differential equation

$$\frac{d^2y}{dx^2} - 2\frac{dy}{dx} + 10y = 0.$$

In this example equation (3) becomes

$$m^2 - 2m + 10 = 0,$$

from which $m_1 = 1 + 3i$, $m_2 = 1 - 3i$. Hence $a = 1$ and $b = 3$ and $y = e^x(A \cos 3x + B \sin 3x)$.

EXERCISES

Solve the following differential equations:

1. $\dfrac{d^2y}{dx^2} + 5\dfrac{dy}{dx} + 6y = 0.$

2. $\dfrac{d^2y}{dx^2} + \dfrac{dy}{dx} - 6y = 0.$

3. $2\dfrac{d^2y}{dx^2} - 5\dfrac{dy}{dx} + 2y = 0.$

4. $\dfrac{d^2y}{dx^2} + 4\dfrac{dy}{dx} + y = 0.$

5. $4\dfrac{d^2y}{dx^2} + 4\dfrac{dy}{dx} + y = 0.$

6. $\dfrac{d^2y}{dx^2} + 6\dfrac{dy}{dx} + 9y = 0.$

7. $\dfrac{d^2y}{dx^2} + 4y = 0.$

8. $\dfrac{d^2y}{dx^2} + \dfrac{dy}{dx} + y = 0.$

9. $\dfrac{d^2y}{dx^2} - k^2 y = 0.$

10. $\dfrac{d^2y}{dx^2} + k^2 y = 0.$

112. Physical applications. In § 105 we solved the problem of simple harmonic motion, using a special method. We now proceed to solve it as a problem in linear differential equations of the second order.

Example 1. Set up and solve the equation of simple harmonic motion.

Let $x =$ displacement. Then

$$\frac{d^2x}{dt^2} = -k^2 x$$

represents the motion. The solution of this equation is

$$x = A \cos kt + B \sin kt.$$

If in this result we set

$$\frac{A}{(A^2 + B^2)^{\frac{1}{2}}} = \sin a, \qquad \frac{B}{(A^2 + B^2)^{\frac{1}{2}}} = \cos a,$$

we have $\quad x = (A^2 + B^2)^{\frac{1}{2}}(\sin a \cos kt + \cos a \sin kt)$

$\qquad\quad = (A^2 + B^2)^{\frac{1}{2}} \sin(kt + a).$

Example 2. A weight on a spring is set in motion and slides along a groove. If the force due to the spring is proportional to the displacement and is toward the position of equilibrium, and if the force due to the friction is proportional to the velocity of the weight, set up and solve the differential equation of motion of the weight.

There are two forces producing acceleration in the weight, the force of the spring and the force due to friction.

Let $x =$ displacement of the weight from the position of equilibrium,
$$\frac{dx}{dt} = \text{velocity of the weight.}$$

Now $-k_2 x =$ force due to the pull of the spring,

$$-k_1 \frac{dx}{dt} = \text{force due to friction,}$$

$$-k_2 x - k_1 \frac{dx}{dt} = \text{total force acting on the weight.}$$

Since, by Newton's law, force equals mass times acceleration, we have
$$-k_1 \frac{dx}{dt} - k_2 x = w \left(\frac{d^2 x}{dt^2}\right),$$

or
$$w \left(\frac{d^2 x}{dt^2}\right) + k_1 \frac{dx}{dt} + k_2 x = 0,$$

where $w =$ mass of the weight.

Equation (3), § 111, for this becomes
$$w m^2 + k_1 m + k_2 = 0.$$

The solutions of this equation are
$$m_1 = \frac{-k_1 + \sqrt{k_1^2 - 4 k_2 w}}{2w}, \quad m_2 = \frac{-k_1 - \sqrt{k_1^2 - 4 k_2 w}}{2w}.$$

The general solution of the differential equation is
$$x = c_1 e^{\left(\frac{-k_1 + \sqrt{k_1^2 - 4 k_2 w}}{2w}\right)t} + c_2 e^{\left(\frac{-k_1 - \sqrt{k_1^2 - 4 k_2 w}}{2w}\right)t}.$$

There are three cases to be considered.

CASE I. $k_1^2 - 4 k_2 w > 0$.

CASE II. $k_1^2 - 4 k_2 w = 0$.

In these two cases the motion is not oscillatory. The weight can go through the position of equilibrium for one value of t, after which it reaches a position of maximum displacement and then continually approaches but never reaches a position of equilibrium.

CASE III. $k_1^2 - 4 k_2 w < 0$.

In this case there are oscillations in equal times.

EXERCISES

1. Study the motion of a body which falls under the influence of gravity through a medium which opposes the motion with a force proportional to the first power of the velocity.

2. A particle moving with simple harmonic motion has a speed of 20 cm. per second when the displacement is 6 cm. and a speed of 10 cm. per second when the displacement is 12 cm. Find the equation of motion.

3. In Example 2 above find the equation of motion of a unit particle if $k_1 = 3$ and $k_2 = 2$. Plot the graph of the equation if $c_1 = c_2 = 1$.

4. Same as Ex. 3 with $k_1 = 2$, $k_2 = 1$.

5. Same as Ex. 3 with $k_1 = k_2 = 1$.

6. A needle is suspended on a cord so that it hangs in a horizontal position. If the needle is twisted through an angle θ from the position of equilibrium, the restoring force in the cord is proportional to the angular displacement. Find the motion of the needle.

7. If in Ex. 6 the retarding force due to the resistance of the fibers of the cord is proportional to the velocity of rotation, find the equation of motion.

113. Method of undetermined coefficients for finding a particular integral, Y, of the nonhomogeneous linear differential equation. When R contains only such expressions as e^{ax}, sin ax, cos ax, x^n (n a positive integer) which have a finite number of distinct derivatives, the method of undetermined coefficients may be applied. This method is not general, but has the advantage of being simple. Moreover, in a large number of problems arising in the elementary applications of physics and engineering R contains only such expressions. The following examples will illustrate the method.

Example 1. Solve the differential equation

$$\frac{d^2y}{dx^2} - 5\frac{dy}{dx} + 6y = 52 \sin 2x. \tag{1}$$

In this example the complementary function is

$$y_0 = c_1 e^{2x} + c_2 e^{3x}.$$

Let Y be the particular integral which must be added to y_0 in order that we may obtain $52 \sin 2x$ in the right member of (1).

Since the right member of (1) contains $\sin 2x$, we are led to take Y in the form
$$Y = a \sin 2x + b \cos 2x, \qquad (2)$$
because both $\sin 2x$ and $\cos 2x$ may be present when we find the first and second derivatives of $\sin 2x$. From (2) we have
$$\frac{dY}{dx} = 2a \cos 2x - 2b \sin 2x. \qquad (3)$$
$$\frac{d^2Y}{dx^2} = -4a \sin 2x - 4b \cos 2x. \qquad (4)$$

A substitution from (2), (3), and (4) in (1) gives
$$(2a + 10b)\sin 2x + (2b - 10a)\cos 2x = 52 \sin 2x. \qquad (5)$$
Since the left member of (5) must be identically the same as the right member, we have
$$2a + 10b = 52,$$
$$10a - 2b = 0,$$
from which we obtain $a = 1, b = 5$. Therefore $Y = \sin 2x + 5 \cos 2x$, and the complete solution is
$$y = c_1 e^{2x} + c_2 e^{3x} + \sin 2x + 5 \cos 2x.$$

Example 2. Solve the differential equation
$$\frac{d^2y}{dx^2} - y = e^x + 2x.$$
The complementary function is
$$y_0 = c_1 e^x + c_2 e^{-x}.$$
We note that one of the terms of the complementary function contains e^x and that e^x also occurs in the right member of the original equation. In such a situation we proceed as follows: If a term in the right member is a term of the complementary function, replace that term and its derivatives by x^n ($n = 1, 2, \cdots$) until the method of undetermined coefficients leads to consistent equations. Thus to solve the above equation we write the particular integral in the form
$$Y = axe^x + bx + c.$$
From this we have
$$\frac{d^2Y}{dx^2} = axe^x + 2ae^x.$$

A substitution in the given equation leads to the equation
$$2ae^x - bx - c = e^x + 2x.$$
Therefore $\quad a = \tfrac{1}{2}, \quad b = -2, \quad c = 0.$
Hence the complete solution is
$$y = c_1 e^x + c_2 e^{-x} + \tfrac{1}{2} x e^x - 2x.$$

EXERCISES

Solve the following differential equations:

1. $\dfrac{d^2y}{dx^2} + 3\dfrac{dy}{dx} - 4y = e^{2x}.$

2. $\dfrac{d^2y}{dx^2} + 9y = x^2.$

3. $\dfrac{d^2y}{dx^2} - (m+n)\dfrac{dy}{dx} + mny = 4.$

4. $\dfrac{d^2y}{dx^2} - (k+1)\dfrac{dy}{dx} + ky = \sin 3x.$

5. $\dfrac{d^2y}{dx^2} + 2\dfrac{dy}{dx} + 3y = \cos x.$

6. $\dfrac{d^2y}{dx^2} - 5\dfrac{dy}{dx} + 6y = e^{2x} + x.$

7. $\dfrac{d^2y}{dx^2} + k^2 y = \sin kx.$

8. $m^2\dfrac{d^2y}{dx^2} - 3m\dfrac{dy}{dx} + 2y = e^x + \sin x.$

9. $\dfrac{d^2y}{dx^2} - 5\dfrac{dy}{dx} - 6x = 0.$

10. $\dfrac{d^2y}{dx^2} - 2\dfrac{dy}{dx} - 3 = 0.$

11. A steel weight suspended on a spring and immersed in a fluid, whose resistance is proportional to the velocity of the weight in the fluid, is set in motion in a vertical line. At the same time an electromagnet is turned on, acting along the line of motion with a constant force E. Find the differential equation of motion.

$$Ans. \quad w\dfrac{d^2x}{dt^2} + k_1 \dfrac{dx}{dt} + k_2 x = E.$$

12. Solve the differential equation obtained in Ex. 11 if

(1) $k_1 = 5, \ k_2 = w = 2$
(2) $k_1 = w = 4, \ k_2 = 1$
(3) $k_1 = k_2 = w = 1.$

13. Solve the differential equation obtained in Ex. 11 under the conditions imposed in Ex. 12 with the further condition that the electromagnet acts with a force $E = E_0 \cos 3t.$

CHAPTER XIII

SUCCESSIVE DIFFERENTIATION AND INTEGRATION

114. Successive differentiation. In § 43 we defined derivatives of higher order. There are some ways of dealing with them which we shall now illustrate by examples.

Example 1. Find d^2y/dx^2 in terms of the derivatives of x with respect to y.

We start from the relation $\dfrac{dy}{dx} = \dfrac{1}{\frac{dx}{dy}}$.

Differentiating both sides with respect to x, we have

$$\frac{d^2y}{dx^2} = \frac{d}{dx}\left(\frac{dy}{dx}\right) = \frac{d}{dx}\left(\frac{1}{\frac{dx}{dy}}\right) = \frac{d}{dy}\left(\frac{1}{\frac{dx}{dy}}\right) \cdot \frac{dy}{dx}.$$

Hence
$$\frac{d^2y}{dx^2} = \frac{-\dfrac{d^2x}{dy^2}}{\left(\dfrac{dx}{dy}\right)^2} \cdot \frac{1}{\dfrac{dx}{dy}} = -\frac{\dfrac{d^2x}{dy^2}}{\left(\dfrac{dx}{dy}\right)^3}.$$

Example 2. Find d^2y/dx^2 when x and y are connected by the relation $b^2x^2 + a^2y^2 = a^2b^2$.

In the usual way we find that

$$\frac{dy}{dx} = -\frac{b^2x}{a^2y}.$$

Differentiating with respect to x, we have

$$\frac{d^2y}{dx^2} = \frac{d}{dx}\left(\frac{dy}{dx}\right) = -\frac{b^2}{a^2}\frac{d}{dx}\left(\frac{x}{y}\right) = -\frac{b^2}{a^2}\frac{y - x\dfrac{dy}{dx}}{y^2}$$

$$= -\frac{b^2}{a^2}\frac{y - x\left(-\dfrac{b^2}{a^2}\dfrac{x}{y}\right)}{y^2} = \frac{-b^2(a^2y^2 + b^2x^2)}{a^4y^3} = \frac{-b^2(a^2b^2)}{a^4y^3} = -\frac{b^4}{a^2y^3}.$$

EXERCISES

Find $\dfrac{d^2y}{dx^2}$ in the case of the following relations:

1. $x^2 - xy + y^2 = c$. \hfill Ans. $\dfrac{6c}{(x-2y)^3}$.

2. $x^3 + y^3 - 3axy = 0$. \hfill Ans. $-\dfrac{2a^3xy}{(y^2 - ax)^3}$.

3. When $b^2x^2 + a^2y^2 = a^2b^2$ show that
$$\frac{d^3y}{dx^3} = -\frac{3b^6x}{a^4y^5}.$$

4. Let y be a function both of x and of z and write $y = f(x)$ and $y = \phi(z)$. Then show that
$$\frac{d^2y}{dx^2} = \phi''(z)\left(\frac{dz}{dx}\right)^2 + \phi'(z)\frac{d^2z}{dx^2}.$$

HINT. Begin with the relation
$$\frac{dy}{dx} = \phi'(z)\frac{dz}{dx}.$$

5. If x and y are both functions of t, show that
$$\frac{d^2y}{dx^2} = \frac{\dfrac{dx}{dt}\dfrac{d^2y}{dt^2} - \dfrac{dy}{dt}\dfrac{d^2x}{dt^2}}{\left(\dfrac{dx}{dt}\right)^3}.$$

(When $y = t$ this reduces to the result given in illustrative example 1.)

6. Show that $\dfrac{d^2}{dx^2}(uv) = \dfrac{d^2u}{dx^2}v + 2\dfrac{du}{dx}\dfrac{dv}{dx} + u\dfrac{d^2v}{dx^2}$.

7. Show that
$$\frac{d^3}{dx^3}(uv) = \frac{d^3u}{dx^3}v + 3\frac{d^2u}{dx^2}\frac{dv}{dx} + 3\frac{du}{dx}\frac{d^2v}{dx^2} + u\frac{d^3v}{dx^3}.$$

8. Beginning from the results in Exs. 6 and 7, prove by mathematical induction the following formula due to Leibnitz:
$$\frac{d^n}{dx^n}(uv) = \frac{d^nu}{dx^n}v + n\frac{d^{n-1}u}{dx^{n-1}}\frac{dv}{dx} + \frac{n(n-1)}{2!}\frac{d^{n-2}u}{dx^{n-2}}\frac{d^2v}{dx^2}$$
$$+ \cdots + n\frac{du}{dx}\frac{d^{n-1}v}{dx^{n-1}} + u\frac{d^nv}{dx^n}.$$

The numerical coefficients here are those of the binomial formula.

9. By Leibnitz's formula (Ex. 8) show that
$$\frac{d^n}{dx^n}(e^x x^2) = e^x\left[x^2 + 2nx + n(n-1)\right].$$

115. Successive integration. Since the indefinite integral of a function of x is itself a function of x, it may also be integrated; the resulting indefinite integral is a function which may again be integrated; and so on indefinitely. Thus we have the successive indefinite integrals of a given function. Since each integration introduces an arbitrary constant, there will be in the result a number of arbitrary constants equal to the number of integrations performed. In particular problems one or more of these constants may be determined by given conditions.

Example 1. Find four successive integrals of $\sin ax$.

The successive integrals are found by continued integrations (which the student may perform without written computations). They are

$$-\frac{1}{a}\cos ax + c_1,$$

$$-\frac{1}{a^2}\sin ax + c_1 x + c_2,$$

$$\frac{1}{a^3}\cos ax + \frac{c_1}{2}x^2 + c_2 x + c_3,$$

$$\frac{1}{a^4}\sin ax + \frac{c_1}{6}x^3 + \frac{c_2}{2}x^2 + c_3 x + c_4.$$

Example 2. Find y when the third derivative of y with respect to x is x.

We have
$$\frac{d^3 y}{dx^3} = x, \quad \frac{d^2 y}{dx^2} = \frac{x^2}{2} + c_1,$$

$$\frac{dy}{dx} = \frac{x^3}{6} + c_1 x + c_2, \quad y = \frac{x^4}{24} + \frac{c_1}{2}x^2 + c_2 x + c_3.$$

Example 3. How far will a body fall from rest in 20 sec.?

It is known that the acceleration due to gravity is a constant. Denoting it by g and the distance fallen in t seconds by s, we have

$$\frac{d^2 s}{dt^2} = g.$$

Then
$$v = \frac{ds}{dt} = gt + c_1.$$

Since the body starts from rest, its velocity at the time $t = 0$ is 0. Hence $c_1 = 0$. Then we have

$$\frac{ds}{dt} = gt, \quad \text{whence} \quad s = \tfrac{1}{2} gt^2 + c_2.$$

Since $s = 0$ when $t = 0$, we have $c_2 = 0$. Therefore the formula for s is

$$s = \tfrac{1}{2} gt^2.$$

Putting 32.2 for g and 20 for t, we have 6440 ft. for the distance fallen in 20 sec.

EXERCISES

1. Find three successive integrals of the following functions:

$$e^{ax}, \quad x^3 + x, \quad \frac{1}{x^2}, \quad 3x^2 - \frac{1}{x^2}, \quad \frac{1}{x}.$$

2. Find y when

(1) $\dfrac{d^4y}{dx^4} = \sin x;$ (2) $\dfrac{d^3y}{dx^3} = \cos^3 x.$

3. The second derivative of the ordinate of a curve with respect to the abscissa is at each point equal to the abscissa decreased by 1. Find the equation of the curve so that it shall pass through the point (1, 0) and have its tangent at that point parallel to the x-axis.

Ans. $y = \tfrac{1}{6}(x-1)^3.$

116. Rolle's theorem and the law of the mean. We are now to give (in this section and the next one) a succession of three connected theorems the last of which affords one of the fundamental applications of higher derivatives to be given in Chapter XV. We begin with a very simple geometric property of a curve. Let $y = f(x)$ be a single-valued function which is continuous on an interval (a, b) and is such that $f(a) = f(b)$, as indicated in the adjacent figure. Moreover, let us suppose that the derivative $f'(x)$ is continuous on the interval (a, b). This means that the tangent to the curve turns continuously (without sudden jumps) as we pass continuously along the curve and that it does not become perpendicular to the x-axis. It is therefore clear that there must

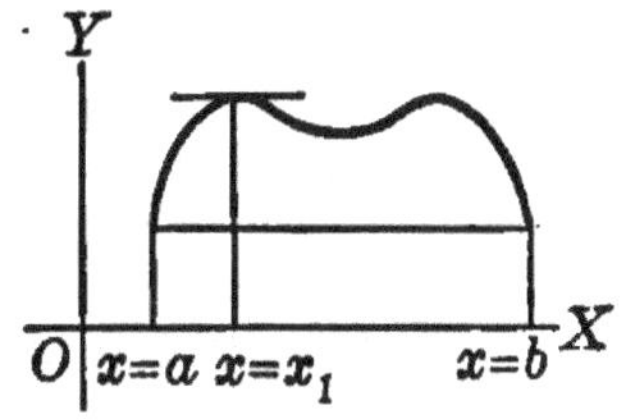

FIG. 51

be at least one point (x_1, y_1) on the curve such that the tangent at this point is parallel to the x-axis, x_1 being between a and b. This means that $f'(x_1)$ is equal to 0. Hence we have the following theorem :

Rolle's theorem. *Let $f(x)$ and its derivative $f'(x)$ be single-valued and continuous for all values of x on an interval (a, b), and let $f(a)$ be equal to $f(b)$. Then there is a value x_1 of x on the interior of the interval (a, b) for which we have $f'(x_1) = 0$.*

This theorem is readily extended. In its proof, to say that the tangent for the point for which $x = x_1$ is parallel to the x-axis is obviously the same as to say that it is parallel to the chord joining the points on the curve for which $x = a$ and $x = b$. Let us remove the restriction that $f(a) = f(b)$. If $f(x)$ and $f'(x)$ are single-valued and continuous, whence the curve and its slope are continuous, it is easy to prove that there is a point (x_1, y_1) on the curve

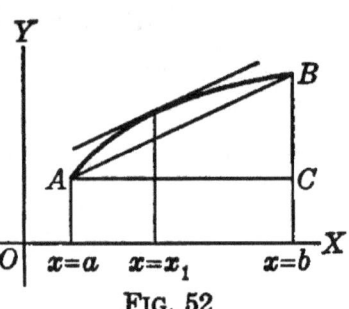

FIG. 52

such that the tangent at this point is parallel to the chord joining the points for which $x = a$ and $x = b$, x_1 being between a and b. For, if the curve lies above the chord (as in the figure), the slope of the tangent near A to the right is greater that that of the chord while the slope of the tangent near B to the left is less than that of the chord, so that at some intermediate point the slope of the tangent must be equal to that of the chord. A similar argument may be made if the curve lies below the chord, and also in the case where the curve lies partly above and partly below the chord. The slope of the chord AB is

$$\frac{BC}{AC} \quad \text{or} \quad \frac{f(b) - f(a)}{b - a}.$$

The slope of the tangent at the point (x_1, y_1) is $f'(x_1)$. Since these two slopes are equal, we have

$$\frac{f(b) - f(a)}{b - a} = f'(x_1). \quad \text{(See § 24.)}$$

This leads us to the following important result:

Law of the mean. *Let $f(x)$ and its derivative $f'(x)$ be single-valued and continuous for all values of x on an interval (a, b). Then there is a value x_1 of x on the interior of (a, b) for which we have*

$$\frac{f(b) - f(a)}{b - a} = f'(x_1).$$

Denote $b - a$ by h. Let θ be a number such that $x_1 = a + \theta h$. Since x_1 lies between a and b it is clear that θ lies between 0 and 1. Putting $a + h$ and $a + \theta h$ for b and x_1 in the foregoing equation, we have

$$\frac{f(a + h) - f(a)}{h} = f'(a + \theta h), \quad (0 < \theta < 1)$$

or
$$f(a + h) = f(a) + h f'(a + \theta h). \quad (0 < \theta < 1)$$

These two forms of the theorem of the mean are of frequent use.

The foregoing proof of the theorem of the mean is the one most readily understood by the learner. It is instructive for him to examine the following more analytical proof as a preparation for the proof of a more extended theorem in the next section.

Let us write
$$\frac{f(b) - f(a)}{b - a} = Q.$$

Then Q is the quantity to be determined. From this equation we have
$$f(b) - f(a) - (b - a)Q = 0.$$

Let $F(x)$ denote the function obtained from the first member of this equation on replacing a by x. Then

$$F(x) = f(b) - f(x) - (b - x)Q.$$

Then $F(a) = 0$ and $F(b) = 0$.

Moreover, $F'(x) = -f'(x) + Q$, so that $F'(x)$ is continuous since $f'(x)$ is continuous. Hence on applying Rolle's theorem we see that a number x_1 exists such that $F'(x_1) = 0$, x_1 lying between a and b. In the relation $F'(x) = -f'(x) + Q$ replace x by x_1 and use the relation $F'(x_1) = 0$. Then we have $0 = -f'(x_1) + Q$, and therefore the desired result

$$Q = f'(x_1).$$

117. Taylor's theorem. Let us now consider the relation

$$f(b) = f(a) + \frac{(b-a)}{1!}f'(a) + \frac{(b-a)^2}{2!}f''(a) + \frac{(b-a)^3}{3!}f'''(a)$$
$$+ \cdots + \frac{(b-a)^{n-1}}{(n-1)!}f^{(n-1)}(a) + \frac{(b-a)^n}{n!}R. \qquad (1)$$

Here a may be either less than or greater than b. It is obvious that this equation defines the number R in terms of the other quantities in the equation. In order to evaluate R in another useful form let us consider the function $F(x)$ defined by the equation

$$F(x) = f(b) - f(x) - \frac{(b-x)}{1!}f'(x) - \frac{(b-x)^2}{2!}f''(x) - \cdots$$
$$- \frac{(b-x)^{n-1}}{(n-1)!}f^{(n-1)}(x) - \frac{(b-x)^n}{n!}R. \qquad (2)$$

Then we have

$$F'(x) = -\frac{(b-x)^{n-1}}{(n-1)!}f^{(n)}(x) + \frac{(b-x)^{n-1}}{(n-1)!}R. \qquad (3)$$

The second member of equation (2) is what one obtains from (1) by transposing all terms to the first member and replacing a by x in that member. Hence $F(a) = 0$. But it is readily seen from (2) that $F(b) = 0$. Moreover, it follows from (3) that $F'(x)$ is continuous if $f^{(n)}(x)$ is continuous, as we shall now suppose. Hence we can apply Rolle's theorem (§ 116) to show that a number x_1 exists (between a and b) such that $F'(x_1) = 0$. Replacing x by x_1 in (3) and remembering that $F'(x_1) = 0$, we find on solving for R that

$$R = f^{(n)}(x_1).$$

Putting this value for R in place of R in equation (1), we have the following equation:

$$f(b) = f(a) + \frac{(b-a)}{1!}f'(a) + \frac{(b-a)^2}{2!}f''(a) + \frac{(b-a)^3}{3!}f'''(a)$$
$$+ \cdots + \frac{(b-a)^{n-1}}{(n-1)!}f^{(n-1)}(a) + \frac{(b-a)^n}{n!}f^{(n)}(x_1), \qquad (4)$$

where x_1 lies between a and b. This equation embodies *Taylor's Theorem*, sometimes called also the extended law of the mean.

In proving it we have assumed that $f(x), f'(x), f''(x), \cdots, f^{(n)}(x)$ are single-valued and continuous on (a, b).

118. Maxima and minima. Taylor's theorem affords a means of developing a useful test for maxima and minima. From equation (4) of the preceding section we write down two equations, one of them obtained by replacing b by $a + h$ and the other by replacing b by $a - h$; they are the following:

$$f(a + h) = f(a) + \frac{h}{1!}f'(a) + \frac{h^2}{2!}f''(a) + \frac{h^3}{3!}f'''(a) + \cdots$$
$$+ \frac{h^{n-1}}{(n-1)!}f^{(n-1)}(a) + \frac{h^n}{n!}f^{(n)}(x_1),$$

$$f(a - h) = f(a) - \frac{h}{1!}f'(a) + \frac{h^2}{2!}f''(a) - \frac{h^3}{3!}f'''(a) + \cdots$$
$$+ (-1)^{n-1}\frac{h^{n-1}}{(n-1)!}f^{(n-1)}(a) + (-1)^n\frac{h^n}{n!}f^{(n)}(x_2),$$

where x_1 is between a and $a + h$ and x_2 is between a and $a - h$. We take h to be positive in both equations.

Let us now consider the condition under which $f(x)$ will have a maximum or a minimum for $x = a$. (The student will do well to review § 69 before proceeding.) We shall suppose that $f'(x)$ is continuous for $x = a$. A first necessary condition for a maximum or a minimum is that $f'(a)$ shall be zero. It may happen that some of the higher derivatives are also equal to zero. We shall consider the case in which at least one of them is different from zero. Let $f^{(k)}(a)$ be the first of the quantities $f'(a), f''(a), \cdots$ which is not equal to zero. Taking $n = k$ in the two foregoing equations, we have

$$f(a + h) - f(a) = \frac{h^k}{k!}f^{(k)}(x_1),$$

$$f(a - h) - f(a) = (-1)^k \frac{h^k}{k!}f^{(k)}(x_2).$$

For a maximum $f(a + h) - f(a)$ and $f(a - h) - f(a)$ must both be negative when h is sufficiently near zero (see § 69). Both x_1 and x_2 approach a as h approaches zero. Hence the condition for a maximum is fulfilled when and only when k is even and $f^{(k)}(a)$ is negative.

For a mimimum both $f(a+h) - f(a)$ and $f(a-h) - f(a)$ must be positive for h sufficiently near zero (see § 69). Hence the condition for a minimum is fulfilled when and only when k is even and $f^{(k)}(a)$ is positive.

Summing up these results, we have the following theorem:

Let $f(x)$ and its derivative $f'(x)$ be continuous for $x = a$ and let $f'(a)$ be equal to zero. Furthermore let $f(x)$ be such that its first k derivatives exist and are continuous for $x = a$ and that the kth derivative is the first one which does not vanish for $x = a$. If k is odd, $f(x)$ has neither a maximum nor a minimum for $x = a$. If k is even, then $f(x)$ has a maximum or a minimum for $x = a$ according as $f^{(k)}(a)$ is negative or positive.

This theorem extends and completes theorem II of § 69. In finding maxima or minima the student may now proceed according to the working rule given in § 69 and employ this theorem as an additional means of testing the critical values in those cases in which a sufficient number of the derivatives of $f(x)$ are continuous for $x = a$ and one of them does not vanish.

Example. Test $e^x + e^{-x} - x^2$ for maxima and minima.

We have $f(x) = e^x + e^{-x} - x^2$, $f'(x) = e^x - e^{-x} - 2x$.

Now $f'(x)$ is continuous for all values of x; hence the only critical values are those for which $f'(x) = 0$. By inspection we see that $x = 0$ is a critical value. To test this critical value we have

$$f''(x) = e^x + e^{-x} - 2, \quad f'''(x) = e^x - e^{-x}, \quad f^{iv}(x) = e^x + e^{-x}.$$

The first of these which does not vanish for $x = 0$ is the fourth derivative and it has the positive value 2. Hence $f(x)$ is a minimum for $x = 0$.

To show that there is no other critical value we observe first that $f''(x)$, or $e^x + e^{-x} - 2$, is never negative; for $e^x + e^{-x}$ is the sum of a positive number e^x and its reciprocal e^{-x}, and such a sum is always equal to or greater than 2 (problem 6 at end of Chapter VI). Since $f''(x)$ is never negative the graph of $f'(x)$ can never descend as x increases. Hence $f'(x)$ cannot be zero for more than one value of x. Hence the only critical value is the value $x = 0$ already treated.

EXERCISES

Examine for maxima and minima the functions in problems 1 to 6.

1. $e^x + 2 \cos x + e^{-x}$. *Ans.* Minimum when $x = 0$.
2. $(x-2)^2(x+2)^3$. *Ans.* Minimum when $x = 2$; maximum when $x = 2/5$.
3. $e^x - 2 \sin x - e^{-x}$. *Ans.* No maxima or minima.
4. $x^4 - 4x^3 + 27$. *Ans.* Minimum when $x = 3$.
5. $\cos x(1 - \sin x)$. *Ans.* Maximum when $x = -\dfrac{\pi}{6}$; minimum when $x = -\dfrac{5\pi}{6}$.
6. $\sin x(2 - \sin x)$. *Ans.* Maximum when $x = \dfrac{\pi}{2}$; minimum when $x = \dfrac{3\pi}{2}$.

7. Suppose that $f(x)$ and its first k derivatives are continuous for $x = a$, k being an odd number, and suppose further that the kth derivative of $f(x)$ is the first one which does not vanish for $x = a$. Show that $f(x)$ is an increasing or a decreasing function for x near a according as $f^{(k)}(a)$ is positive or negative.

8. The intensity of light varies inversely as the square of the distance from the source. If one of two lights is k times as intense as the other $(k > 1)$, show that the point of least illumination on the line joining them is $\sqrt[3]{k}$ times as far from the more intense source as from the less intense.

9. The velocity v of waves of length λ in deep water is given by the formula
$$v = k\sqrt{\frac{\lambda}{a} + \frac{a}{\lambda}},$$
where a and k are positive constants. For what value of λ is the velocity a minimum? *Ans.* a.

10. The illumination of a flat surface by a given light varies inversely as the square of the distance from the light and directly as the sine of the inclination of the ray to the surface. At what height should a light be placed on a wall so as to afford the best illumination at a point on the floor a feet from the wall? *Ans.* $\tfrac{1}{2}a\sqrt{2}$ ft.

119. Indeterminate forms. We shall now give the promised systematic discussion of the evaluation of indeterminate forms.

The student is advised to review § 15 before proceeding with the reading and study of this section.

If as x approaches a the functions $f(x)$, $F(x)$, $\phi(x)$, $\psi(x)$, $g(x)$ approach $0, 0, \infty, \infty, 1$, respectively, then the functions

$$\frac{f(x)}{F(x)}, \quad \frac{\phi(x)}{\psi(x)}, \quad f(x)\cdot\phi(x), \quad \phi(x)-\psi(x),$$

$$(f(x))^{F(x)}, \qquad (\phi(x))^{f(x)}, \qquad (g(x))^{\phi(x)} \tag{1}$$

are said to take the respective indeterminate forms

$$\frac{0}{0}, \quad \frac{\infty}{\infty}, \quad 0\cdot\infty, \quad \infty-\infty, \quad 0^0, \quad \infty^0, \quad 1^\infty \tag{2}$$

as x approaches a. But it may happen that a function in (1) approaches a definite limit as x approaches a even though it takes one of these forms. On finding this limit one is said to evaluate the corresponding indeterminate form. The indeterminate form $0/0$ is the most important one for the learner; the others may be reduced to it, as we shall see later. It has been frequently encountered by the student in his study of differentiation.

I. *If as x approaches a the functions $f(x)$ and $F(x)$ both approach 0 and if $\lim_{x=a} F'(x) \neq 0$, we have*

$$\lim_{x=a} \frac{f(x)}{F(x)} = \lim_{x=a} \frac{f'(x)}{F'(x)}.$$

To prove this employ the law of the mean (§ 116). From it we have

$$f(a+h) = f(a) + hf'(a+\theta_1 h), \quad (0 < \theta_1 < 1)$$
$$F(a+h) = F(a) + hF'(a+\theta_2 h). \quad (0 < \theta_2 < 1)$$

But $f(a) = F(a) = 0$ by hypothesis. Hence we have

$$\frac{f(a+h)}{F(a+h)} = \frac{f'(a+\theta_1 h)}{F'(a+\theta_2 h)}.$$

As h approaches 0 both $a+h$, $a+\theta_1 h$, and $a+\theta_2 h$ approach a. Therefore

$$\lim_{x=a} \frac{f(x)}{F(x)} = \lim_{h=0} \frac{f(a+h)}{F(a+h)} = \lim_{h=0} \frac{f'(a+\theta_1 h)}{F'(a+\theta_2 h)} = \frac{f'(a)}{F'(a)}.$$

II. *If as x approaches a the functions $f(x)$ and $F(x)$, together with their first $n-1$ derivatives, approach 0, and if $F^{(n)}(a) \neq 0$, we have*
$$\lim_{x \to a} \frac{f(x)}{F(x)} = \frac{f^{(n)}(a)}{F^{(n)}(a)}.$$

To prove this we employ Taylor's theorem (§ 117), writing x for b and omitting the terms which vanish on account of the vanishing of $f(a), f'(a), \cdots, f^{(n-1)}(a), F(a), F'(a), \cdots, F^{(n-1)}(a)$; thus we have
$$f(x) = \frac{(x-a)^n}{n!} f^{(n)}(x_1), \quad F(x) = \frac{(x-a)^n}{n!} F^{(n)}(x_2)$$
where x_1 and x_2 lie between a and x. Thence we have
$$\frac{f(x)}{F(x)} = \frac{f^{(n)}(x_1)}{F^{(n)}(x_2)}.$$

As x approaches a so do x_1 and x_2. Hence, on taking the limits as x approaches a in the last equation, we have the result stated in the theorem.

These theorems yield the following rule for evaluating the indeterminate form $0/0$: *Differentiate the numerator for a new numerator and the denominator for a new denominator. The resulting fraction has the same limit as the original one when x approaches a. If it still has the indeterminate form $0/0$, repeat the process.* In applying the rule it is sometimes desirable to modify the *form* of the fraction before performing some of the differentiations.

It may be shown that the form ∞ / ∞ may be evaluated by the same rule; but the proof lies beyond the scope of this work. This form ∞ / ∞ may also be reduced to the form $0/0$ by means of the identity
$$\frac{\phi(x)}{\psi(x)} = \frac{\dfrac{1}{\psi(x)}}{\dfrac{1}{\phi(x)}}.$$

The form $0 \cdot \infty$ may be reduced to the form $0/0$ by means of the identity
$$f(x) \cdot \phi(x) = \frac{f(x)}{\dfrac{1}{\phi(x)}}.$$

The form $\infty - \infty$ may be reduced to the form $0/0$ by means of the identity
$$\phi(x) - \psi(x) = \frac{\dfrac{1}{\psi(x)} - \dfrac{1}{\phi(x)}}{\dfrac{1}{\psi(x)\phi(x)}}.$$

To evaluate a function which assumes one of the forms 0^0, ∞^0, 1^∞ we take its logarithm and evaluate the resulting function by means of one of the previous methods, giving the logarithm of the limit desired. Thence we pass to the limit itself.

In a particular case it may be possible to reduce one of the forms ∞/∞, $0 \cdot \infty$ and $\infty - \infty$ to the form $0/0$ by inspection. This method will be illustrated in Examples 3 and 4.

Example 1. Evaluate $\displaystyle\lim_{x \to 0} \frac{e^x - e^{-x}}{\sin x}$.

This takes the form $0/0$. Applying the rule, we have
$$\lim_{x \to 0} \frac{e^x - e^{-x}}{\sin x} = \lim_{x \to 0} \frac{e^x + e^{-x}}{\cos x} = \frac{2}{1} = 2.$$

Example 2. Evaluate $\displaystyle\lim_{x \to 0} \frac{\log x}{\cot x}$.

This takes the form ∞/∞. Applying the rule, we have
$$\lim_{x \to 0} \frac{\log x}{\cot x} = \lim_{x \to 0} \frac{\dfrac{1}{x}}{-\csc^2 x} = \lim_{x \to 0} \frac{-\sin^2 x}{x}$$
$$= - \lim_{x \to 0} \frac{\sin x}{x} \cdot \lim_{x \to 0} \sin x = - 1 \cdot 0 = 0.$$

Example 3. Evaluate $\displaystyle\lim_{x \to \pi/2} (\sec x - \tan x)$.

This has the form $\infty - \infty$. But
$$\sec x - \tan x = \frac{1}{\cos x} - \frac{\sin x}{\cos x} = \frac{1 - \sin x}{\cos x}.$$

As x approaches $\pi/2$ the last fraction takes the form $0/0$. Hence
$$\lim_{x \to \pi/2} (\sec x - \tan x) = \lim_{x \to \pi/2} \frac{1 - \sin x}{\cos x} = \lim_{x \to \pi/2} \frac{-\cos x}{-\sin x} = 0.$$

Example 4. Evaluate $\lim_{x \to \pi/2} \cos 3x \sec x$.

This takes the form $0 \cdot \infty$. But
$$\cos 3x \sec x = \frac{\cos 3x}{\cos x}.$$

As x approaches $\pi/2$ this last fraction takes the form $0/0$. Hence
$$\lim_{x \to \pi/2} \cos 3x \sec x = \lim_{x \to \pi/2} \frac{\cos 3x}{\cos x} = \lim_{x \to \pi/2} \frac{-3 \sin 3x}{-\sin x} = -3.$$

Example 5. Evaluate $\lim_{x \to 0} \frac{\sin x - x}{x^3}$.

This takes the form $0/0$. The student will observe from the following work that it maintains this form for two differentiations of the terms of the fraction, so that three differentiations are necessary to evaluate it:
$$\lim_{x \to 0} \frac{\sin x - x}{x^3} = \lim_{x \to 0} \frac{\cos x - 1}{3x^2} = \lim_{x \to 0} \frac{-\sin x}{6x} = \lim_{x \to 0} \frac{-\cos x}{6} = -\frac{1}{6}.$$

Example 6. Evaluate $\lim_{x \to \infty} \frac{\log x}{x}$.

Writing $1/t$ for x, we have
$$\lim_{x \to \infty} \frac{\log x}{x} = \lim_{t \to 0} \frac{-\log t}{\frac{1}{t}} = \lim_{t \to 0} \frac{-\frac{1}{t}}{-\frac{1}{t^2}} = \lim_{t \to 0} t = 0.$$

Example 7. Evaluate $\lim_{x \to 0} (e^x + x)^{\frac{1}{x}}$.

Putting
$$y = (e^x + x)^{\frac{1}{x}}$$

we have
$$\log y = \frac{\log(e^x + x)}{x}.$$

Now
$$\lim_{x \to 0} \frac{\log(e^x + x)}{x} = \lim_{x \to 0} \frac{\frac{e^x + 1}{e^x + x}}{1} = 2.$$

Hence
$$\lim_{x \to 0} \log y = 2.$$

Therefore
$$\lim_{x \to 0} (e^x + x)^{\frac{1}{x}} = \lim_{x \to 0} y = e^2.$$

EXERCISES

Evaluate the following limits:

1. $\lim\limits_{x=1} \dfrac{x^n - 1}{x - 1} = n.$

2. $\lim\limits_{x=1} \dfrac{x - 1}{\log x} = 1.$

3. $\lim\limits_{x=0} \dfrac{e^x - e^{-x} - 2x}{x - \sin x} = 2.$

4. $\lim\limits_{x=0} \dfrac{e^x + e^{-x} - 2}{x^2} = 1.$

5. $\lim\limits_{x=0} \dfrac{\sin x - x}{x - \tan x} = \dfrac{1}{2}.$

6. $\lim\limits_{x=0} \dfrac{x^2}{1 - \cos x} = 2.$

7. $\lim\limits_{x=a} \dfrac{x - a}{\sin x - \sin a} = \sec a.$

8. $\lim\limits_{x=\infty} \dfrac{x}{e^x} = 0.$

9. $\lim\limits_{x=0} x \log \sin x = 0.$

10. $\lim\limits_{x=0} \left(\dfrac{1}{\sin x} - \dfrac{1}{x}\right) = 0.$

11. $\lim\limits_{x=0} \left(\dfrac{1}{x} - \cot x\right) = 0.$

12. $\lim\limits_{x=0} x^x = 1.$

13. $\lim\limits_{x=1} x^{\frac{1}{x-1}} = e.$

14. $\lim\limits_{x=0} (\cot x)^{\sin x} = 1.$

15. $\lim\limits_{x=0} (1 + ax)^{\frac{1}{x}} = e^a.$

16. $\lim\limits_{x=\infty} \left(1 + \dfrac{1}{x}\right)^x = e.$

17. $\lim\limits_{x=\infty} \left(1 + \dfrac{a}{x}\right)^x = e^a.$

18. $\lim\limits_{x=1} \left(\dfrac{x}{\log x} - \dfrac{1}{\log x}\right) = 1.$

19. $\lim\limits_{x=0} x^n \log x = 0$ if $n > 0.$

20. $\lim\limits_{x=0} \dfrac{\log(1+x) - x}{x^2} = -\dfrac{1}{2}.$

21. $\lim\limits_{x=0} (\tan x)^x = 1.$

22. $\lim\limits_{x=0} \dfrac{\arctan x - x}{x^2} = 0.$

23. $\lim\limits_{x=0} \dfrac{\arctan x - x}{x^3} = -\dfrac{1}{3}.$

24. $\lim\limits_{x=0} \dfrac{e^{\sin x} - 1}{x} = 1.$

25. $\lim\limits_{x=0} \dfrac{x^2 - \sin^2 x}{x^4} = \dfrac{1}{3}.$

26. $\lim\limits_{x=0} \dfrac{\arcsin x - x}{x^3} = \dfrac{1}{6}.$

27. $\lim\limits_{x=0} \dfrac{\log(e^x + 1) - \log 2}{x} = \dfrac{1}{2}.$

28. $\lim\limits_{x=0} (1 + x^3)^{\frac{1}{x}} = 1.$

29. $\lim\limits_{x=\pi} \left(1 - \tan \dfrac{x}{4}\right) \sec \dfrac{x}{2} = 1.$

30. $\lim\limits_{x=1} \left(\dfrac{x}{x-1} - \dfrac{1}{\log x}\right) = \dfrac{1}{2}.$

CHAPTER XIV

INFINITE SERIES

120. Introduction. The student has already become acquainted with the important notions of sequence and limiting value of a sequence (§ 30). We shall first consider the particularly simple example of the sequence whose terms are the terms, t_n, of the familiar geometric progression for which the first term is unity and the common ratio is $\frac{1}{2}$. Here

$$\{t_n\} = \left\{\frac{1}{2^{n-1}}\right\}. \qquad (n = 1, 2, 3, \cdots) \qquad (1)$$

From the terms t_n of the sequence (1) a second sequence $\{s_n\}$ can be constructed in the following manner. Define $s_1 = 1$, $s_2 = 1 + \frac{1}{2}$, and, in general,

$$s_n = 1 + \tfrac{1}{2} + \tfrac{1}{4} + \cdots + \frac{1}{2^{n-1}}. \qquad (n = 1, 2, \cdots) \qquad (2)$$

The new sequence $\{s_n\}$ thus associated with the original geometric progression is called the *partial sum sequence* of the geometric progression. It is customary to represent this partial sum sequence by the new symbol

$$1 + \tfrac{1}{2} + \tfrac{1}{4} + \cdots + \frac{1}{2^{n-1}} + \cdots, \qquad (3)$$

which is spoken of as the *geometric series* determined by the geometric progression (1). The nth partial sum for any fixed value of n, no matter how large, is a perfectly definite number, since the sum of a finite number of terms t_n is always well defined. On the contrary, the value of the new symbol (3) itself is not well defined until we have agreed upon what is to be meant by the sum of an infinite number of the terms t_n. Now in paragraph 5 of § 8 it was shown that

$$s_n = 2 - \frac{1}{2^{n-1}} \qquad (4)$$

and hence that the limiting value of the partial sum sequence is 2. Since it is natural to define the value of the symbol (3) as the limiting value of its nth partial sum as n tends to infinity, we are led to assign the number 2 as the value of the symbol (3). More briefly, we simply say the sum of the geometric series (3) is 2, and write symbolically

$$1 + \tfrac{1}{2} + \tfrac{1}{4} + \cdots + \frac{1}{2^{n-1}} + \cdots = 2. \tag{5}$$

In paragraph 6 of § 8, considerations precisely like those just presented led us to define the number $\dfrac{a}{1-r}$ as the sum of the series

$$a + ar + ar^2 + \cdots + ar^{n-1} + \cdots \tag{6}$$

associated with the geometric progression

$$\{ar^{n-1}\}, \quad |r| < 1. \quad (n = 1, 2, 3, \cdots) \tag{7}$$

Consider now any sequence whatever of real numbers

$$\{u_n\} \quad (n = 1, 2, 3, \cdots) \tag{8}$$

and the associated sequence of partial sums $\{s_n\}$ where

$$s_n = u_1 + u_2 + \cdots + u_n. \quad (n = 1, 2, 3, \cdots) \tag{9}$$

It is customary to represent this partial sum sequence $\{s_n\}$ by the new symbol

$$u_1 + u_2 + u_3 + \cdots + u_n + \cdots. \tag{10}$$

This symbol (10) we shall call an *infinite series*.

If for some fixed n all the terms of (10) after the nth have the value zero, in which case the series is known as a *terminating series*, the value of the symbol (10), that is, the sum of the series, is simply the sum (9) of a finite number of terms and is therefore always well defined. If, on the contrary, no matter how large n is chosen there are always terms beyond u_n which are different from zero, then, as in the special cases (3) and (6), the notion of the value of the new symbol (10), that is, the idea of the sum of the infinite series, is initially quite meaningless. In order to attach significance to the notion of the sum of an infinite series (10) we shall agree, as in the special cases treated above, to give a definition of sum in terms of the behavior of the partial sum sequence $\{s_n\}$. The limiting values of this sequence

$$\lim_{n=\infty} s_n \tag{11}$$

may or may not exist. If this limit does not exist, the infinite series (10) is said to *diverge* and we make no further attempt to assign a meaning to the notion of its sum. If, however, this limit does exist, the series (10) is said to *converge* and in particular to converge to the limit s or have the sum s in case the limiting value of s_n is s.

In this latter case we write symbolically

$$u_1 + u_2 + u_3 + \cdots + u_n + \cdots = s, \qquad (12)$$

but the student should realize that this equation, like the equation $\tan \pi/2 = \infty$, is based on a definite convention as to the meaning of the sum of an infinite series, and is in fact merely a convenient abbreviation for the lengthy statement necessary for the complete expression of this convention.

In the applications it is usually the infinite series (10) itself which arises for study rather than either the basic sequence $\{u_n\}$ or the associated partial sum sequence $\{s_n\}$, but the student should always bear in mind that if we are to be concerned with the *value* of such a series it must be regarded merely as a convenient symbol for the partial sum sequence $\{s_n\}$ deducible from it and in terms of which the notion of its sum (as we have agreed) alone has meaning.

In this text our interest in a series (10) ceases as soon as we have shown that it diverges. On the contrary, we shall find that convergent infinite series are of much use in the calculus. Evidently two main problems arise in connection with any infinite series. The first is to decide whether or not the given series converges, and the second is to find to what value a convergent series really converges, that is, to actually compute its sum. In general only the first problem will be considered in what follows, as the second problem does not admit of an elementary treatment. In attempting to solve the first problem for a given series it is most helpful to observe that in view of the definition of convergence, which depends on the behavior of the partial sum sequence as n tends to infinity, it is permissible to omit any finite number of the terms of the given series, for such omission does not alter the *behavior* of the partial sum sequence although, of course, it does alter its *limiting value* in case it converges. This observation will considerably simplify the application of certain of the convergence tests developed in the next sections.

As a first example of a divergent series consider the alternating series
$$1 - 1 + 1 - 1 + \cdots.$$

Evidently as n tends to infinity the nth partial sum takes on alternately the values 1 and 0, and hence does not approach a limiting value. When a series (10) exhibits such oscillatory behavior it is said to be *improperly divergent*.

The so-called *harmonic* series
$$1 + \tfrac{1}{2} + \tfrac{1}{3} + \tfrac{1}{4} + \cdots$$
is an example of the second type of divergent series, for in this case, no matter how large a number $G > 0$ may be chosen, the nth partial sum ultimately exceeds G. In fact, if we choose $n_0 > 2G$, then for all $n > 2^{n_0}$ we have

$$s_n > (1 + \tfrac{1}{2}) + (\tfrac{1}{3} + \tfrac{1}{4}) + (\tfrac{1}{5} + \cdots + \tfrac{1}{8}) + \cdots + \left(\frac{1}{2^{n_0-1}+1} + \cdots + \frac{1}{2^{n_0}}\right)$$
$$> \tfrac{1}{2} + 2 \cdot \tfrac{1}{4} + 4 \cdot \tfrac{1}{8} + \cdots + 2^{n_0-1} \cdot \frac{1}{2^{n_0}} = \frac{n_0}{2} > G.$$

When a series (10) behaves like the harmonic series, it is said to be *properly divergent*.

As an example of an important general class of convergent series we may cite the geometric series (6) on page 241.

Alternating series of the form
$$u_1 - u_2 + u_3 - u_4 + \cdots, \tag{13}$$
where $0 < u_n < u_{n-1}$ for $n = 1, 2, \cdots$, and $\lim_{n=\infty} u_n = 0$, constitute another important general class of convergent series. We shall prove that such series are actually convergent in the next section.

The special alternating series
$$1 - \tfrac{1}{2} + \tfrac{1}{3} - \tfrac{1}{4} + \cdots, \tag{14}$$
$$1 - \tfrac{1}{2} + \tfrac{1}{4} - \tfrac{1}{8} + \cdots, \tag{15}$$
satisfy the conditions placed on the series (13) and hence are convergent. The series which we obtain from the first of these

special series by making the signs of all the terms positive is the harmonic series and is therefore divergent. But the series $1+\frac{1}{2}+\frac{1}{4}+\cdots$ obtained in the same manner from the second of these alternating series is one of the geometric series (6) and is convergent. On the basis of these simple examples we are led to make the following important distinction between types of convergent infinite series. A convergent infinite series (10) in which $u_n \geqq 0$, for $n = 1, 2, 3, \cdots$, may remain convergent when u_n is replaced by $|u_n|$ for $n = 1, 2, 3, \cdots$. Such a series will hereafter be called *absolutely convergent*. On the contrary, the resulting series of absolute values may diverge. In this case the original convergent series will be called *conditionally convergent*.

We state without proof that we can operate with absolutely convergent infinite series as if they were finite sums. But for conditionally convergent series this statement is not true. For example, it can be shown that a mere change in the order of the terms of an absolutely convergent series does not affect its convergence or the value of its sum. In this respect absolutely convergent series behave like finite sums. Conditionally convergent series no longer possess this property, as the following example shows.

Let us denote by s the sum of the conditionally convergent series (14) so that we have

$$s = 1 - \tfrac{1}{2} + \tfrac{1}{3} - \tfrac{1}{4} + \tfrac{1}{5} - \tfrac{1}{6} + \cdots. \tag{16}$$

Then
$$\frac{s}{2} = \quad\ \tfrac{1}{2}\ \quad -\tfrac{1}{4}\ \quad + \tfrac{1}{6} - \cdots. \tag{17}$$

On adding, by combining terms in the same column, we obtain

$$\frac{3s}{2} = 1 + \tfrac{1}{3} - \tfrac{1}{2} + \tfrac{1}{5} + \tfrac{1}{7} - \tfrac{1}{4} + \tfrac{1}{9} + \tfrac{1}{11} - \tfrac{1}{6} + \cdots. \tag{18}$$

The series on the right in (18) is quite obviously a simple rearrangement of the series (16) in the sense that every term in the first series occurs once and only once in the second, and conversely. Nevertheless the sum of the second series is $\tfrac{3}{2}$ of the sum of the first. With this striking example of an apparent paradox before him the student should never fail to use great caution in manipulating conditionally convergent series.

121. Tests for convergence. In practice we cannot usually find the actual limit of the sum S_n of the first n terms of a series as n becomes infinite. But it is of great importance to us to know that *a limit exists*; for, if it does not exist, the notion of *sum* of the series is not defined. We shall now give some tests for convergence suited for use when we cannot evaluate the limit of S_n. The series to be tested we shall write in the form

$$u_1 + u_2 + u_3 + \cdots . \tag{1}$$

We denote by S_n the sum of its first n terms.

Let us suppose that (1) converges and let s denote its sum. Then we have
$$\lim_{n=\infty} S_n = s, \quad \lim_{n=\infty} S_{n+1} = s.$$
Hence
$$\lim_{n=\infty} (S_{n+1} - S_n) = 0.$$

But $S_{n+1} - S_n = u_{n+1}$. Therefore $\lim_{n=\infty} u_n = 0$. Hence

I. *A necessary condition that a series shall converge is that the nth term shall approach zero as n becomes infinite.*

The series
$$\frac{1}{1+1} + \frac{2}{2+1} + \frac{3}{3+1} + \cdots + \frac{n}{n+1} + \cdots$$
is divergent since its nth term approaches 1 as n becomes infinite.

That the condition in theorem I is not *sufficient* is shown by series (3) of the previous section. This series diverges although its nth term $1/n$ approaches 0 as n becomes infinite.

II. *Let $u_1 + u_2 + u_3 + \cdots$ be a series of positive terms to be tested for convergence, and let $a_1 + a_2 + a_3 + \cdots$ be a convergent series of positive terms whose sum is A. Suppose, moreover, that each term of the latter is equal to or greater than the corresponding term of the former (so that $a_n \geq u_n$ for $n = 1, 2, 3, \cdots$). Then the series $u_1 + u_2 + u_3 + \cdots$ is convergent and its sum is not greater than A.*

Let $S_n = u_1 + u_2 + \cdots + u_n$, $s_n = a_1 + a_2 + \cdots + a_n$.

Then $\lim_{n=\infty} s_n = A$ and $S_n \leq s_n < A$. Hence S_n increases with

n and never becomes greater than A. Then from theorem V of § 10 it follows that $\lim_{n \to \infty} S_n$ exists and is not greater than A. This is the result to be proved.

Example 1. Show that the series $1 + \dfrac{1}{2^2} + \dfrac{1}{3^3} + \dfrac{1}{4^4} + \cdots$ is convergent.

The terms are never greater than the corresponding terms of the convergent series
$$1 + \frac{1}{2^2} + \frac{1}{2^3} + \frac{1}{2^4} + \cdots.$$

Hence the given series is convergent.

By a method similar to that employed in the last proof the following theorem may be demonstrated:

III. If $u_1 + u_2 + u_3 + \cdots$ is a series of positive terms, and if $b_1 + b_2 + b_3 + \cdots$ is a divergent series of positive terms such that $b_n \leqq u_n$ for $n = 1, 2, 3, \cdots$, then the series $u_1 + u_2 + u_3 + \cdots$ is divergent.

Example 2. Show that the series $2 + \tfrac{2}{3} + \tfrac{2}{5} + \tfrac{2}{7} + \cdots$ is divergent.

This follows from the fact that it is term by term greater than the known divergent series
$$1 + \tfrac{1}{2} + \tfrac{1}{3} + \tfrac{1}{4} + \cdots.$$

Example 3. Test the following series for convergence:
$$1 + \frac{1}{2^p} + \frac{1}{3^p} + \frac{1}{4^p} + \cdots. \tag{2}$$

For $p = 1$ this is the harmonic series (3) of the previous section, and the latter series has been proved divergent. If $p < 1$ the terms of the given series are equal to or greater than the terms of the harmonic series; it follows therefore from theorem III that it is divergent. Hence the given series diverges when $p \leqq 1$.

We shall now show that it converges when $p > 1$. For this purpose we observe that
$$\frac{1}{2^p} + \frac{1}{3^p} < \frac{1}{2^p} + \frac{1}{2^p} = \frac{2}{2^p} = \frac{1}{2^{p-1}},$$
$$\frac{1}{4^p} + \frac{1}{5^p} + \frac{1}{6^p} + \frac{1}{7^p} < \frac{1}{4^p} + \frac{1}{4^p} + \frac{1}{4^p} + \frac{1}{4^p} = \frac{4}{4^p} = \left(\frac{1}{2^{p-1}}\right)^2,$$
$$\frac{1}{8^p} + \cdots + \frac{1}{15^p} < \frac{8}{8^p} = \left(\frac{1}{2^{p-1}}\right)^3,$$
$$\cdots\cdots\cdots\cdots\cdots\cdots\cdots\cdots\cdots\cdots$$

Now the series
$$1 + \frac{1}{2^{p-1}} + \left(\frac{1}{2^{p-1}}\right)^2 + \left(\frac{1}{2^{p-1}}\right)^3 + \cdots \qquad (3)$$

is a geometric series (like (6) of § 120) with ratio less than 1; it is therefore convergent. The sum S_n of n terms of (2) is less than the sum of (3). But S_n increases as n increases. From theorem V of § 10 it follows therefore that the limit of S_n exists and hence that (2) is convergent when $p > 1$.

This example is of sufficient importance as a comparison series for use with theorems II and III to justify us in stating the result as a theorem:

IV. *The p-series* $1 + \dfrac{1}{2^p} + \dfrac{1}{3^p} + \dfrac{1}{4^p} + \cdots$

is convergent when $p > 1$ and is divergent when $p \leq 1$.

Example 4. Show that the series $\dfrac{1}{1 \cdot 2} + \dfrac{1}{2 \cdot 3} + \dfrac{1}{3 \cdot 4} + \cdots$ is convergent.

Its terms are less than the corresponding terms of the p-series for $p = 2$. From theorems IV and II it follows therefore that it is convergent.

Example 5. Show that the following series is divergent:
$$\frac{1}{(1 \cdot 2)^{\frac{1}{3}}} + \frac{1}{(2 \cdot 3)^{\frac{1}{3}}} + \frac{1}{(3 \cdot 4)^{\frac{1}{3}}} + \cdots.$$

This series is term by term greater than the series
$$\frac{1}{2^{\frac{2}{3}}} + \frac{1}{3^{\frac{2}{3}}} + \frac{1}{4^{\frac{2}{3}}} + \cdots.$$

From theorem IV it follows that the latter is divergent. Hence it follows from theorem III that the given series is divergent.

V. Ratio test. *If $u_1 + u_2 + u_3 + \cdots$ is a series of positive terms such that*
$$\lim_{n \to \infty} \frac{u_{n+1}}{u_n}$$

exists and has the value ρ, then the given series is convergent if $\rho < 1$ and is divergent if $\rho > 1$. If $\rho = 1$ there is no test.

For the case of the p-series we have $\rho = 1$, since in that case
$$\lim_{n=\infty}\frac{u_{n+1}}{u_n} = \lim_{n=\infty}\frac{1}{(n+1)^p} \div \frac{1}{n^p} = \lim_{n=\infty}\left(\frac{n}{n+1}\right)^p = 1^p = 1.$$
But the series diverges when $p \leqq 1$ and converges when $p > 1$. Hence we get no general test when $\rho = 1$.

In case $\rho > 1$ we let R be a number lying between 1 and ρ. When m is sufficiently large we have $u_{m+1}/u_m > R > 1$. Hence $u_{m+1} > u_m$ for m sufficiently large. Hence u_n cannot approach zero as n increases indefinitely. Hence it follows from theorem I that the series is divergent.

In case $\rho < 1$ we let r be a number lying between ρ and 1. For m sufficiently large we have
$$u_{m+1} < ru_m, \quad u_{m+2} < ru_{m+1} < r^2 u_m,$$
$$u_{m+3} < ru_{m+2} < r^3 u_m, \quad \cdots.$$
Therefore, after the term u_m, we have the terms of the given series less than the terms of the series
$$ru_m + r^2 u_m + r^3 u_m + \cdots.$$
But the latter is a geometrical series with the ratio r less than 1. It is therefore convergent. From theorem II it follows then that the given series is convergent.

Example 6. Show that the following series is convergent:
$$1 + \frac{a}{1!} + \frac{a^2}{2!} + \frac{a^3}{3!} + \cdots, \quad (a > 0)$$

We have
$$\frac{u_{n+1}}{u_n} = \frac{a^n}{n!} \div \frac{a^{n-1}}{(n-1)!} = \frac{a}{n}.$$

But the limit of this quantity, as n becomes infinite, is zero. Hence the series converges.

So far (with the exception of theorem I) we have given tests only for series with positive terms. These can be carried over to certain other series by means of the following general principle, which we shall state without proof:

VI. A series whose terms have different signs is convergent if the series resulting from it by making all the signs positive is convergent.

To this we must add the following theorem, so frequently applicable in practice as to be of considerable importance.

VII. *If* $u_1 - u_2 + u_3 - u_4 + \cdots$ $(u_i > 0)$ *is an alternating series such that no term is numerically greater than the preceding term, and if* $\lim_{n=\infty} u_n = 0$, *then the given series is convergent.*

We have
$$S_{2m} = (u_1 - u_2) + (u_3 - u_4) + \cdots + (u_{2m-1} - u_{2m}),$$
$$S_{2m+1} = u_1 - (u_2 - u_3) - (u_4 - u_5) - \cdots - (u_{2m} - u_{2m+1}),$$
$$S_{2m} = S_{2m+1} - u_{2m+1}.$$

From the second of these relations it follows that $S_{2m+1} \leq u_1$. Then from the third we see that $S_{2m} \leq u_1$. But from the first it follows that S_{2m} increases with m. From theorem V of § 10 we then conclude that $\lim_{m=\infty} S_{2m}$ exists. Thence from the third of the given relations it follows that $\lim_{m=\infty} S_{2m+1}$ exists and has the same value. Therefore $\lim_{n=\infty} S_n$ exists, as was to be proved.

Example 7. Show that the series $1 - \frac{1}{3} + \frac{1}{5} - \frac{1}{7} + \cdots$ is convergent.

It is an alternating series whose terms decrease numerically, and the nth term approaches zero as n approaches infinity. Therefore it follows from theorem VII that the series is convergent.

EXERCISES

Show that the following eight series are convergent:

1. $1 + \dfrac{1}{2^3} + \dfrac{1}{3^3} + \dfrac{1}{4^3} + \cdots$.

2. $1 - \dfrac{1}{2^2} + \dfrac{1}{3^2} - \dfrac{1}{4^2} + \cdots$.

3. $\dfrac{1}{1 \cdot 2 \cdot 3} + \dfrac{1}{4 \cdot 5 \cdot 6} + \dfrac{1}{7 \cdot 8 \cdot 9} + \cdots$.

4. $\dfrac{1^2}{2^4} + \dfrac{2^2}{3^4} + \dfrac{3^2}{4^4} + \dfrac{4^2}{5^4} + \cdots$.

5. $\dfrac{1}{\log 2} - \dfrac{1}{\log 3} + \dfrac{1}{\log 4} - \dfrac{1}{\log 5} + \cdots$.

6. $\dfrac{a^3}{3!} + \dfrac{a^5}{5!} + \dfrac{a^7}{7!} + \dfrac{a^9}{9!} + \cdots$.

7. $1 + \dfrac{1}{2 \cdot 2^2} + \dfrac{1}{3 \cdot 2^3} + \dfrac{1}{4 \cdot 2^4} + \cdots$.

8. $\dfrac{1}{2 \cdot 3 \cdot 4} + \dfrac{2}{3 \cdot 4 \cdot 5} + \dfrac{3}{4 \cdot 5 \cdot 6} + \dfrac{4}{5 \cdot 6 \cdot 7} + \cdots$.

Show that the following six series are divergent:

9. $\tfrac{1}{2} + \tfrac{1}{4} + \tfrac{1}{6} + \cdots$.

10. $1 + \tfrac{1}{3} + \tfrac{1}{5} + \tfrac{1}{7} + \cdots$.

11. $1 + \tfrac{1}{2} + \tfrac{2}{3} + \tfrac{3}{4} + \tfrac{4}{5} + \cdots$.

12. $1 - \tfrac{1}{2} + \tfrac{2}{3} - \tfrac{3}{4} + \tfrac{4}{5} - \cdots$.

13. $\dfrac{2!}{a^2} + \dfrac{3!}{a^3} + \dfrac{4!}{a^4} + \cdots$.

14. $\dfrac{1 \cdot 2}{3 \cdot 4 \cdot 5} + \dfrac{2 \cdot 3}{4 \cdot 5 \cdot 6} + \dfrac{3 \cdot 4}{5 \cdot 6 \cdot 7} + \dfrac{4 \cdot 5}{6 \cdot 7 \cdot 8} + \cdots$.

Test the following six series for convergence and divergence:

15. $\dfrac{2^2 + 1}{2^3 + 1} + \dfrac{3^2 + 1}{3^3 + 1} + \dfrac{4^2 + 1}{4^3 + 1} + \cdots$.

16. $\dfrac{2^2 + 1}{2^4 - 1} + \dfrac{3^2 + 1}{3^4 - 1} + \dfrac{4^2 + 1}{4^4 - 1} + \cdots$.

17. $\dfrac{2^2 + 1}{2^4 + 1} + \dfrac{3^2 + 1}{3^4 + 1} + \dfrac{4^2 + 1}{4^4 + 1} + \cdots$.

18. $\dfrac{1}{2\sqrt{2}} + \dfrac{1}{3\sqrt{3}} + \dfrac{1}{4\sqrt{4}} + \cdots$.

19. $\dfrac{1}{\sqrt{2}} - \dfrac{1}{\sqrt{3}} + \dfrac{1}{\sqrt{4}} - \dfrac{1}{\sqrt{5}} + \cdots$.

20. $\dfrac{1}{\sqrt{2}} + \dfrac{1}{\sqrt{3}} + \dfrac{1}{\sqrt{4}} + \dfrac{1}{\sqrt{5}} + \cdots$.

122. Power series. A series of ascending integral powers of x, with coefficients independent of x, is called a *power series* in x. Such a series may be written in the form

$$a_0 + a_1 x + a_2 x^2 + a_3 x^3 + \cdots.$$

To test such a series for convergence we first apply the ratio test in theorem V of the preceding section. This sometimes yields full information. But sometimes we encounter cases where the test of that theorem fails in one respect to give complete information. Then we must resort to the other theorems. The student will see the method from the following examples:

Example 1. Test the following series for convergence:

$$1 - \frac{x^2}{2!} + \frac{x^4}{4!} - \frac{x^6}{6!} + \cdots. \qquad (1)$$

We first examine the corresponding series of absolute values, namely,
$$1 + \left|\frac{x^2}{2!}\right| + \left|\frac{x^4}{4!}\right| + \left|\frac{x^6}{6!}\right| + \cdots. \qquad (2)$$

The $(n+1)$th term of this series is
$$u_{n+1} = \left|\frac{x^{2n}}{(2n)!}\right|.$$

Hence $\quad \dfrac{u_{n+1}}{u_n} = \left|\dfrac{x^{2n}}{(2n)!}\right| \div \left|\dfrac{x^{2n-2}}{(2n-2)!}\right| = \left|\dfrac{x^2}{2n(2n-1)}\right|.$

This approaches 0 as n becomes infinite. Hence series (2) converges for all values of x (theorem V of § 121). Therefore series (1) converges for all values of x (theorem VI of § 121). We express this fact by saying that the interval of convergence of (1) is that for which $-\infty < x < +\infty$.

Example 2. Test the following series for convergence:
$$x - \frac{x^2}{2} + \frac{x^3}{3} - \frac{x^4}{4} + \cdots. \qquad (3)$$

Let u_n denote the nth term of the corresponding series of absolute values, namely,
$$|x| + \left|\frac{x^2}{2}\right| + \left|\frac{x^3}{3}\right| + \left|\frac{x^4}{4}\right| + \cdots. \qquad (4)$$

Then $\quad \dfrac{u_{n+1}}{u_n} = \left|\dfrac{x^{n+1}}{n+1}\right| \div \left|\dfrac{x^n}{n}\right| = \left|x \cdot \dfrac{n}{n+1}\right| = |x| \cdot \dfrac{n}{n+1}.$

Hence $\quad \lim\limits_{n=\infty} \dfrac{u_{n+1}}{u_n} = |x|.$

Therefore series (4) converges if $|x| < 1$; that is, if $-1 < x < 1$. When $|x| > 1$ we have u_{n+1} greater than u_n if n is sufficiently large; hence u_n fails to approach 0 when n becomes infinite; therefore the nth term of (3) fails to approach 0 when n becomes infinite; and hence series (3) diverges when $|x| > 1$. The ratio test fails for the case when $|x| = 1$; that is, when $x = 1$ or $x = -1$. But for $x = 1$ the original series converges, as we saw in § 120, while for $x = -1$ the original series diverges. Summing up our results we say that series (3) converges on the interval $-1 < x \leq 1$.

EXERCISES

Find the intervals on which the following power series converge:

1. $1 + x + x^2 + x^3 + x^4 + \cdots$.   Ans. $-1 < x < 1$.

2. $x - \dfrac{x^2}{2^2} + \dfrac{x^3}{3^2} - \dfrac{x^4}{4^2} + \dfrac{x^5}{5^2} \cdots$.   Ans. $-1 \leqq x \leqq 1$.

3. $x + \dfrac{x^2}{2} + \dfrac{x^3}{3} + \dfrac{x^4}{4} + \cdots$.   Ans. $-1 \leqq x < 1$.

4. $1 + x + \dfrac{x^2}{2!} + \dfrac{x^3}{3!} + \dfrac{x^4}{4!} + \cdots$.   Ans. $-\infty < x < +\infty$.

5. $x - \dfrac{x^3}{3!} + \dfrac{x^5}{5!} - \dfrac{x^7}{7!} + \cdots$.   Ans. $-\infty < x < +\infty$.

6. $1 + x + 2x^2 + 3x^3 + 4x^4 + \cdots$.   Ans. $-1 < x < 1$.

7. $1 + \dfrac{x}{a} + \dfrac{x^2}{2a^2} + \dfrac{x^3}{3a^3} + \dfrac{x^4}{4a^4} + \cdots$.   Ans. $-a \leqq x < a$.

8. $x - \dfrac{x^3}{3} + \dfrac{x^5}{5} - \dfrac{x^7}{7} + \cdots$.   Ans. $-1 \leqq x \leqq 1$.

9. $x + x^4 + x^9 + x^{16} + x^{25} + \cdots$.   Ans. $-1 < x < 1$.

10. $x + \tfrac{1}{2} x^4 + \tfrac{1}{3} x^9 + \tfrac{1}{4} x^{16} + \tfrac{1}{5} x^{25} + \cdots$.   Ans. $-1 \leqq x < 1$.

CHAPTER XV

EXPANSIONS OF FUNCTIONS

123. Expansions in power series. On replacing r by x in the series in paragraph 6 of § 8, we have

$$\frac{a}{1-x} = a + ax + ax^2 + ax^3 + \cdots.$$

We say that the infinite series in the second member of this equation is the power-series expansion of the function in the first member, and that it is valid when $|x| < 1$ since the sum of the series in this interval is the given function. In this chapter we shall study the expansion of various functions in power series both in powers of x (as in the foregoing example) and in powers of $x - a$. The latter reduces to the former on taking $a = 0$.

If a function $f(x)$ has an expansion in the form

$$f(x) = c_0 + c_1(x-a) + c_2(x-a)^2 + c_3(x-a)^3 + \cdots,$$

valid on a certain interval for x, the coefficients $c_0, c_1, c_2, c_3, \cdots$ being constants, the derivative of that function has the expansion

$$f'(x) = c_1 + 2\,c_2(x-a) + 3\,c_3(x-a)^2 + 4\,c_4(x-a)^3 + \cdots,$$

obtained by differentiating term by term the series for $f(x)$, and this expansion of $f'(x)$ is valid in the interior of the interval in which the expansion of $f(x)$ is valid. This important theorem is proved in more advanced treatises on the calculus. We shall employ it freely but shall not give a proof of it.

Another important theorem which we shall use without proof is the following: A particular indefinite integral $g(x)$ of the function $f(x)$ has the expansion

$$g(x) = c + c_0(x-a) + \frac{c_1}{2}(x-a)^2 + \frac{c_2}{3}(x-a)^3 + \frac{c_3}{4}(x-a)^4 + \cdots,$$

obtained by integrating term by term the series for $f(x)$, where

$c = g(a)$, and this expansion is valid in the interior of the interval in which the expansion of $f(x)$ is valid.

These two theorems may be summarized briefly (but incompletely) by saying that a power series may be differentiated term by term and that it may be integrated term by term. There are many types of infinite series which do not have these useful properties.

124. Taylor's series. Let $f(x)$ be a function which is continuous for values of x near a and let its derivatives (with respect to x) of various orders be all continuous for x near a. If it has an expansion of the form

$$f(x) = c_0 + c_1(x-a) + c_2(x-a)^2 + c_3(x-a)^3 + \cdots,$$

where $c_0, c_1, c_2, \cdots$ are constants, the coefficients c may be readily evaluated in terms of the function and its derivatives. On differentiating the series repeatedly term by term we have

$$f'(x) = c_1 + 2\,c_2(x-a) + 3\,c_3(x-a)^2 + 4\,c_4(x-a)^4 + \cdots,$$
$$f''(x) = 1\cdot 2\,c_2 + 2\cdot 3\,c_3(x-a) + 3\cdot 4\,c_4(x-a)^2 + \cdots,$$
$$f'''(x) = 1\cdot 2\cdot 3\,c_3 + 2\cdot 3\cdot 4\,c_4(x-a) + 3\cdot 4\cdot 5\,c_5(x-a)^2 + \cdots,$$
$$f^{\text{iv}}(x) = 1\cdot 2\cdot 3\cdot 4\,c_4 + 2\cdot 3\cdot 4\cdot 5\,c_5(x-a) + \cdots,$$
$$\cdots\cdots\cdots\cdots\cdots\cdots\cdots\cdots\cdots\cdots\cdots\cdots\cdots\cdots$$

Putting a for x in the original expansion and these various derived expansions, we have readily

$$c_0 = f(a),\quad c_1 = \frac{f'(a)}{1!},\quad c_2 = \frac{f''(a)}{2!},\quad c_3 = \frac{f'''(a)}{3!},\quad c_4 = \frac{f^{\text{iv}}(a)}{4!},\cdots.$$

Substituting these values of the coefficients into the original expansion, we have the *Taylor's series* for the function $f(x)$, namely,

$$f(x) = f(a) + \frac{f'(a)}{1!}(x-a) + \frac{f''(a)}{2!}(x-a)^2$$
$$+ \frac{f'''(a)}{3!}(x-a)^3 + \cdots.$$

125. Maclaurin's series. Maclaurin's series is the special case of Taylor's series in which $a = 0$. It may be written in the form

$$f(x) = f(0) + \frac{f'(0)}{1!}x + \frac{f''(0)}{2!}x^2 + \frac{f'''(0)}{3!}x^3 + \cdots.$$

126. Taylor's theorem and Taylor's expansion of a given function. We have already given Taylor's theorem in § 117. Writing x for b in equation (4) of that section, we have for $f(x)$ the finite expansion

$$f(x) = f(a) + \frac{f'(a)}{1!}(x-a) + \frac{f''(a)}{2!}(x-a)^2 + \frac{f'''(a)}{3!}(x-a)^3 + \cdots$$
$$+ \frac{f^{(n-1)}(a)}{(n-1)!}(x-a)^{n-1} + \frac{(x-a)^n}{n!}f^{(n)}(x_1),$$

where x_1 lies between x and a. The last term in the second member is called the remainder term in Taylor's theorem. It will be observed that the first n terms in this finite expansion are the same as the first n terms in the infinite expansion given at the end of § 124. The actual validity of the finite expansion has already been established in § 117. If we let n become infinite in the finite expansion and if the remainder term approaches zero, then this finite expansion yields the infinite expansion given at the end of § 124. Hence

Taylor's series for $f(x)$ yields a valid expansion for $f(x)$ when and only when the remainder term in Taylor's theorem approaches zero when n becomes infinite.

For the functions which appear in this book the Taylor series for $f(x)$ yields a valid expansion for $f(x)$ throughout the interval of convergence of the series for $f(x)$. We shall not prove this statement in general but shall verify it for each case as the occasion arises.

Since the $(k+1)$th derivative of a polynomial of the kth degree is zero, it follows that Taylor's series for every polynomial is valid for all values of x.

Since Taylor's theorem reduces to Maclaurin's when $a=0$ we need not treat the latter series separately. It is the Maclaurin series which the student will usually meet in this course.

Example 1. Expand e^x in powers of x.

The derivative of every order is e^x. For $x=0$, e^x has the value 1. Hence the Maclaurin series for e^x is to be found from the formula in § 125 by replacing $f(0), f'(0), f''(0), \ldots$ each by 1. Hence the expansion for e^x is

$$e^x = 1 + x + \frac{x^2}{2!} + \frac{x^3}{3!} + \frac{x^4}{4!} + \cdots.$$

It remains to determine the range of validity of this expansion. The remainder term in Taylor's theorem after n terms is

$$\frac{x^n}{n!} e^{x_1}.$$

When x is positive, x_1 is positive and less than x. Hence

$$\frac{x^n}{n!} e^{x_1} \leq \frac{x^n}{n!} e^x \quad \text{when } x \geq 0.$$

We shall show that the second member approaches zero as n becomes infinite. On changing n to $n+1$ this function is replaced by its former value multiplied by $x/(n+1)$. Since this factor approaches zero as n becomes infinite, it follows that

$$\frac{x^n}{n!} e^x, \text{ and hence } \frac{x^n}{n!} e^{x_1},$$

approaches zero when $x \geq 0$.

If $x < 0$, then x_1 is negative and

$$\left| \frac{x^n}{n!} e^{x_1} \right| < \left| \frac{x^n}{n!} \right|.$$

We can now proceed as before to show that the remainder term approaches zero.

Since the remainder term approaches zero for all values of x, it follows that the foregoing expansion of e^x is valid for all values of x.

Example 2. Expand $\log (1-x)$ in powers of x.

This expansion could be found by forming the successive derivatives and substituting into the series in § 125. But in this case it can be found more readily by the following frequently useful method. We have

$$\frac{d}{dx} \log(1-x) = -\frac{1}{1-x} = -(1 + x + x^2 + x^3 + \cdots),$$

this series (which is given at the beginning of the chapter) being valid for $|x| < 1$. Integrating term by term, we have

$$\log(1-x) = c - x - \frac{x^2}{2} - \frac{x^3}{3} - \frac{x^4}{4} - \cdots.$$

Setting $x = 0$, we find that $c = 0$. Hence, finally,

$$\log(1-x) = -x - \frac{x^2}{2} - \frac{x^3}{3} - \frac{x^4}{4} - \cdots.$$

This is valid when $|x| < 1$.

It can be shown that this is also valid for $x = -1$, a value for which the series is convergent; but we shall usually omit the discussion of validity at the ends of the intervals.

EXERCISES

Compute the following five expansions and prove their validity in the ranges indicated:

1. $\sin x = x - \dfrac{x^3}{3!} + \dfrac{x^5}{5!} - \dfrac{x^7}{7!} + \cdots.$ $\quad(-\infty < x < \infty)$

2. $\cos x = 1 - \dfrac{x^2}{2!} + \dfrac{x^4}{4!} - \dfrac{x^6}{6!} + \cdots.$ $\quad(-\infty < x < \infty)$

3. $\log(1+x) = x - \dfrac{x^2}{2} + \dfrac{x^3}{3} - \dfrac{x^4}{4} + \cdots.$ $\quad(-1 < x < 1)$

4. $e^{-x} = 1 - x + \dfrac{x^2}{2!} - \dfrac{x^3}{3!} + \dfrac{x^4}{4!} - \cdots.$ $\quad(-\infty < x < \infty)$

5. $\arctan x = x - \dfrac{x^3}{3} + \dfrac{x^5}{5} - \dfrac{x^7}{7} + \cdots.$ $\quad(-1 < x < 1)$

Compute the following nine expansions as far as indicated:

6. $(1-x)^{-\frac{1}{2}} = 1 + \tfrac{1}{2}x + \dfrac{1 \cdot 3}{2 \cdot 4}x^2 + \dfrac{1 \cdot 3 \cdot 5}{2 \cdot 4 \cdot 6}x^3 + \cdots.$

7. $(1-x^2)^{-\frac{1}{2}} = 1 + \tfrac{1}{2}x^2 + \dfrac{1 \cdot 3}{2 \cdot 4}x^4 + \dfrac{1 \cdot 3 \cdot 5}{2 \cdot 4 \cdot 6}x^6 + \cdots.$

8. $\arcsin x = x + \tfrac{1}{2} \cdot \dfrac{x^3}{3} + \dfrac{1 \cdot 3}{2 \cdot 4} \cdot \dfrac{x^5}{5} + \dfrac{1 \cdot 3 \cdot 5}{2 \cdot 4 \cdot 6} \cdot \dfrac{x^7}{7} + \cdots.$

9. $(1-x)^{-m} = 1 + mx + \dfrac{m(m+1)}{2!}x^2 + \dfrac{m(m+1)(m+2)}{3!}x^3 + \cdots.$

10. $(1+x)^m = 1 + mx + \dfrac{m(m-1)}{2!}x^2 + \dfrac{m(m-1)(m-2)}{3!}x^3 + \cdots.$

11. $e^{\sin x} = 1 + x + \dfrac{x^2}{2} - \dfrac{x^4}{8} + \cdots.$

12. $\tan x = x + \dfrac{x^3}{3} + \dfrac{2x^5}{15} + \cdots.$

13. $\sec x = 1 + \dfrac{x^2}{2} + \dfrac{5x^4}{24} + \cdots$.

14. $\log \cos x = -\dfrac{x^2}{2} - \dfrac{x^4}{12} - \dfrac{x^6}{45} - \cdots$.

15. Obtain the series for $\cos x$ (1) by differentiating that for $\sin x$ (2) by integrating that for $\sin x$.

16. Show that
$$\log x = (x-1) - \frac{(x-1)^2}{2} + \frac{(x-1)^3}{3} - \frac{(x-1)^4}{4} + \cdots. \quad (0 < x < 2)$$

17. Expand x^3 in powers of $x - 2$ and verify the result by simplifying

18. Expand $x^4 + 7x^3 - 8$ in powers of $x + 2$ and verify the result by simplifying.

19. Show that
$$\sin(a + x) = \sin a + \frac{x}{1!}\cos a - \frac{x^2}{2!}\sin a - \frac{x^3}{3!}\cos a + \cdots.$$

Thence show that $\sin(a + x) = \sin a \cos x + \cos a \sin x$.

20. Show that
$$\sin x = \sin a + \frac{(x-a)}{1!}\cos a - \frac{(x-a)^2}{2!}\sin a - \frac{(x-a)^3}{3!}\cos a + \cdots.$$

127. Computation by means of series. Infinite series are useful in numerical computation. We shall illustrate this remark by means of examples.

Example 1. Compute the value of e.

Putting 1 for x in the series for e^x in Example 1 of § 126, we have
$$e = 1 + 1 + \frac{1}{2!} + \frac{1}{3!} + \frac{1}{4!} + \cdots.$$

In the discussion of the series for e^x (in § 126) we saw that the remainder after n terms is less than
$$\frac{x^n}{n!}e^x.$$

Hence for $x = 1$ this remainder is less than $e/n!$. Thus if we take 10 terms of the foregoing series for e the error is less than $e/10!$. We can thus determine an upper bound to the error made by taking any given number of terms of the series for e. By taking a sufficient number of terms we can thus show that
$$e = 2.7182818 \cdots.$$

Example 2. Compute the value of π.

For this purpose we use the series of problem 8 in the foregoing section. Setting $x = \frac{1}{2}$ and remembering that arc $\sin \frac{1}{2} = \frac{\pi}{6}$, we have

$$\frac{\pi}{6} = \frac{1}{2} + \frac{1}{2} \cdot \frac{1}{2^3 \cdot 3} + \frac{1 \cdot 3}{2 \cdot 4} \cdot \frac{1}{2^5 \cdot 5} + \cdots,$$

whence
$$\pi = 3.14159 \cdots.$$

It is sometimes very difficult to determine the percentage of error in a numerical computation. The *actual error* is of course equal to the sum of the neglected terms. But often this cannot be found. There is, however, one class of series in which it is easy to find an upper bound to the error. These are the alternating series whose terms decrease in numerical value. From the proof of theorem VII of §121 it follows that the sum of such a series lies between the sum of n terms and the sum of $n+1$ terms for every value of n. Hence the error made in taking the sum of the first n terms of the series for the sum of the series is numerically less than the first term neglected. Hence the alternating series are particularly useful for purposes of computation, owing to the ease with which one may determine an upper bound to the error made in summing only a given number of terms of the series.

Example 3. Compute sin 1.

From the series for $\sin x$ in problem 1 of the preceding section we have
$$\sin 1 = 1 - \frac{1}{3!} + \frac{1}{5!} - \frac{1}{7!} + \cdots = .84147 \cdots.$$

(In computing the values of the trigonometric functions by means of the power-series expansions which we have developed we must always express the angle in radian measure. Why?)

Natural logarithms might be computed by the aid of the power-series expansion for $\log(1 + x)$. But it is often convenient to use another formula for this purpose. We start from the following power series (given in § 126):

$$\log(1-x) = -x - \frac{x^2}{2} - \frac{x^3}{3} - \frac{x^4}{4} - \cdots, \qquad (1)$$

$$\log(1+x) = x - \frac{x^2}{2} + \frac{x^3}{3} - \frac{x^4}{4} + \cdots. \qquad (2)$$

By subtraction we have
$$\log(1+x) - \log(1-x) = \log\frac{1+x}{1-x} \tag{3}$$
$$= 2\left(x + \frac{x^3}{3} + \frac{x^5}{5} + \cdots\right).$$

Putting
$$\frac{1+x}{1-x} = \frac{y+1}{y}$$
we have
$$x = \frac{1}{2y+1}. \tag{4}$$

Substituting for x in (3) its value in terms of y, and remembering that
$$\log\frac{y+1}{y} = \log(y+1) - \log y$$
we have
$$\log(y+1) = \log y + 2\left[\frac{1}{2y+1} + \frac{1}{3}\left(\frac{1}{2y+1}\right)^3 \right.$$
$$\left. + \frac{1}{5}\left(\frac{1}{2y+1}\right)^5 + \cdots\right]. \tag{5}$$

Since series (1) and (2) are valid when x is positive and less than 1, and since x meets this condition when y is positive (as one sees from (4)), it follows that equation (5) is valid for all positive values of y. It is a convenient series for the computation of logarithms.

Now $\log 1 = 0$. Hence if we take $y = 1$ we have
$$\log 2 = 2[\tfrac{1}{3} + \tfrac{1}{3}(\tfrac{1}{3})^3 + \tfrac{1}{5}(\tfrac{1}{3})^5 + \cdots]$$
$$= 0.693147\cdots.$$
Then $\log 3 = \log 2 + 2[\tfrac{1}{5} + \tfrac{1}{3}(\tfrac{1}{5})^3 + \tfrac{1}{5}(\tfrac{1}{5})^5 + \cdots]$
$$= 1.098612\cdots.$$
$\log 4 = 2\log 2 = 1.386294\cdots.$
$\log 5 = \log 4 + 2[\tfrac{1}{9} + \tfrac{1}{3}(\tfrac{1}{9})^3 + \tfrac{1}{5}(\tfrac{1}{9})^5 + \cdots]$
$$= 1.609438\cdots.$$
$\log 6 = \log 2 + \log 3 = 1.791759\cdots.$

Logarithms to the base 10 are found by means of the formula (compare a result at the end of § 17)
$$\log_{10} a = \frac{\log_e a}{\log_e 10} = (0.4342945\cdots)\log_e a.$$

128. Approximate formulas.
Let us consider the series
$$\sin x = x - \frac{x^3}{3!} + \frac{x^5}{5!} - \frac{x^7}{7!} + \cdots.$$

This is an alternating series. If we assign to x a certain range and hold that range fixed, then we can go far enough out into the series to insure that the following terms decrease numerically for every value of x in the range. Hence we can choose a certain finite number of terms whose sum will approximate to the value of $\sin x$ for x in the given range, and the difference between the true value and the approximate value will be numerically less than the first term omitted from the series. Thus if x ranges between -1 and 1 the terms decrease numerically from the beginning. Hence $\sin x$ differs from x by less than $x^3/6$. If x is a very small angle the difference between $\sin x$ and x is numerically much smaller. Again, the function $\sin x$ is represented by $x - x^3/6$ with an error numerically less than $x^5/5!$ if x is in the range -1 to 1.

Many functions whose expansions are in alternating series may be approximated in this way.

Whenever for a given function a convenient upper bound may be obtained for the value of the remainder term in Taylor's theorem, the omission of the remainder term leads to an approximate formula which is often useful. But the detailed discussion of these approximations lies outside the scope of this book.

EXERCISES

Compute the results given in the following eight problems correct to four decimal places:

1. $\cos 1 = .5403$.
2. $\tan 12° = .2126$.
3. $\sin 15° = .2588$.
4. $\sin \dfrac{\pi}{4} = .7071$.
5. $e^2 = 7.3891$.
6. $\sqrt{e} = 1.6487$.
7. $\sin 19° 20' = .3311$.
8. $\cos 30° 14' = .8640$.

Compute the logarithms in the following four problems correct to five places:

9. $\log 7 = 1.94591$.
10. $\log 84 = 4.43082$.
11. $\log 73 = 4.29046$.
12. $\log_{10} 5 = .69897$.

13. Using the value log 4 = 1.3863, compute log 4.01 by aid of the expansion of log (1 + x) in powers of x. *Ans.* 1.3888.

14. Find log 5.005. *Ans.* 1.6104.

15. Discuss the limits of accuracy of the formula
$$\cos x = 1 - \frac{x^2}{2} + \frac{x^4}{24}$$
for x between $-.1$ and $.1$.

16. Compare the graphs of
$$x, \quad x - \frac{x^3}{6}, \quad x - \frac{x^3}{6} + \frac{x^5}{120}$$
with the graph of sin x.

17. Find the approximate value of $\sqrt[3]{1333}$ by writing it in the form
$$\sqrt[3]{1333} = (1331 + 2)^{\frac{1}{3}} = 11(1 + \tfrac{2}{1331})^{\frac{1}{3}}$$
and employing problem 10 of § 126 to evaluate the last cube root.
Ans. 11.005507.

18. Find the fifth root of 100,001 correct to five decimal places.
Ans. 10.00002.

19. Find the approximate values of the reciprocals of 98 and 101, using problem 9 of § 126, and verify the results by a direct computation.

CHAPTER XVI

PROPERTIES OF PLANE CURVES

129. Concavity. In treating the concavity of a plane curve it is convenient to think of the x-axis as being horizontal and the y-axis as being vertical with the positive direction upward. Then a curve is said to be *concave upward* at a point if an arc of a curve containing the point in its interior lies above the tangent at that point. If such an arc lies below the tangent at a point the curve is said to be *concave downward* at the point. The

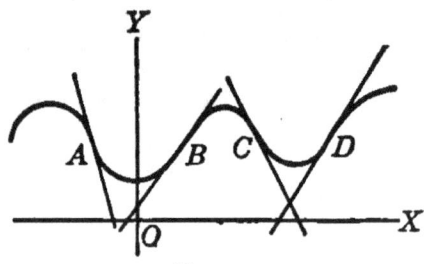

FIG. 53

adjacent curve is concave upward between A and B and also between C and D; it is concave downward between B and C.

As x increases for points along an arc which is concave upward the slope is increasing; for an arc concave downward the slope is decreasing. Hence

I. A curve $y = f(x)$ is concave upward or downward at a given point on it according as $f'(x)$ increases or decreases with increase in x.

Where $f'(x)$ is increasing, its derivative $f''(x)$ is positive, and where $f'(x)$ is decreasing, its derivative is negative. Hence

II. A curve $y = f(x)$ is concave upward at a point if the second derivative $f''(x)$ is positive at that point; it is concave downward if the second derivative is negative.

130. Points of inflection. If the inclination of a curve is continuous at a given point, and if this point separates an arc concave upward from one concave downward, it is called a *point of inflection*. In the figure of the previous section the curve has points of inflection at A, B, C, D.

A point of inflection of the curve $y = f(x)$ is a point at which $f'(x)$ changes from an increasing to a decreasing function, or vice versa, as we see from the first theorem in the preceding section. A special examination should be made of those points at which the tangent to the curve is perpendicular to the x-axis. At all other points the conditions on $f'(x)$ for a point of inflection of the curve $y = f(x)$ are the same as those for a maximum or a minimum of $f'(x)$. Hence the totality of points to be examined for points of inflection are those at which $f''(x)$ is zero or becomes infinite and those at which $f'(x)$ becomes infinite. We have a point of inflection at one of these points if and only if $f''(x)$ changes sign there.

Example 1. Test the curve $y = x^4 - 4x^3 + 12$ for points of inflection and direction of bending.

We have

$$f(x) = x^4 - 4x^3 + 12, \quad f'(x) = 4x^3 - 12x^2, \quad f''(x) = 12x(x-2).$$

Since $f'(x)$ and $f''(x)$ are everywhere continuous, the only points to be examined are those at which $f''(x) = 0$, namely, the points for which $x = 0$ and $x = 2$. At each of these points $f''(x)$ changes sign. Hence the curve has a point of inflection at $(0, 12)$ and at $(2, -4)$. The curve is concave downward between these two points. To the left of $(0, 12)$ and to the right of $(2, -4)$ it is concave upward.

Example 2. Test the curve $y = 1 + (x-2)^{\frac{1}{3}}$ for points of inflection and direction of bending.

We have

$$f(x) = 1 + (x-2)^{\frac{1}{3}}, \quad f'(x) = \tfrac{1}{3}(x-2)^{-\frac{2}{3}}, \quad f''(x) = -\tfrac{2}{9}(x-2)^{-\frac{5}{3}}.$$

At the point for which $x = 2$ both $f'(x)$ and $f''(x)$ become infinite. Moreover $f''(x)$ is positive for $x < 2$ and negative for $x > 2$. Hence there is a point of inflection at $(2, 1)$; to the left of this point the curve is concave upward; to the right it is concave downward.

EXERCISES

Test the following curves for points of inflection and direction of bending:

1. $y = x^3 - x^2 - x + 1$.

   Ans. Concave downward to the left and concave upward to the right of the point of inflection $(\tfrac{1}{3}, \tfrac{16}{27})$.

2. $y = 2 + (x-1)^{\frac{1}{3}}$.

Ans. Concave upward to the left and concave downward to the right of the point of inflection (1, 2).

3. $y = x^2 + 2x + 17$.   *Ans.* Concave upward everywhere.
4. $y = x^4 - 6x^3 + 12x^2 - 20$.
5. $y = a + (x-c)^5$.
6. $y = \sin x$ and $y = \tan x$.
7. Show that the following curves are concave upward everywhere:

$$y = x^2, \quad y = x^4, \quad y = e^x, \quad y = e^{x^2}.$$

8. Show that the following curves are concave downward everywhere:

$$y = -7 - x - 3x^2, \quad y = 5 + x - x^4, \quad y = \log x.$$

9. Show that a conic section does not have a point of inflection.

131. Singular points. Certain points at which a plane curve has peculiarities are called *singular points*. These are of different types as follows:

1. A *multiple point* is a point at which two or more branches of a curve intersect.

2. A *tacnode* is a point at which two branches of a curve meet and have a common tangent but do not stop.

3. A *cusp* is a point at which two branches of a curve meet and stop and have a common tangent. It is of the *first species* if the two branches lie on opposite sides of the tangent; it is of the *second species* if they lie on the same side.

4. A *conjugate point*, or an *isolated point*, is a point which satisfies the equation of a curve while no other point in its neighborhood satisfies the equation of the curve.

5. A *salient point* is a point at which two branches of a curve meet and stop but do not have a common tangent.

6. An *end point* is a point at which a single branch of a curve stops.

Example 1. The curve $y = x \log x$ has a single branch for positive values of x. It stops at the point (0, 0), since y is not defined when x is negative. The point (0, 0) is therefore an end point of the curve.

Example 2. The curve $y = e^{-\frac{1}{x}}$ has two branches, one for x positive and one for x negative. The former has an end point at the origin.

Example 3. The curve
$$y = \frac{x}{1 + e^{\frac{1}{x}}}$$
has two branches, one for x positive and one for x negative. They meet at the origin. They do not have a common tangent, since the derivative
$$\frac{dy}{dx} = \frac{1}{1 + e^{\frac{1}{x}}} + \frac{e^{\frac{1}{x}}}{x(1 + e^{\frac{1}{x}})^2}$$
approaches 1 if x approaches 0 from the left and 0 if x approaches zero from the right. Hence the origin is a salient point. One branch is tangent to the x-axis; the other makes an angle of 45° with the x-axis.

Example 4. The curve $y^2 = x^2(x^2 - 1)$ has the origin for an isolated point.

Example 5. The curve $y^2 = x^2(1 - x^2)$ has two branches which come together at the origin, these being given by the equations
$$y = x\sqrt{1 - x^2}, \quad y = -x\sqrt{1 - x^2}.$$
For these two we have, respectively,
$$\frac{dy}{dx} = \frac{1 - 2x^2}{\sqrt{1 - x^2}}, \quad \frac{dy}{dx} = -\frac{1 - 2x^2}{\sqrt{1 - x^2}}.$$
As x approaches 0 these derivatives approach 1 and -1 respectively. Hence the two branches have the tangents $y = x$ and $y = -x$ at $(0, 0)$. Hence the origin is a multiple point at which two branches of the curve cross each other at right angles.

Example 6. The curve $y^2 = x^4(1 - x^2)$ has two branches which meet at the origin and proceed in both directions from the origin. By working as in the foregoing example it may be shown that the slope of each branch at the origin is 0. Hence the two branches have the common tangent $y = 0$ at the origin. Therefore the origin is a tacnode.

Example 7. The curve $y^2 = (x - 1)^5$ has two branches meeting at the point $(1, 0)$; they stop there, since y is imaginary for x less than 1. It can be shown that the slope of each branch at $(1, 0)$ is 0. Hence the x-axis is a common tangent to the two branches at (1, 0). The curve therefore has a cusp at this point. The two branches lie on opposite sides of the tangent; hence the cusp is of the first species.

Example 8. What is the character of the curve
$$x^4 + 2x^2y - 3xy^2 + y^3 = 0$$
at the point (0, 0)?

The equation $y = mx$, by the proper choice of m, can be made to represent any line through the point (0, 0) except the y-axis. We shall think of it as a secant of the given curve. We solve the given equation and the equation $y = mx$ simultaneously. Putting mx for y in the given equation, we have
$$x^4 + 2mx^3 - 3m^2x^3 + m^3x^3 = 0.$$

For the point of intersection (of the curve and secant) other than the origin we have $x \neq 0$; hence, for such a point, we may divide the last equation through by x^3 and obtain
$$x + 2m - 3m^2 + m^3 = 0.$$

For small values of x this equation, as an equation in m, has three real roots. As x approaches zero these approach the roots of the equation
$$2m - 3m^2 + m^3 = 0,$$
namely 0, 1, 2. But as x approaches zero m approaches the slope of the given curve at the origin. Hence this curve has three branches with slopes 0, 1, 2 at the origin and hence with the tangents
$$y = 0, \quad y = x, \quad y = 2x.$$

This discussion leaves unsettled the relation of the y-axis to the curve. To test that we use the equation $x = \mu y$ and proceed as before, this enabling us to deal with every possible tangent except the x-axis. We have
$$\mu^4 y^4 + 2\mu^2 y^3 - 3\mu y^3 + y^3 = 0.$$
Dividing by y^3, we have
$$\mu^4 y + 2\mu^2 - 3\mu + 1 = 0.$$

When y approaches zero we have $2\mu^2 - 3\mu + 1 = 0$, an equation with roots $\frac{1}{2}$ and 1. This leads to the tangents $x = \frac{1}{2} y$ and $x = 1 \, y$, but not to the tangent $x = 0$. Hence the y-axis is not tangent to a branch of the curve at the origin. The other two equations are equations of two of the tangents already found.

Therefore we conclude finally that three branches of the curve intersect at (0, 0) having there the tangents $y = 0$, $y = x$ and $y = 2x$. Hence the origin is a *triple* point of the curve, that is, a multiple point where three branches meet.

If we were testing the point (a, b) on the curve, we would use in a similar way the more general equations

$$y - b = m(x - a) \quad \text{and} \quad x - a = \mu(y - b)$$

instead of the equations $y = mx$ and $x = \mu y$.

If it should turn out that m and μ have no real values, we should conclude that no branches pass through the point in question. Then the point on the curve which is being tested is a conjugate or isolated point.

EXERCISES

1. Show that the curve $y^2 = x^3$ has a cusp of the first species at the origin.

2. Show that the curve $(y - x^2)^2 = x^5$ has a cusp of the second species at the origin.

3. Show that the origin is an isolated point of the curve $y^4 - y^2 - x^2 = 0$.

4. Show that the curve $(x - y)^2 = (x - 1)^3$ has a cusp of the first species at the point (1, 1), the tangent being $y = x$.

5. What is the character of the curve $x^3 + y^3 = 3\,axy$ at the point (0, 0)?   *Ans.* Double point with tangents $x = 0$, $y = 0$.

6. Show that the curve $y^2 = x^3/(2a - x)$ has a cusp of the first species at (0, 0).

7. Show that the origin is an isolated point for the curve $y^2(x^2 - 1) = x^4$.

8. What is the character of the curve $x^3 = y^2(y^2 + x)$ at the point (0, 0)?

Ans. It has a triple point with the tangents $y = x$, $y = -x$, $x = 0$.

9. Show that the curve $y = x \arctan 1/x$ has a salient point at the origin if $\arctan 1/x$ lies between $-\tfrac{1}{2}\pi$ and $+\tfrac{1}{2}\pi$.

10. Show that the curve $(y - 2)^2 = (x - 1)^3$ has a cusp of the first species at the point (1, 2).

11. In the case of the lemniscate $(x^2 + y^2)^2 = a^2(x^2 - y^2)$ what is the character of the point (0, 0)?

Ans. Double point with tangents $y = \pm x$.

12. Show that the curve $x^{\frac{2}{3}} + y^{\frac{2}{3}} = a^{\frac{2}{3}}$ has a cusp of the first species at each of the points where it cuts a coördinate axis.

13. What is the character of the point (0, 0) in the case of the cardioid
$$x^2 + y^2 + ax = a\sqrt{x^2 + y^2}?$$

14. What is the character of the point (0, 0) in the case of the strophoid
$$y^2 = x^2 \frac{a+x}{a-x}?$$

Discuss the character of the point (0, 0) in the case of each of the following curves:

15. $\rho = a \sin 2\theta$.
16. $\rho = a \cos 2\theta$.
17. $\rho = a \sin 4\theta$.
18. $\rho^2 = a^2 \sin 2\theta$.

132. Asymptotes. An *asymptote* to a plane curve is the limiting position of a tangent whose point of contact with the curve moves off to infinity. It is understood that this limiting line approached must lie partly in a finite portion of the plane. Hence an asymptote must pass within a finite distance of the origin. It must therefore have a finite intercept on at least one of the coördinate axes. It is evident that a curve cannot have an asymptote without having at least one infinite branch. The student has already met asymptotes in the case of the hyperbola; and he has also learned that the parabola is a curve with an infinite branch but without an asymptote.

There are two principal methods of finding asymptotes in the case of rectangular axes.

The method of limiting intercepts. The equation of the tangent to the curve $y = f(x)$ at the point (x_1, y_1) is (see §54)
$$y - y_1 = f'(x_1)(x - x_1).$$

Its intercepts on the x-axis and the y-axis are, respectively,
$$x_1 - \frac{y_1}{f'(x_1)}, \quad y_1 - x_1 f'(x_1).$$

One finds the limiting values of these expressions as the point (x_1, y_1) moves off to infinity on the curve. One of the limits must be finite. If the other is infinite the asymptote is parallel to one of the axes and its equation is readily determined. If both are finite the asymptote is determined by its intercepts except when both are zero. In this latter case it is necessary to find also the limiting value of $f'(x_1)$ in order to know the slope of the asymptote; and then its equation is readily found. (This method, though simple in theory, is frequently too complicated for practical use.)

Example 1. Show that the hyperbola $b^2x^2 - a^2y^2 = a^2b^2$ has the two asymptotes $ay = \pm bx$.

We have $dy/dx = b^2x/a^2y$. Hence the equation of the tangent is

$$y - y_1 = \frac{b^2 x_1}{a^2 y_1}(x - x_1), \quad \text{or} \quad \frac{xx_1}{a^2} - \frac{yy_1}{b^2} = 1.$$

The intercepts of the tangent are therefore a^2/x_1 and $-b^2/y_1$. Both of these approach zero as (x_1, y_1) moves off to infinity on the curve. Hence we have to find the limiting value of the slope $f'(x_1)$. Now

$$f'(x_1) = \pm \frac{b}{a} \frac{1}{\sqrt{1 - \frac{a^2}{x_1^2}}};$$

and this approaches $\pm b/a$ as x_1 becomes infinite. Hence the asymptotes pass through the origin and have the slopes $\pm b/a$. Hence their equations are $ay = \pm bx$.

The method of substitution for algebraic curves. We take the equation of the curve in the form $f(x, y) = 0$, where $f(x, y)$ is a polynomial of degree n in x and y. Every tangent can be represented by the equation $y = mx + b$ except those tangents which are parallel to the y-axis. Now the tangent is the limiting position of a secant line as two points of intersection of the secant with the curve come into coincidence. For a secant approaching an asymptote these points of intersection must approach coincidence as each recedes to infinity. Hence, for an asymptote, the equation

$$f(x, mx + b) = 0$$

must have two infinite roots. In algebra it is shown that the condition for this is that the coefficient of x^n and that of x^{n-1} shall be zero. On equating them to zero we have two equations for determining m and b. For every pair of values of m and b satisfying these equations we have in the corresponding equation $y = mx + b$ an asymptote which is not parallel to the y-axis.

For determining the asymptotes parallel to the y-axis we employ the equation $x = c$ and proceed in a similar manner by equating to zero the coefficients of y^n and y^{n-1}.

Example 2. Examine the curve $x^3 + y^3 - 3\,axy = 0$ for asymptotes.

Substituting $mx + b$ for y in the equation, we have

$$(1 + m^3)x^3 + (3\,m^2b - 3\,am)x^2 + (3\,mb^2 - 3\,ab)x + b^3 = 0.$$

Equating to zero the coefficients of x^3 and x^2, we have

$$1 + m^3 = 0, \quad 3\,m^2b - 3\,am = 0; \quad \text{or,} \quad m = -1, \quad b = -a.$$

Hence we have the asymptote $y + x + a = 0$.

If we put $x = c$ we have $y^3 - 3\,acy + c^3 = 0$. The coefficient of y^3 cannot be made zero. Hence we get no asymptote parallel to the y-axis.

Example 3. Show that the curve $y^2(2\,a - x) = x^3$ has the asymptote $x = 2\,a$.

Putting $x = c$ in the equation, we have

$$(2\,a - c)y^2 - c^3 = 0.$$

The coefficient of y^3 is zero. Setting the coefficient of y^2 equal to zero, we have $2\,a - c = 0$, or $c = 2\,a$. Hence the line $x = 2\,a$ is an asymptote to the curve.

Asymptote for polar coördinates. Let $f(\rho, \theta) = 0$ be the equation of a curve in polar coördinates. Since an asymptote must pass within a finite distance of the origin, it must be parallel to the limiting position of the radius vector from the origin to a point P on the curve when P itself recedes to infinity. Hence the inclination of the asymptote to the polar axis is the same as the value of θ for which ρ becomes infinite. (If there is more than one such asymptote there will be a value of θ for each asymptote.) To determine the distance from the origin to the asymptote we find the distance from the origin to the tangent at the general point (ρ_1, θ_1) and obtain the limit of this distance as the point recedes to infinity. The inclination of the asymptote and its distance from the origin are sufficient to determine it.

In order to apply this method it is necessary to have a means of finding the distance from the origin to the tangent. For this purpose we first derive an auxiliary formula which is useful also for other purposes.

Let (ρ, θ) denote any fixed point P on the curve $\rho = g(\theta)$. Let ψ be the angle between the tangent at this point and the radius vector to this point. Let τ be the angle between the tangent PT and the polar axis. Then $\tau = \theta + \psi$, by geometry. Let Q be the point $(\rho + \Delta\rho, \theta + \Delta\theta)$ near the point P, and draw PA perpendicular to OQ. Then

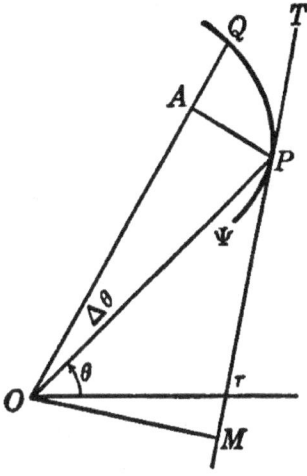

FIG. 54

$$\tan PQA = \frac{PA}{AQ} = \frac{PA}{OQ - OA}$$
$$= \frac{\rho \sin \Delta\theta}{\rho + \Delta\rho - \rho \cos \Delta\theta}.$$

As Q approaches P the angle PQA approaches ψ and the increments $\Delta\theta$ and $\Delta\rho$ approach zero. Hence

$$\tan \psi = \lim_{\Delta\theta \to 0} \frac{\rho \sin \Delta\theta}{\rho + \Delta\rho - \rho \cos \Delta\theta} = \lim_{\Delta\theta \to 0} \frac{\rho \frac{\sin \Delta\theta}{\Delta\theta}}{\frac{\Delta\rho}{\Delta\theta} + \rho \frac{1 - \cos \Delta\theta}{\Delta\theta}};$$

or
$$\tan \psi = \frac{\rho}{\frac{d\rho}{d\theta}} = \rho \frac{d\theta}{d\rho},$$

by § 16 and Ex. 3 in that section. But if OM is perpendicular to PT, then
$$OM = \rho \sin \psi.$$

Hence we have a means of determining the distance OM from the origin to the tangent.

Incidentally we have for the slope of the tangent the important formula
$$\tan \tau = \tan(\theta + \psi) = \frac{\tan \theta + \tan \psi}{1 - \tan \theta \tan \psi}.$$

In the case when ρ becomes infinite on a branch having an asymptote the angle ψ approaches 0. Hence the distance from the origin to the asymptote is

$$\lim_{\rho=\infty} OM = \lim_{\rho=\infty} \rho \sin \psi = \lim_{\rho=\infty} \rho \tan \psi \cdot \lim_{\rho=\infty} \cos \psi = \lim_{\rho=\infty} \rho^2 \frac{d\theta}{d\rho}.$$

Example 4. Show that the line at distance a from the origin and parallel to the polar axis is an asymptote to the hyperbolic spiral $\rho\theta = a$.

That the asymptote (if there is one) is parallel to the polar axis follows from the fact that ρ becomes infinite only when θ approaches 0. The distance of the asymptote from the origin is

$$\lim_{\rho=\infty} \rho^2 \frac{d\theta}{d\rho} = \lim_{\rho=\infty} \rho^2 \left(-\frac{a}{\rho^2}\right) = -a.$$

(The negative sign corresponds to the fact that the asymptote is to the left of the origin when one looks from the origin in the direction of the infinite radius vector.)

EXERCISES

Find the asymptotes of the curves with the following equations:

1. $y = e^x$.   Ans. $y = 0$.
2. $y = \log x$.   Ans. $x = 0$.
3. $x^2 y = 4 a^2(2a - y)$.   Ans. $y = 0$.
4. $x^2 y^2 = (y + a)^2(b^2 - y^2)$.   Ans. $y = 0$.
5. $y^2 = x^2 \dfrac{a + x}{a - x}$.   Ans. $x = a$.
6. $y = e^{-x^2}$.   Ans. $y = 0$.
7. $xy = a$.   Ans. $x = 0$, $y = 0$.
8. $x^3 + y^3 = c$.   Ans. $x + y = 0$.
9. $xy^2 + x^2 y + c = 0$.   Ans. $x = 0$, $y = 0$, $x + y = 0$.
10. $y = \tan x$.   Ans. $x = (k + \tfrac{1}{2})\pi$, $k =$ integer.
11. $\rho = a \csc \theta + b$.
    Ans. Line parallel to the polar axis at a distance a from it.
12. $\rho^2 \theta = a^2$.   Ans. The polar axis.
13. $\rho = a \sec 2\theta$.
    Ans. Four asymptotes at distance $a/2$ from origin.

133. Curve tracing. The general form of the curve can often be determined rapidly by means of properties of plane curves already treated. In carrying out such work in the case of rectangular axes the student will do well to take some or all of the following steps:

1. Test the curve for symmetry with respect to the axes. When only even powers of x are present in the equation the curve is symmetrical with respect to the y-axis; when only even powers of y are present the curve is symmetrical with respect to the x-axis.
2. Find where the curve cuts the axes.
3. Determine the maximum and the minimum points.
4. Determine the points of inflection and thence the intervals of concavity upward or downward.
5. Examine the curve for infinite branches and the asymptotes to them (if such exist).
6. Examine the curve for singular points.
7. Plot such additional points as may seem necessary and then sketch in the curve.

EXERCISES

Sketch the following curves:

1. $y = x^4 - 2x^2$.
2. $y(1 + x^2) = x$.
3. $y^2 = x^2 \dfrac{a + x}{a - x}$.
4. $y^2 = x^3$.
5. $x^2 y = 4 a^2 (2a - y)$.
6. $y^2 (2a - x) = x^3$.
7. $(x^2 + y^2)^2 = a^2(x^2 - y^2)$.
8. $\rho^2 = a^2 \cos 2\theta$.
9. $x^2 y^2 = (y + a)^2 (b^2 - y^2)$.
10. $\rho = a \csc \theta + b$.
11. $\rho = a \sin 3\theta$.
12. $y = \tfrac{1}{2} a\left(e^{\frac{x}{a}} + e^{-\frac{x}{a}}\right)$.
13. $y = e^{-x^2}$.
14. $x^{\frac{1}{2}} + y^{\frac{1}{2}} = a^{\frac{1}{2}}$.
15. $x^{\frac{2}{3}} + y^{\frac{2}{3}} = a^{\frac{2}{3}}$.
16. $x^3 + y^3 - 3axy = 0$.
17. $\rho = a(1 - \cos \theta)$.
18. $\rho = b - a \cos \theta$.
19. $\rho \theta = a$.
20. $\rho^2 \theta = a^2$.
21. $y = x \log x$.
22. $y\left(1 + e^{\frac{1}{x}}\right) = x$.

134. Curvature. The direction of a plane curve at a given point is the same as the direction of its tangent at that point. As a point moves along a curve the tangent continually changes its direction. Many properties of the curve depend on the rate at which this change takes place. To obtain a convenient analytical expression for this rate let us consider an increment of arc Δs from the fixed point P to the movable point Q and the corresponding increment $\Delta \phi$ of inclination of the tangent line at Q over that of the tangent line at P. Then $\Delta \phi$ denotes the total change in direction of the curve in passing from P to Q over the arc length Δs. The ratio $\Delta \phi / \Delta s$ is called the mean curvature of the arc PQ. Its limit as Δs approaches zero is called the *curvature* of the curve at the point P. If we denote it by K, we have

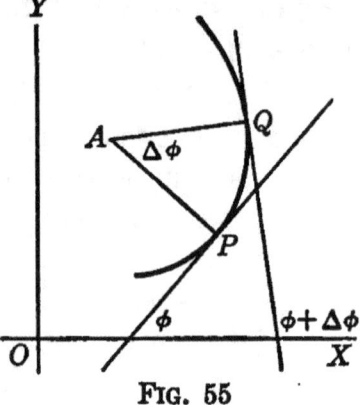

Fig. 55

$$K = \lim_{\Delta s \to 0} \frac{\Delta \phi}{\Delta s} = \frac{d\phi}{ds}.$$

If we represent the curve by the equation $y = f(x)$, then we have
$$\tan \phi = f'(x), \quad \text{or} \quad \phi = \arctan f'(x).$$

Differentiating both members of the last equation with respect to s, we have
$$\frac{d\phi}{ds} = \frac{d}{dx}(\arctan f'(x)) \cdot \frac{dx}{ds}$$
$$= \frac{f''(x)}{1 + [f'(x)]^2} \cdot \frac{1}{\sqrt{1 + [f'(x)]^2}}$$

by §§ 76 and 62. Hence the curvature K is given by the formula

$$K = \frac{f''(x)}{\{1 + [f'(x)]^2\}^{\frac{3}{2}}} = \frac{\dfrac{d^2y}{dx^2}}{\left[1 + \left(\dfrac{dy}{dx}\right)^2\right]^{\frac{3}{2}}}.$$

In the case of a circle the normals PA and QA are radii. Therefore, if the radius of the circle is r, we have

$$r\Delta\phi = \Delta s, \quad \text{or} \quad \frac{\Delta\phi}{\Delta s} = \frac{1}{r}.$$

Hence the curvature of a circle is constant and is equal to the reciprocal of the radius.

Let a circle be drawn tangent to the curve $y = f(x)$ at a point P and with radius equal to the reciprocal of the curvature of the given curve at P. This circle is called the *circle of curvature* of the given curve at the point P, its center is the *center of curvature* of that curve at P; and its radius is the *radius of curvature* of that curve at P. The radius of curvature R is therefore given by the formula

$$R = \frac{ds}{d\phi} = \frac{\left[1 + \left(\frac{dy}{dx}\right)^2\right]^{\frac{3}{2}}}{\frac{d^2y}{dx^2}}.$$

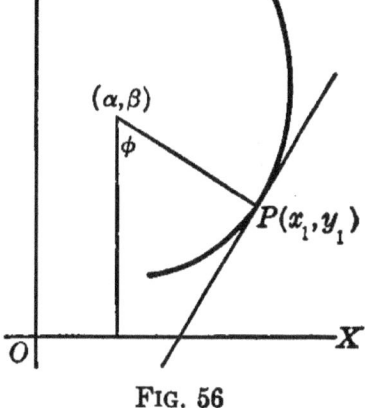

FIG. 56

Since the curve is tangent to its circle of curvature at P, it follows that the normal at P lies along the radius of curvature.

The sign of R is the same as that of $f''(x)$. Hence it is positive or negative according as the curve is concave upward or concave downward at P. In many cases we are concerned only with the numerical value of R and not with its algebraic sign.

Let the point P be denoted by (x_1, y_1) and the center of curvature by (α, β). To find the formulas for α and β we note from the figure that

$$x_1 - \alpha = R \sin\phi, \quad y_1 - \beta = -R \cos\phi.$$

But $\tan\phi = f'(x_1)$; whence

$$\cos\phi = \frac{1}{\sqrt{1 + [f'(x_1)]^2}}, \quad \sin\phi = \frac{f'(x_1)}{\sqrt{1 + [f'(x_1)]^2}}.$$

Putting these values of $\cos \phi$ and $\sin \phi$ in the foregoing equations and solving for the coördinates α and β of the center of curvature, we have the formulas

$$\alpha = x_1 - \frac{f'(x_1)\{1 + [f'(x_1)]^2\}}{f''(x_1)},$$

$$\beta = y_1 + \frac{1 + [f'(x_1)]^2}{f''(x_1)}.$$

If the curve is given by the parametric equations

$$x = f(\theta), \quad y = \phi(\theta)$$

we have (compare problem 5 in § 114)

$$\frac{dy}{dx} = \frac{\frac{dy}{d\theta}}{\frac{dx}{d\theta}}, \quad \frac{d^2y}{dx^2} = \frac{\frac{dx}{d\theta}\frac{d^2y}{d\theta^2} - \frac{dy}{d\theta}\frac{d^2x}{d\theta^2}}{\left(\frac{dx}{d\theta}\right)^3}.$$

Substituting these into the formula for R and simplifying, we have

$$R = \frac{\left[\left(\frac{dx}{d\theta}\right)^2 + \left(\frac{dy}{d\theta}\right)^2\right]^{\frac{3}{2}}}{\frac{dx}{d\theta}\frac{d^2y}{d\theta^2} - \frac{dy}{d\theta}\frac{d^2x}{d\theta^2}}.$$

If the curve is given in polar coördinates we may take the parametric representation in the special form $x = \rho \cos \theta$ and $y = \rho \sin \theta$, where ρ is a function of θ. In this case the formula for R reduces to the following form:

$$R = \frac{\left[\rho^2 + \left(\frac{d\rho}{d\theta}\right)^2\right]^{\frac{3}{2}}}{\rho^2 + 2\left(\frac{d\rho}{d\theta}\right)^2 - \rho\frac{d^2\rho}{d\theta^2}}.$$

The student should verify this reduction.

The computation of curvature, radius of curvature, and center of curvature of a given curve is to be effected by a direct substitution into the formulas developed in this section.

EXERCISES

Find the radius of curvature of the following curves at any point:

1. $y^2 = 4\,px$.   Ans. $-\dfrac{(y^2 + 4\,p^2)^{\frac{3}{2}}}{4\,p^2}$.

2. $\rho = e^{a\theta}$.   Ans. $\rho\sqrt{1 + a^2}$.

3. $y = \tfrac{1}{2} a(e^{\frac{x}{a}} + e^{-\frac{x}{a}})$.   Ans. $\dfrac{y^2}{a}$.

4. $x = a(\theta - \sin\theta)$, $y = a(1 - \cos\theta)$.   Ans. $-2\,a\sqrt{2 - 2\cos\theta}$.

5. $y = \log(x + a)$.   Ans. $-\dfrac{[(x + a)^2 + 1]^{\frac{3}{2}}}{x + a}$.

6. $y = \log \sec x$.   Ans. $\sec x$.

7. $x^{\frac{1}{2}} + y^{\frac{1}{2}} = 2\,c^{\frac{1}{2}}$.   Ans. $c^{-\frac{1}{2}}(x + y)^{\frac{3}{2}}$.

8. What is the curvature of the probability curve $y = ke^{-ax^2}$ at the point $(0, k)$?   Ans. $-2\,ak$.

9. At what point has the exponential curve $y = e^x$ the greatest curvature?   Ans. $(\tfrac{1}{2}\log\tfrac{1}{2},\ 2^{-\frac{1}{2}})$.

10. Show that the parabola has its greatest curvature at the vertex.

11. Determine the points at which the curvature of an ellipse is (1) the greatest, (2) the least.

12. Show that the radius of curvature of the curve $xy^2 = a^2(a - x)$ at the point $(a, 0)$ is numerically equal to $a/2$.

13. Show that the center of the circle of curvature of the hyperbola $b^2x^2 - a^2y^2 = a^2b^2$ at the point (x, y) has the coördinates
$$\alpha = \frac{(a^2 + b^2)x^3}{a^4}, \quad \beta = -\frac{(a^2 + b^2)y^3}{b^4}.$$

14. For the parabola $x^{\frac{1}{2}} + y^{\frac{1}{2}} = a^{\frac{1}{2}}$ show that we have the relation $\alpha + \beta = 3(x + y)$.

15. Find the coördinates of the center of curvature of the ellipse $b^2x^2 + a^2y^2 = a^2b^2$.

16. For the equilateral hyperbola $2\,xy = 1$ show that
$$\alpha + \beta = (y + x)^3, \quad \alpha - \beta = (y - x)^3.$$

135. Roulettes, involutes, and evolutes. Let the curve C be held fixed and a second curve C' be rolled on C without slipping. Then any point P on C' will describe a curve Γ. Such a curve Γ is called a *roulette*.

In the special case when the rolling curve C' is a straight line the roulette is called an *involute*. A given curve has an infinitude of involutes varying with the choice of the point P on the revolving straight line by means of which they are generated. If a flexible cord is wrapped on C and is then unwrapped, being kept taut and in the plane of C, it is clear that a fixed point on the free part of the cord will generate an arc of an involute of C.

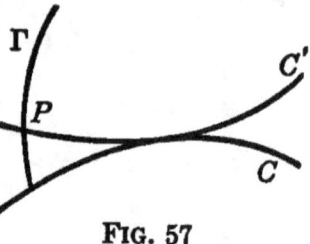

FIG. 57

Let (x, y) denote the general point A on the fixed curve C and (x_1, y_1) denote the corresponding point P on the involute. Let B be the point at which P is in contact with the curve C and let arc length s on C be measured from B. Then $AB = AP = s$. Hence

$$y - y_1 = s \sin \phi, \quad x - x_1 = s \cos \phi, \quad (1)$$

where $\phi = \angle DPA$. Therefore

$$y_1 = y - s \sin \phi, \quad x_1 = x - s \cos \phi.$$

Giving to x the increment Δx and to the other variables corresponding increments, we have

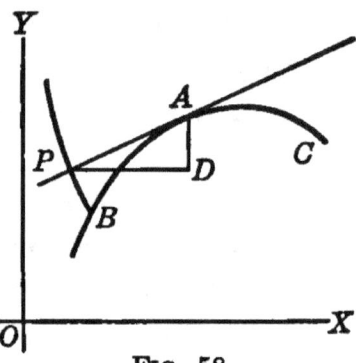

FIG. 58

$$y_1 + \Delta y_1 = y + \Delta y - (s + \Delta s) \sin (\phi + \Delta \phi),$$
$$x_1 + \Delta x_1 = x + \Delta x - (s + \Delta s) \cos (\phi + \Delta \phi).$$

Therefore

$$\Delta y_1 = \Delta y - s\{\sin (\phi + \Delta \phi) - \sin \phi\} - \Delta s \sin (\phi + \Delta \phi),$$
$$\Delta x_1 = \Delta x - s\{\cos (\phi + \Delta \phi) - \cos \phi\} - \Delta s \cos (\phi + \Delta \phi).$$

Taking the quotients of these equations member by member and dividing the terms of the fraction in the resulting second member by Δx, we have

$$\frac{\Delta y_1}{\Delta x_1} = \frac{\dfrac{\Delta y}{\Delta x} - s \dfrac{\sin (\phi + \Delta \phi) - \sin \phi}{\Delta \phi} \cdot \dfrac{\Delta \phi}{\Delta x} - \dfrac{\Delta s}{\Delta x} \sin (\phi + \Delta \phi)}{1 - s \dfrac{\cos (\phi + \Delta \phi) - \cos \phi}{\Delta \phi} \cdot \dfrac{\Delta \phi}{\Delta x} - \dfrac{\Delta s}{\Delta x} \cos (\phi + \Delta \phi)}.$$

Taking the limits as the increments approach zero, we have

$$\frac{dy_1}{dx_1} = \frac{\dfrac{dy}{dx} - s\cos\phi\,\dfrac{d\phi}{dx} - \dfrac{ds}{dx}\sin\phi}{1 + s\sin\phi\,\dfrac{d\phi}{dx} - \dfrac{ds}{dx}\cos\phi}.$$

But we have seen (§§ 59 and 62) that

$$\sin\phi = \frac{dy}{ds}, \quad \cos\phi = \frac{dx}{ds}, \quad \frac{dy}{dx} = \tan\phi.$$

Hence $\quad \dfrac{dy}{dx} - \dfrac{ds}{dx}\sin\phi = 0, \quad 1 - \dfrac{ds}{dx}\cos\phi = 0.$

Therefore the previous value for dy_1/dx_1 reduces to

$$\frac{dy_1}{dx_1} = \frac{-s\cos\phi\,\dfrac{d\phi}{dx}}{s\sin\phi\,\dfrac{d\phi}{dx}} = -\frac{1}{\tan\phi} = -\frac{1}{\dfrac{dy}{dx}}. \tag{2}$$

Hence the slope of the involute is the negative reciprocal of the slope of the original curve C at the corresponding point. Therefore *the generating line AP is perpendicular to the involute at the point P*. In other words,

Any normal to an involute is tangent to the fixed curve.

From the manner in which an involute is generated it follows that

Any two involutes of the same curve are parallel in the sense that they have common normals and intercept a common distance on these normals.

From equations (1) and (2) we have

$$\frac{y - y_1}{x - x_1} = \tan\phi = \frac{dy}{dx} = -\frac{dx_1}{dy_1}.$$

Hence

$$(x - x_1) + (y - y_1)\frac{dy_1}{dx_1} = 0. \tag{3}$$

Differentiating with respect to x_1, we have
$$\frac{dx}{dx_1} - 1 + \left(\frac{dy}{dx_1} - \frac{dy_1}{dx_1}\right)\frac{dy_1}{dx_1} + (y - y_1)\frac{d^2y_1}{dx_1^2} = 0.$$

Employing from (2) the fact that $dx/dy = - dy_1/dx_1$ we may reduce this to the following:
$$-1 - \left(\frac{dy_1}{dx_1}\right)^2 + (y - y_1)\frac{d^2y_1}{dx_1^2} = 0.$$

Hence
$$y = y_1 + \frac{1 + \left(\frac{dy_1}{dx_1}\right)^2}{\frac{d^2y_1}{dx_1^2}}$$

Thence from (3) we have
$$x = x_1 - \frac{\frac{dy_1}{dx_1}\left[1 + \left(\frac{dy_1}{dx_1}\right)^2\right]}{\frac{d^2y_1}{dx_1^2}}.$$

Comparing the last two equations with the formulas for the center of curvature of the curve whose general point is (x_1, y_1), we see that this center of curvature is the corresponding point (x, y) on the given curve C. Hence *the given curve C is the locus of centers of curvature of any one of its involutes.*

The locus of centers of curvature of any given curve is called the *evolute* of that curve. In § 134 we gave formulas for the coördinates α and β of the center of curvature of a given curve in terms of the coördinates x_1 and y_1 of the point on the curve. If y_1 is replaced in these formulas by its value in terms of x_1, the formulas become the parametric equations of the evolute, the parameter being x_1. Then the running coördinates are α and β. Or, by eliminating x_1 and y_1 from the equation $y_1 = f(x_1)$ and the equations giving the values of α and β we may obtain the equation of the evolute in terms of α and β.

We have seen that a given curve has an infinitude of involutes, and that the given curve is itself the evolute of each of the involutes. It can also be shown that every given curve is an involute of its evolute.

EXERCISES

Find the equations of the evolutes of the following curves:

1. $\dfrac{x^2}{a^2} - \dfrac{y^2}{b^2} = 1.$ Ans. $(a\alpha)^{\frac{2}{3}} - (b\beta)^{\frac{2}{3}} = (a^2 + b^2)^{\frac{2}{3}}.$

2. $\dfrac{x^2}{a^2} + \dfrac{y^2}{b^2} = 1.$ Ans. $(a\alpha)^{\frac{2}{3}} + (b\beta)^{\frac{2}{3}} = (a^2 - b^2)^{\frac{2}{3}}.$

3. $y^2 = 4\,px.$ Ans. $4(\alpha - 2\,p)^3 = 27\,p\beta^2.$

4. The *cycloid* is the roulette described by a point P on the circumference of a circle which rolls on a straight line. Take this line as the x-axis and a line perpendicular to it at O as the y-axis. Let the circle start from a position in which P coincides with O. Let θ denote the angle PCM through which the circle has turned when P reaches the point (x, y), as in the figure.

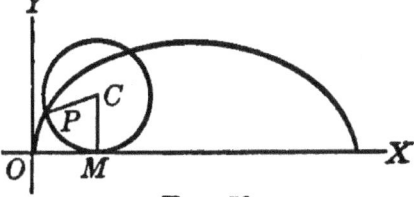

Fig. 59

Thence derive the following parametric equations for the cycloid:
$$x = a(\theta - \sin \theta), \quad y = a(1 - \cos \theta),$$
a being the radius of the rolling circle.

5. Show that the parametric equations of the evolute of the cycloid are
$$\alpha = a(\theta + \sin \theta), \quad \beta = -a(1 - \cos \theta).$$

6. Show that the evolute of a cycloid is itself a cycloid whose generating circle is equal to that of the given cycloid.

7. Show that the normal at any point on a cycloid is equal to half the radius of curvature at the same point.

8. The hypocycloid of four cusps is the roulette described by a point on a circle of radius $\frac{1}{4}a$ rolling on the interior of a fixed circle of radius a. Taking the axes through the cusps so that they intersect at the center of the fixed circle, show that the equation of the hypocycloid is
$$x^{\frac{2}{3}} + y^{\frac{2}{3}} = a^{\frac{2}{3}}.$$

CHAPTER XVII

APPLICATIONS TO GEOMETRY AND MECHANICS

136. Plane areas in polar coördinates. If the equation of a plane curve is written in polar coördinates, the area considered is the area swept over by a radius vector in passing from an initial to a final position. A formula for finding this area is developed as follows.

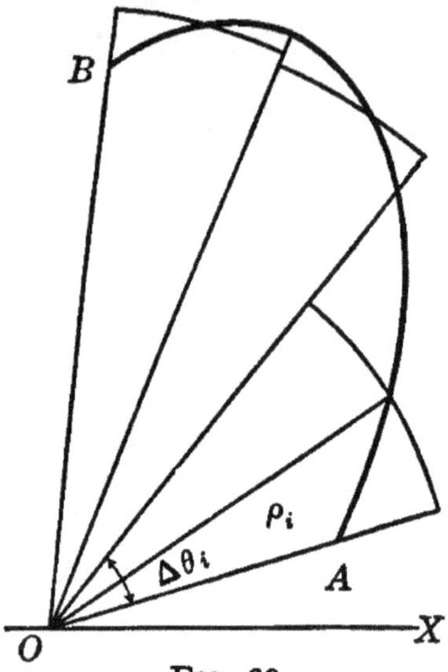

Let $\rho = f(\theta)$ be the equation of the curve in the adjoining figure. Let AOB be the area A swept over by the radius vector from its initial position OA to its final position OB. Let $\angle XOA = \theta_1$ and $\angle XOB = \theta_2$. Let $\angle AOB$ be divided into n parts $\Delta \theta_i$ and let ρ_i be a radius vector lying in the angle $\Delta \theta_i$. Then with O as center and radius ρ_i draw arcs of circles cutting the sides of the angles $\Delta \theta_i$. There will thus be formed n sectors of circles. The area of each sector will be $\frac{1}{2}\rho_i^2 \Delta \theta_i$.* The sum of the areas thus formed is

$$\sum_1^n \tfrac{1}{2} \rho_i^2 \Delta \theta_i.$$

Fig. 60

Now let n increase indefinitely in such a way that every $\Delta \theta_i$ approaches zero as a limit. Then, by the method of § 31, we have

$$\text{Area } AOB = \lim_{n=\infty} \sum_1^n \tfrac{1}{2} \rho_i^2 \Delta \theta_i.$$

---

*It is shown in plane geometry that the area of a sector of a circle equals one half the radius times the arc.

Also, by § 32, we have
$$\lim_{n=\infty} \sum_{1}^{n} \tfrac{1}{2}\rho_i{}^2 \Delta\theta_i = \int_{\theta_1}^{\theta_2} \tfrac{1}{2}\rho^2\,d\theta.$$

Therefore
$$A = \tfrac{1}{2}\int_{\theta_1}^{\theta_2} \rho^2\,d\theta.$$

Example 1. Find the area A swept out by the radius vector of the spiral of Archimedes $\rho = a\theta$ from $\theta = 0$ to $\theta = \pi$.

$$A = \tfrac{1}{2}\int_0^{\pi} \rho^2\,d\theta = \tfrac{1}{2}\int_0^{\pi} a^2\theta^2\,d\theta = \tfrac{1}{6}\pi^3 a^2.$$

Example 2. Find the area A swept out by the radius vector of the curve $\rho = \sec\theta$ from $\theta = 0$ to $\theta = \pi/4$.

$$A = \tfrac{1}{2}\int_0^{\frac{\pi}{4}} \sec^2\theta\,d\theta = \tfrac{1}{2}.$$

REMARK. The student now has at his command two formulas for finding areas bounded by curves, the one of the present section and the one given in § 99. If the equation of the curve is taken in rectangular coördinates the formula of § 99 is used. If the equation is in polar coördinates the formula of the present section applies. In certain problems, however, the student must choose his own coordinate system, and a proper choice will very materially simplify his labor. The following example will illustrate this fact.

Example 3. Find the area of a circle of radius r.

1. Let the center of the circle be the origin of coördinates. The equation of the circle in rectangular coördinates is then $x^2 + y^2 = r^2$. Its area is given by the formula

$$A = 4\int_0^r y\,dx = 4\int_0^r \sqrt{r^2 - x^2}\,dx = 4\left[\frac{x}{2}\sqrt{r^2 - x^2} + \frac{r^2}{2}\sin^{-1}\frac{x}{r}\right]_0^r = \pi r^2.$$

2. If the center of the circle is taken as the origin of coördinates, the equation of the circle in polar coördinates is $\rho = r$.

The formula for area in this case yields the relations

$$A = \tfrac{1}{2}\int_0^{2\pi} \rho^2\,d\theta = \tfrac{1}{2}\int_0^{2\pi} r^2\,d\theta = \pi r^2.$$

The integration in 2 is much simpler than the integration in 1. Hence it is preferable in this case to use polar coördinates. In other examples it might be better to use rectangular coördinates. The student should practice using both formulas on such examples, in order to develop his judgment as to which one would likely be the better coördinate system for a particular problem.

137. Lengths of curves in polar coördinates. It has been shown (§ 63) that the differential of arc length in polar coördinates is

$$ds = \sqrt{\rho^2 + \left(\frac{d\rho}{d\theta}\right)^2}\, d\theta.$$

Therefore
$$s = \int_{\theta_1}^{\theta_2} \sqrt{\rho^2 + \left(\frac{d\rho}{d\theta}\right)^2}\, d\theta,$$

which is the required formula.

Example. Find the entire length of the curve $\rho = a \sin^3(\theta/3)$.

Here $\dfrac{d\rho}{d\theta} = a \sin^2 \dfrac{\theta}{3} \cos \dfrac{\theta}{3}.$

Hence
$$s = 2a \int_0^{\frac{3\pi}{2}} \sqrt{\sin^6 \frac{\theta}{3} + \sin^4 \frac{\theta}{3} \cos^2 \frac{\theta}{3}}\, d\theta,$$

$$= 2a \int_0^{\frac{3\pi}{2}} \sqrt{\sin^4 \frac{\theta}{3}}\, d\theta,$$

$$= 2a \int_0^{\frac{3\pi}{2}} \sin^2 \frac{\theta}{3}\, d\theta,$$

$$= \frac{3\pi a}{2}.$$

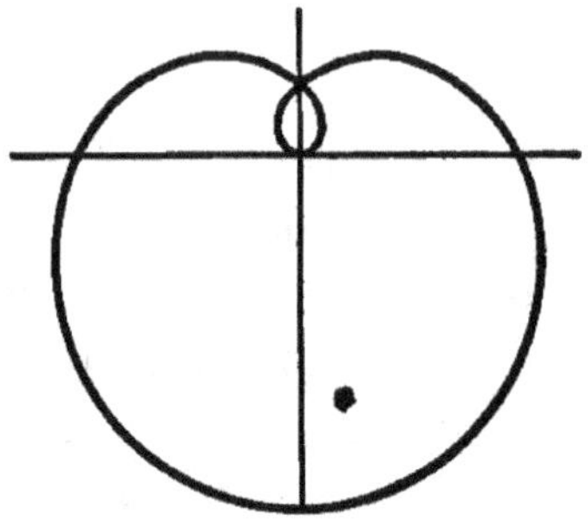

FIG. 61

EXERCISES

1. Find the area bounded by the lemniscate $\rho^2 = a^2 \cos 2\theta$.

   Ans. a^2.

2. Find the area swept out by the radius vector of the logarithmic spiral $\log \rho = a\theta$ from $\theta = 0$ to $\theta = \pi$.

   Ans. $\dfrac{e^{2\pi a} - 1}{4a}$.

3. Find the area bounded by the cardioid $\rho = a(1 + \cos \theta)$.

   Ans. $\tfrac{3}{2}\pi a^2$.

4. Find the area of one loop of the curve $\rho = a \sin 2\theta$.

   Ans. $\dfrac{\pi a^2}{8}$.

5. Find the area swept out by the radius vector of the parabola $\rho = a \sec^2 \theta/2$ from $\theta = 0$ to $\theta = \pi/2$.

Ans. $\dfrac{4 a^2}{3}$.

6. Find the length of the cardioid $\rho = 2 a(1 + \cos \theta)$.

Ans. $16 a$.

7. Find the length of the spiral $\rho = e^{a\theta}$ from $\theta = 0$ to $\theta = \pi/2$.

Ans. $\dfrac{1}{a}\sqrt{1 + a^2}\left(e^{\frac{a\pi}{2}} - 1\right)$.

8. Find the length of the curve $\rho = a \sec \theta$ from $\theta = 0$ to $\theta = \pi/4$.

Ans. a.

9. Find the length of the circumference of a circle with radius a, using (1) rectangular coördinates, (2) polar coördinates.

Ans. $2 \pi a$.

10. The equations $x = \rho \cos \theta$, $y = \rho \sin \theta$, and $x^2 + y^2 = \rho^2$ express the relations existing between polar and rectangular coördinates. Using these relations show that the equation of the parabola in Ex. 5 in rectangular coördinates is $y^2 = 4 a^2 - 4 ax$, and from this equation find the required area.

11. The equation of a straight line in rectangular coördinates is $x + y = 2 a$. Show that its equation in polar coördinates is

$$\rho = \frac{2 a \sec \theta}{1 + \tan \theta}.$$

Find the area bounded by the line and the rectangular coördinate axes, using (1) polar coördinates, (2) rectangular coördinates.

Ans. $2 a^2$.

12. Find the area of the loop of the folium of Descartes

$$\rho = \frac{3 a \tan \theta \sec \theta}{1 + \tan^3 \theta}. \qquad Ans.\ \frac{3 a^2}{2}.$$

138. Volumes of solids of revolution. Let $y = f(x)$ be the equation of a plane curve continuous on the interval from $x = a$ to $x = b$. The area bounded by $y = f(x)$ and the ordinates $x = a$ and $x = b$ will, if revolved about the x-axis, generate a solid of revolution. A formula for finding this volume may be derived as follows:

Let the interval between $x = a$ and $x = b$ be divided into n parts Δx_i. Let y_i be an ordinate of $y = f(x)$ on the interval Δx_i, and let the rectangles having y_i for altitudes and Δx_i for bases be formed. Each of these rectangles will generate a cylinder of radius y_i and altitude Δx_i. Its volume is therefore

$$\pi y_i^2 \Delta x_i,$$

and the sum $\sum_1^n \pi y_i^2 \Delta x_i$

is an approximation to the required volume.

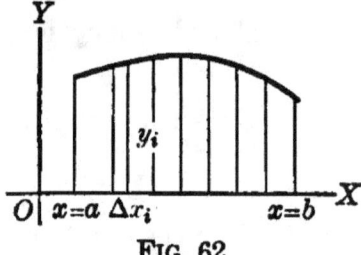

FIG. 62

If now n is allowed to increase indefinitely in such a way that every Δx_i approaches zero as a limit, the required volume V (§ 31) is given by the relation

$$V = \lim_{n=\infty} \sum_1^n \pi y_i^2 \Delta x_i.$$

Also, by § 32, we have

$$\lim_{n=\infty} \sum_1^n \pi y_i^2 \Delta x_i = \pi \int_a^b y^2 \, dx.$$

Therefore the required formula for V is the following:

$$V = \pi \int_a^b y^2 \, dx.$$

Similarly it may be shown that if the area is revolved about the y-axis the volume is given by the formula

$$V = \pi \int_c^d x^2 \, dy,$$

where c and d are the values of y corresponding respectively to the values a and b of x.

Example. Find the volume generated by revolving the parabola $y^2 = 4x$ about the x-axis from $x = 0$ to $x = 3$.

$$V = \pi \int_0^3 4x \, dx = \pi \Big[2 x^2\Big]_0^3 = 18 \pi.$$

EXERCISES

1. Find the volume generated by revolving the area bounded by the line $2x + y = 8$ and the coördinate axes (1) about the x-axis; (2) about the y-axis.

Ans. (1) $\dfrac{256 \pi}{3}$; (2) $\dfrac{128 \pi}{3}$.

2. Find the volume of a sphere as a volume of revolution.

Ans. $\frac{4}{3}\pi r^3$.

3. Find the volume generated by revolving about the x-axis the plane area bounded by the x-axis, the line $x = a$, and the cissoid

$$y^2 = \frac{x^3}{2a - x}.$$

Ans. $\pi\left(8a^3 \log 2 - \frac{16a^3}{3}\right)$.

4. Find the volume of the solid generated by revolving about the x-axis the part of the parabola $x^{\frac{1}{2}} + y^{\frac{1}{2}} = a^{\frac{1}{2}}$ intercepted by the axes.

Ans. $\dfrac{\pi a^3}{15}$.

5. Find the volume generated by revolving the ellipse $\dfrac{x^2}{a^2} + \dfrac{y^2}{b^2} = 1$ (1) about the x-axis; (2) about the y-axis.

Ans. (1) $\frac{4}{3}\pi ab^2$; (2) $\frac{4}{3}\pi a^2 b$.

6. The segment of the parabola $x^2 - 3x + 2y = 0$ above the x-axis is revolved about the x-axis. Find the volume generated.

Ans. $\dfrac{81\pi}{40}$.

7. Find the entire volume generated by revolving about the x-axis the hypocycloid $x^{\frac{2}{3}} + y^{\frac{2}{3}} = a^{\frac{2}{3}}$.   Ans. $\frac{32}{105}\pi a^3$.

8. The area under one arch of the sine curve is revolved about the x-axis. Find the volume generated.

Ans. $\dfrac{\pi^2}{2}$.

9. The curve $y^2 = x(x-1)(x-2)$ rotates about the x-axis. Find the closed volume generated.

Ans. $\dfrac{\pi}{4}$.

10. A round hole of radius a is bored through the center of a sphere of radius $2a$. Find the volume of the hole cut out.

Ans. $\frac{4}{3}\pi a^3(8 - 3\sqrt{3})$.

11. Find the volume generated by revolving about the x-axis one loop of the curve $a^4 y^2 = a^2 x^4 - x^6$.

Ans. $\dfrac{2\pi a^3}{35}$.

139. Surfaces of revolution, rectangular coördinates. Let $y = f(x)$ be the equation of a plane curve. If the portion of this curve between the ordinates $x = a$ and $x = b$ is revolved about the x-axis it will generate a surface. To find the area of this surface let the arc between $x = a$ and $x = b$ be divided into n parts and let the chords of these n arcs be drawn. If the ith chord is revolved about the x-axis it will generate the convex surface of the frustum of a cone. Let the length of the ith chord be Δc_i and the length of the ordinate to its midpoint be y_i. The area of the surface generated will be

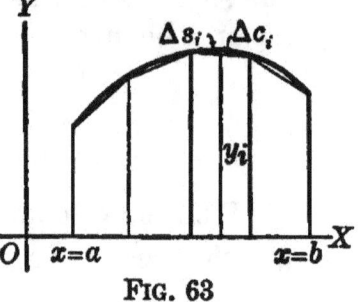

FIG. 63

$$2 \pi y_i \Delta c_i,$$

and the total surface generated by the n chords will be

$$\sum_1^n 2 \pi y_i \Delta c_i = \sum_1^n 2 \pi y_i \frac{\Delta c_i}{\Delta s_i} \cdot \Delta s_i,$$

where Δs_i is the arc subtended by the chord Δc_i.

If n is allowed to increase indefinitely in such a way that every Δc_i approaches zero, then, by § 31,

$$\lim_{n \to \infty} \sum_1^n 2 \pi y_i \frac{\Delta c_i}{\Delta s_i} \cdot \Delta s_i = \text{surface generated by the curve } y = f(x)$$

between $x = a$ and $x = b$.

Obviously,
$$\lim_{\Delta c_i \to 0} \frac{\Delta c_i}{\Delta s_i} = 1.$$

Therefore, by § 32,

$$\lim_{n \to \infty} \sum_1^n 2 \pi y_i \frac{\Delta c_i}{\Delta s_i} \cdot \Delta s_i = 2 \pi \int y \, ds,$$

with the proper limits inserted.

Hence the surface S is given by

$$S = 2 \pi \int_a^b y \sqrt{1 + \left(\frac{dy}{dx}\right)^2} \, dx; \tag{1}$$

since
$$ds = \sqrt{1 + \left(\frac{dy}{dx}\right)^2} \, dx.$$

Similarly, if the curve is revolved about the y-axis the surface is given by the formula

$$S = 2 \pi \int_a^b x \sqrt{1 + \left(\frac{dy}{dx}\right)^2} \, dx. \tag{2}$$

The student should observe that there are two forms for ds, namely,
$$ds = \sqrt{1 + \left(\frac{dy}{dx}\right)^2}\, dx \quad \text{and} \quad ds = \sqrt{1 + \left(\frac{dx}{dy}\right)^2}\, dy.$$
The proper choice of the form to use in a particular problem will often simplify the integration.

Example. Find the area generated by revolving the circle $x^2 + y^2 = a^2$ about the x-axis.

We have
$$\frac{dy}{dx} = -\frac{x}{y}.$$

Therefore $S = 4\pi \int_0^a y\sqrt{1 + \frac{x^2}{y^2}}\, dx = 4\pi \int_0^a a\, dx = 4\pi a^2.$

140. Surfaces of revolution, polar coördinates. From §139 we have
$$S = 2\pi \int y\, ds. \tag{1}$$
with the proper limits inserted.

In polar coördinates $y = \rho \sin\theta$, and $ds = \sqrt{\rho^2 + \left(\frac{d\rho}{d\theta}\right)^2}\, d\theta$. Substituting these values in (1), we obtain
$$S = 2\pi \int_{\theta_1}^{\theta_2} \rho \sin\theta \sqrt{\rho^2 + \left(\frac{d\rho}{d\theta}\right)^2}\, d\theta, \tag{2}$$
which is the required formula when the curve $\rho = f(\theta)$ is revolved about the initial line.

Example. Find the surface generated by revolving the curve $\rho = e^\theta$ about the initial line from $\theta = 0$ to $\theta = \pi/2$.

Here $\qquad \rho = e^\theta \quad \text{and} \quad \dfrac{d\rho}{d\theta} = e^\theta.$

Hence (2) becomes
$$S = 2\pi \int_0^{\frac{\pi}{2}} e^\theta \sin\theta \sqrt{e^{2\theta} + e^{2\theta}}\, d\theta$$
$$= 2\sqrt{2}\,\pi \int_0^{\frac{\pi}{2}} e^{2\theta} \sin\theta\, d\theta. \tag{3}$$

This form must be integrated by parts.

Let $\qquad u = e^{2\theta}, \qquad dv = \sin\theta\, d\theta.$
Then $\qquad du = 2e^{2\theta}\, d\theta, \qquad v = -\cos\theta.$
Therefore
$$S = 2\sqrt{2}\,\pi \left[\left(-e^{2\theta}\cos\theta\right)_0^{\frac{\pi}{2}} + 2\int_0^{\frac{\pi}{2}} e^{2\theta}\cos\theta\, d\theta\right].$$

Integrating by parts again, we have

$$S = 2\sqrt{2}\,\pi\left[\left(-e^{2\theta}\cos\theta + 2\,e^{2\theta}\sin\theta\right)\Big|_0^{\frac{\pi}{2}} - 4\int_0^{\frac{\pi}{2}} e^{2\theta}\sin\theta\, d\theta\right]$$

$$= 2\sqrt{2}\,\pi\left[-e^{2\theta}\cos\theta + 2\,e^{2\theta}\sin\theta\right]_0^{\frac{\pi}{2}} - 2\sqrt{2}\,\pi\left(4\int_0^{\frac{\pi}{2}} e^{2\theta}\sin\theta\, d\theta\right).$$

But $\quad\quad\quad 4(2\sqrt{2}\,\pi)\int_0^{\frac{\pi}{2}} e^{2\theta}\sin\theta\, d\theta = 4\,S.\quad\quad$ [See (3).]

Therefore

$$S = \frac{2\sqrt{2}\,\pi}{5}\left[-e^{2\theta}\cos\theta + 2\,e^{2\theta}\sin\theta\right]_0^{\frac{\pi}{2}} = \frac{2\sqrt{2}\,\pi}{5}(2\,e^{\pi} + 1).$$

EXERCISES

1. Find the area of the surface generated by revolving about the x-axis the portion of the line $3x + y = 6$ cut off by the coördinate axes. *Ans.* $12\,\pi\sqrt{10}$.

2. Find the area if the line in the foregoing exercise is revolved about the y-axis. *Ans.* $4\,\pi\sqrt{10}$.

3. Find the area of the surface generated by revolving the circle $x^2 + y^2 = a^2$ about the x-axis. *Ans.* $4\,\pi a^2$.

4. Find the area generated by revolving the curve $x^{\frac{2}{3}} + y^{\frac{2}{3}} = a^{\frac{2}{3}}$ about the y-axis. *Ans.* $\frac{12}{5}\,\pi a^2$.

5. Find the area of the surface generated if the parabola $2x = y^2$ is revolved about the x-axis between the limits $y = 0$ and $y = 1$.

Ans. $\dfrac{2\,\pi}{3}(2\sqrt{2}-1)$.

6. Find the area generated by revolving about the x-axis one arch of the cycloid $x = a(\theta - \sin\theta)$, $y = a(1 - \cos\theta)$. *Ans.* $\frac{64}{3}\,\pi a^2$.

7. Find the area generated by revolving the cubical parabola $a^2 y = x^3$ about the x-axis from $x = 0$ to $x = a$.

Ans. $\dfrac{\pi a^2}{27}(10\sqrt{10}-1)$.

8. Find the area which is generated by revolving the ellipse $b^2 x^2 + a^2 y^2 = a^2 b^2$ about the x-axis.

Ans. $\dfrac{2\,\pi a^2 b}{\sqrt{a^2-b^2}}\sin^{-1}\dfrac{\sqrt{a^2-b^2}}{a} + 2\,\pi b^2$.

9. Find the area generated by revolving the curve $\rho = a\cos\theta$ about the initial line. *Ans.* πa^2.

10. Find the area generated by revolving the curve $\rho = \sin\theta$ about the initial line from $\theta = 0$ to $\theta = \pi/2$.

Ans. $\dfrac{\pi^2}{2}$.

11. Find the area generated by revolving the curve $\rho = \dfrac{a\sec\theta}{1+\tan\theta}$ about the initial line from $\theta = 0$ to $\theta = \pi/2$.

Ans. $\pi a^2\sqrt{2}$.

12. Find the area which is generated by revolving the cardioid $\rho = a(1 - \cos\theta)$ about the initial line.

Ans. $\dfrac{32\,\pi a^2}{5}$.

13. Find the area if the ellipse in Problem 8 is revolved about the y-axis.

14. Find the area if the curve in Problem 10 is revolved about the line $\theta = \pi/2$.

15. Find the area generated by revolving the curve $y = e^x$ about the x-axis from $x = 1$ to $x = 2$.

141. Force. It has been shown in physics that the force on one side of a horizontal surface of area A, at a depth h beneath the surface of a liquid, is WAh, where W is the weight of a cubic unit of the liquid. Since force is transmitted equally in all directions, this fact may be used to calculate the force on a vertical surface.

Let ABC be such a surface (see Fig. 64). Let ABC be divided into horizontal strips of width Δh_i and let h_i be the distance from the surface of the liquid to a point P of the strip Δh_i. Also let l_i be the length of a horizontal line through P and terminated by the boundary of ABC.

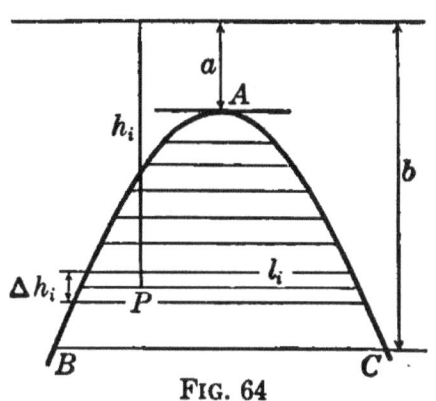

Fig. 64

With this notation the force on the strip Δh_i will be approximately
$$Wl_i h_i \Delta h_i,$$

and the total force on ABC will be approximately

$$\sum_{1}^{n} W l_i h_i \Delta h_i.$$

Whence, by applying the principles of §§ 31 and 32, it is seen that the force P is given by the formula

$$P = \int_a^b Wlh\, dh,$$

where l is a constant or a function of h.

Example 1. Find the force on a vertical rectangular flood gate 4 ft. long and 2 ft. wide, the top of which is 3 ft. below the surface of the water.

In this case

$W = 62.5, \quad l = 4, \quad a = 3, \quad b = 5.$

Hence $P = 62.5 \int_3^5 4\, h\, dh = 2000$ lb.

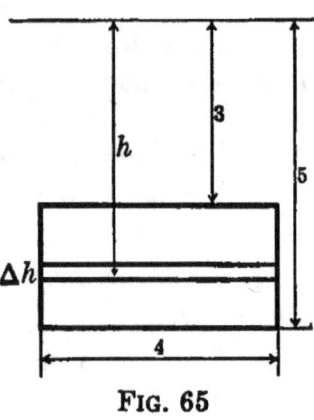

FIG. 65

Example 2. Find the force on a vertical parabolic segment which is 8 ft. across the top and 4 ft. long, with the top at the surface of the water.

Take O as the origin, with the axes as shown in the figure. The equation of the parabola is then $x^2 = 4y$. Consider one half of the area. Then $h = 4 - y$, $l = 2\sqrt{y}$. Therefore the force on one half of the area is given by

$$P = 125 \int_0^4 \sqrt{y}(4-y)dy = 1066\tfrac{2}{3} \text{ lb.,}$$

and the total force is double the foregoing amount.

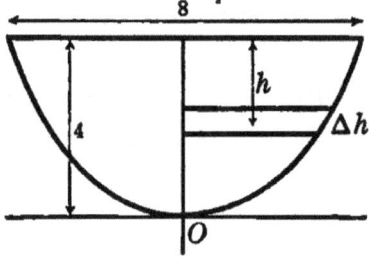

FIG. 66

EXERCISES

1. A vertical rectangular flood gate is 12 ft. long and 8 ft. high. Find the force on it if the upper edge of the gate is submerged 4 ft.
   Ans. 24 tons.

2. A vertical storage dam is in the form of a trapezoid 100 ft. long on the top, 60 ft. long on the bottom, and 20 ft. high. Find the force it must withstand. *Ans.* $458\tfrac{1}{3}$ tons.

3. Find the force on a vertical triangle with base 4 ft. and altitude 2 ft.

   (1) When the base is in the surface of the water.

   (2) When the vertex is at the surface of the water and the base horizontal.

(3) When the base is horizontal and 3 ft. from the surface, and the vertex downward. *Ans.* (1) $166\frac{2}{3}$ lb.; (2) $333\frac{1}{3}$ lb.; (3) $916\frac{2}{3}$ lb.

4. A vessel is filled with equal volumes of water and oil. Its end is a vertical rectangle 4 ft. long and 3 ft. high. The specific gravity of the oil is .5. Find the force on the end. *Ans.* $703\frac{1}{8}$ lb.

5. A cross section of a trough is a parabola. The trough is 5 ft. across the top and 2 ft. deep. Find the force on the end if the trough is full of water and the end is vertical. *Ans.* $333\frac{1}{3}$ lb.

6. Find the force on the vertical ends of a horizontal cylindrical tank, radius a, if it is half full of water. *Ans.* $\dfrac{125\,a^3}{3}$ lb.

7. A vertical circular flood gate is 4 ft. in diameter and has its center 30 ft. below the surface of the water. Find the force on it.

8. A parabolic board 6 ft. across and 3 ft. high is immersed vertically in water so that its vertex lies in the surface. Find the force on it.

142. Work. When a particle is moved through a distance d by a constant force F, work is said to be performed on the particle and is defined quantitatively by the relation

$$W = Fd.$$

Suppose now that the force is variable over an interval from $x = a$ to $x = b$, and that the force is a function of x, as $f(x)$. Separate the interval from $x = a$ to $x = b$ into n parts Δx_i and let x_i be a value of x on the interval Δx_i. The work done in moving the particle over the interval Δx_i will be approximately

$$f(x_i)\Delta x_i,$$

and the sum $\qquad \displaystyle\sum_{1}^{n} f(x_i)\Delta x_i$

will be approximately the work done in moving the particle from $x = a$ to $x = b$. Consequently (§§ 31 and 32) the exact amount of work will be given by the formula

$$W = \int_a^b f(x)dx.$$

Example 1. Find the work done in stretching a spring 6 in., if it takes a force of 15 lb. to stretch it one inch.

By Hooke's law, when a spring is stretched, the stretch is directly proportional to the force applied. Hence if x is the amount that the spring is stretched, $F = kx$. And since $F = 15$ when $x = 1$, $k = 15$. Therefore $F = 15\, x$. Consequently the work done in stretching the spring is given by the formula

$$W = \int_0^6 15\, x\, dx = 270 \text{ inch-pounds.}$$

Example 2. Find the amount of work required to pump the water out of a cylindrical vertical reservoir 10 ft. deep and 4 ft. in diameter.

Consider a thin slice of the cylinder of thickness Δx_i made by planes perpendicular to the axis of the cylinder. Let x_i be the distance from the top of the cylinder to a point within the slice. The work done in lifting this slice to the top of the cylinder is approximately

$$4\, \pi k x_i\, \Delta x_i.$$

If now the cylinder is considered as being made up of n such slices, the work done in lifting all such slices to the top of the reservoir is approximately

$$\sum_1^n 4\, \pi k x_i\, \Delta x_i.$$

Hence the total work done is given by the formula

$$W = 4\, \pi k \int_0^{10} x\, dx = 12{,}500\, \pi \text{ foot-pounds,}$$

where $k = 62.5$.

Example 3. A gas expands in a cylinder against a piston from the volume v_0 to the volume v_1. Find the work done.

By Boyle's law $pv = k$, where p is the pressure per unit of area, v is the volume, and k is a constant.

Let A be the area of a cross section of the cylinder and let the piston be moved a distance Δx_i. The force on the cross section is Ap_i and the work done in moving the piston a distance Δx_i is

$$Ap_i\, \Delta x_i.$$

But $A\, \Delta x_i$ is the volume of expansion of the gas. Hence the work done is given by

$$p_i\, \Delta v_i = k \frac{\Delta v_i}{v_i}.$$

Consequently the total work is given by the formula

$$W = k \int_{v_0}^{v_1} \frac{dv}{v} = k \log \frac{v_1}{v_0} = p_0 v_0 \log \frac{v_1}{v_0}.$$

143. Summary. In this chapter the following topics have been discussed: plane areas in polar coördinates; lengths of curves in polar coördinates; volumes of solids of revolution; surfaces of revolution in rectangular and polar coördinates; force; and work. Special emphasis has been placed on observing when polar and when rectangular coördinates may be used to best advantage in a particular problem.

EXERCISES

1. Find the work done in pumping the water from a rectangular tank 15 ft. square and 10 ft. deep. *Ans.* 703,125 foot-pounds.

2. It takes 5 lb. to stretch a spring one inch. Calculate the work to stretch it 8 in. *Ans.* 160 inch-pounds.

3. A tank is in the form of a frustrum of a pyramid with square bases. The upper base is 10 ft. square, the lower base is 6 ft. square, and the depth is 5 ft. Find the amount of work required to empty it when filled with water.

4. Air confined under a force of 20 lb. per square inch has a volume of 75 cu. in. Find the amount of work done if the air expands to a volume of 200 cu. in. *Ans.* 1500 log $\frac{8}{3}$ inch-pounds.

5. If it takes a force of 50 lb. to stretch a spring 2 in., find the work done in stretching the spring from a length of 24 in. to a length of 36 in.

6. If force doing work on an object varies inversely as the square of the distance of the object from the point of application of the force, derive a formula for the work done in moving the object from a distance a feet to a distance b feet from the point of application of the force. *Ans.* $\dfrac{k(b-a)}{ab}$.

7. Air is confined in a vessel containing 30 cu. ft. under a force of 80 lb. per square inch. Calculate the work done when the air expands so that the pressure is 15 lb. per square inch.

8. Water is pumped from a cylindrical tank 20 ft. in diameter and 10 ft. deep to a point 30 ft. above the top of the tank. Find the work required to empty the tank.

9. When an electric current flows a distance x through a homogeneous conductor of cross section A, the resistance is kx/A, where

k is a constant depending on the material. Find the resistance when the current flows from the inner to the outer surface of a hollow sphere, the two radii being a and b.

Ans. $\dfrac{k(b-a)}{4\pi ab}$.

10. When an electric current i flows through an arc AB (see Fig. 65) it produces at any point O a magnetic force equal to

$$\frac{i\,d\theta}{r},$$

where r is the distance from AB to O. Find the force at the center of a circle due to a current i flowing once around it.

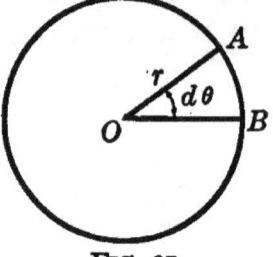

Fig. 67

CHAPTER XVIII

INTEGRATION OF SPECIAL CLASSES OF FUNCTIONS

144. Introduction. In Chapter X a number of methods were used to transform certain functions into functions which could be readily integrated. In the present chapter a further discussion of such transformations will be given.

The first functions to be considered will be in the form of rational fractions. By a rational fraction is meant the quotient of two polynomials, as

$$\frac{x^2 + 3x + 7}{x^4 - 4x^3 + 6x + 7}.$$

If the numerator is of degree equal to or higher than the denominator, the fraction may be reduced to a mixed expression by dividing the numerator by the denominator. For example,

$$\frac{x^3 + 4}{x^2 - 3x + 2} = x + 3 + \frac{7x - 2}{x^2 - 3x + 2}.$$

Then

$$\int \frac{x^3 + 4}{x^2 - 3x + 2} dx = \int (x + 3) dx + \int \frac{7x - 2}{x^2 - 3x + 2} dx.$$

The integral $\int (x+3) dx$ is easily calculated, and attention may be centered on the integral

$$\int \frac{7x - 2}{x^2 - 3x + 2} dx.$$

Such fractions are integrated by separating them into partial fractions, whose denominators are factors of the original denominator. The general discussion of partial fractions belongs to algebra. Only the form of these partial fractions and the methods of determining them will be considered here. Four cases of such fractions will be discussed.

1. *The prime factors of the denominator are all linear and none is repeated.*
2. *The prime factors of the denominator are all linear and some are repeated.*
3. *The denominator contains prime factors of the second degree, but none of them is repeated.*
4. *The denominator contains prime factors of the second degree and some of them are repeated.*

145. Case I. The prime factors of the denominator are all linear and none is repeated. The following example will illustrate this case:

Example. Evaluate the integral

$$\int \frac{x^3 + 6x^2 - 2x - 41}{x^3 + 4x^2 + x - 6} \, dx.$$

Since the numerator of the fraction in the integrand is a polynomial of the same degree as the denominator, the fraction is reduced to a mixed expression, giving

$$\int \left(1 + \frac{2x^2 - 3x - 35}{x^3 + 4x^2 + x - 6}\right) dx.$$

Attention may now be centered on the fraction

$$\frac{2x^2 - 3x - 35}{x^3 + 4x^2 + x - 6}.$$

Its denominator has the factors $x + 2$, $x + 3$, and $x - 1$. These are linear and are all different. When this is the case it is shown in algebra that we have a relation of the form

$$\frac{2x^2 - 3x - 35}{(x+2)(x+3)(x-1)} = \frac{A}{x+2} + \frac{B}{x+3} + \frac{C}{x-1},$$

where A, B, and C are constants.

Clearing of fractions, we obtain

$$2x^2 - 3x - 35 = A(x+3)(x-1) + B(x+2)(x-1) \quad (1)$$
$$+ C(x+2)(x+3).$$

Equating coefficients of like powers of x, we have

$$2 = A + B + C,$$
$$-3 = 2A + B + 5C,$$
$$-35 = -3A - 2B + 6C.$$

Solving this system of equations, we get
$$A = 7, \quad B = -2, \quad C = -3.$$
Therefore
$$\int \frac{x^3 + 6x^2 - 2x - 41}{x^3 + 4x^2 + x - 6} \, dx = \int \left(1 + \frac{7}{x+2} - \frac{2}{x+3} - \frac{3}{x-1}\right) dx$$
$$= x + 7 \log(x+2) - 2\log(x+3) - 3\log(x-1) + C$$
$$= x + \log \frac{(x+2)^7}{(x+3)^2(x-1)^3} + C.$$

REMARK. The values of A, B, and C may be found more readily in this case by the following method. Since equation (1) is an identity it will be true for all values of x. If in this equation we replace x by -2 we obtain $-21 = -3A$ or $A = 7$. Similarly, a replacement of x by -3 gives $-8 = 4B$, or $B = -2$, and a replacement of x by 1 gives $-36 = 12C$, or $C = -3$.

Although this method is simpler in Case I, it is not so readily applied to problems in Cases II, III, and IV, while the first method explained is of general application. It would probably be wise for the student to master thoroughly the first method before attempting the second.

EXERCISES

Integrate the following:

1. $\int \dfrac{3x+2}{x(x-1)} \, dx.$
   Ans. $\log \dfrac{(x-1)^5}{x^2} + C.$

2. $\int \dfrac{x+7}{(x+2)(x+3)} \, dx.$
   Ans. $\log \dfrac{(x+2)^5}{(x+3)^4} + C.$

3. $\int \dfrac{2 \, dx}{x^2 - 1}.$
   Ans. $\log \dfrac{x-1}{x+1} + C.$

4. $\int \dfrac{3x^3 + 8}{x^2 - 5x + 6} \, dx.$
   Ans. $\dfrac{3x^2}{2} + 15x + \log \dfrac{(x-3)^{89}}{(x-2)^{32}} + C.$

5. $\int \dfrac{-x \, dx}{2x^2 + 5x + 2}.$
   Ans. $\log \dfrac{(2x+1)^{\frac{1}{6}}}{(x+2)^{\frac{2}{3}}} + C.$

6. $\int \dfrac{1 - 2x}{2 - x - 3x^2} \, dx.$
   Ans. $\log (1+x)^{\frac{3}{5}}(2-3x)^{\frac{1}{15}} + C.$

7. $\int \dfrac{3x+4}{x^3 - x} \, dx.$
   Ans. $\log \dfrac{(x-1)^{\frac{7}{2}}(x+1)^{\frac{1}{2}}}{x^4} + C.$

8. $\int \dfrac{5\,dx}{x^3 - 3x^2 - 4x + 12}.$ Ans. $\log \dfrac{(x+2)^{\frac{1}{4}}(x-3)}{(x-2)^{\frac{5}{4}}} + C.$

9. $\int \dfrac{x^2 - 2}{x^3 - 2x^2 - x + 2}\,dx.$

10. $\int \dfrac{5x - 6}{(x-1)(x-2)(x-7)}\,dx.$

11. $\int \dfrac{ax + b}{(cx + d)(ex + f)}\,dx.$

12. $\int \dfrac{dx}{a^2 - x^2}.$

146. CASE II. The prime factors of the denominator are all linear and some are repeated. The following example will illustrate the method of handling such fractions:

Example. Evaluate the integral
$$\int \frac{x\,dx}{(x-1)^2(x+1)(x+2)}.$$

The prime factors of the denominator are $x-1$, $x+1$, and $x+2$. The factor $x-1$ is repeated. When such is the case it has been shown in algebra that we have a relation of the form

$$\frac{x}{(x-1)^2(x+1)(x+2)} = \frac{A}{(x-1)^2} + \frac{B}{x-1} + \frac{C}{x+1} + \frac{D}{x+2}.$$

Clearing the above equation of fractions, we have

$$x = A(x+1)(x+2) + B(x-1)(x+1)(x+2) + C(x-1)^2(x+2) + D(x-1)^2(x+1).$$

Equating coefficients of like powers of x, we obtain

$$\begin{aligned} B + C + D &= 0, \\ A + 2B - D &= 0, \\ 3A - B - 3C - D &= 1, \\ 2A - 2B + 2C + D &= 0. \end{aligned} \qquad (1)$$

Solving equations (1), we obtain,
$$A = \tfrac{1}{6};\quad B = \tfrac{1}{36};\quad C = -\tfrac{1}{4};\quad D = \tfrac{2}{9}.$$

Consequently

$$\int \frac{x\,dx}{(x-1)^2(x+1)(x+2)} = \frac{1}{6}\int \frac{dx}{(x-1)^2} + \frac{1}{36}\int \frac{dx}{x-1} - \frac{1}{4}\int \frac{dx}{x+1} + \frac{2}{9}\int \frac{dx}{x+2}$$

$$= -\frac{1}{6(x-1)} + \frac{1}{36}\log \frac{(x-1)(x+2)^8}{(x+1)^9} + C.$$

EXERCISES

Evaluate the following integrals:

1. $\int \dfrac{3x\,dx}{(x-1)^2(x+1)}.$ Ans. $-\dfrac{3}{2(x-1)}+\dfrac{3}{4}\log\dfrac{x-1}{x+1}+C.$

2. $\int \dfrac{x^2-4x}{x^2(x^2-1)}\,dx.$ Ans. $\log\dfrac{x^4}{(x-1)^{\frac{3}{2}}(x+1)^{\frac{5}{2}}}+C.$

3. $\int \dfrac{3x^2+6x+2}{x^3+4x^2}\,dx.$ Ans. $-\dfrac{1}{2x}+\dfrac{1}{8}\log x^{11}(x+4)^{13}+C.$

4. $\int \dfrac{x^2-2x+7}{x^3-3x^2+3x-1}\,dx.$ Ans. $-\dfrac{3}{(x-1)^2}+\log(x-1)+C.$

5. $\int \dfrac{6\,dx}{x^3-3x+2}.$ Ans. $-\dfrac{2}{x-1}+\dfrac{2}{3}\log\dfrac{x+2}{x-1}+C.$

6. $\int \dfrac{4x^2+5}{x^3+x^2-8x-12}\,dx.$

   Ans. $\dfrac{21}{5(x+2)}+\dfrac{1}{25}\log(x+2)^{59}(x-3)^{41}+C.$

7. $\int \dfrac{x^3+3}{x^4+4x^2}\,dx.$

8. $\int \dfrac{3x+7}{(x-2)^2(x+3)}\,dx.$

9. $\int \dfrac{-4x+2}{x^3-4x}\,dx.$

10. $\int \dfrac{3x-7}{4x^3+4x^2+x}\,dx.$

11. $\int \dfrac{11x^2-3}{x^4-2x^2+1}\,dx.$

12. $\int \dfrac{8x^3-3}{x^3-4x^2-3x+18}\,dx.$

147. CASE III. The denominator contains prime factors of the second degree none of which is repeated. Consider the following example:

Example 1. Evaluate the integral

$$\int \dfrac{2\,dx}{(x-1)(x^2+1)}.$$

The factors of the denominator are $x-1$ and x^2+1. One of these factors is of the second degree, and the partial fraction corresponding to this factor is of the form

$$\dfrac{Bx+C}{x^2+1},$$

where B and C are constants. Consequently we write

$$\frac{2}{(x-1)(x^2+1)} = \frac{A}{x-1} + \frac{Bx+C}{x^2+1}.$$

Clearing this equation of fractions, we have

$$2 = A(x^2+1) + (Bx+C)(x-1).$$

Equating coefficients of like powers of x, we obtain

$$A+B=0, \quad -B+C=0, \quad A-C=2.$$

Hence $\quad A=1, \quad B=-1, \quad C=-1.$

Therefore $\displaystyle\int \frac{2\,dx}{(x-1)(x^2+1)} = \int \frac{dx}{x-1} - \int \frac{x+1}{x^2+1}\,dx$

$= \log(x-1) - \tfrac{1}{2}\log(x^2+1) - \tan^{-1}x + C.$

Example 2. Evaluate the integral

$$\int \frac{x^2-2x}{x^3-1}\,dx.$$

The factors of the denominator are $x-1$ and x^2+x+1.

Hence $\displaystyle\frac{x^2-2x}{x^3-1} = \frac{A}{x-1} + \frac{Bx+C}{x^2+x+1}.$

Clearing of fractions, we obtain

$$x^2 - 2x = A(x^2+x+1) + (Bx+C)(x-1).$$

Equating coefficients of like powers of x, we have

$$A+B=1, \quad A-B+C=-2, \quad A-C=0.$$

Therefore $\quad A = -\tfrac{1}{3}, \quad B = \tfrac{4}{3}, \quad C = -\tfrac{1}{3},$
and

$\displaystyle\int\frac{x^2-2x}{x^3-1}\,dx = -\frac{1}{3}\int\frac{dx}{x-1} + \frac{1}{3}\int\frac{4x-1}{x^2+x+1}\,dx$

$\displaystyle = -\frac{1}{3}\int\frac{dx}{x-1} + \frac{1}{3}\int\frac{4x+2}{x^2+x+1}\,dx - \int\frac{dx}{x^2+x+1}$

$\displaystyle = -\frac{1}{3}\log(x-1) + \frac{2}{3}\log(x^2+x+1)$

$\displaystyle \qquad - \frac{2}{\sqrt{3}}\tan^{-1}\left(\frac{2x+1}{\sqrt{3}}\right) + C.$

EXERCISES

Evaluate the following integrals:

1. $\int \dfrac{2x+1}{x^3+x}\, dx.$ Ans. $-\log \dfrac{(x^2+1)^{\frac{1}{2}}}{x} + 2\tan^{-1} x + C.$

2. $\int \dfrac{x^2-7}{x^4-1}\, dx.$ Ans. $\dfrac{3}{2}\log \dfrac{x+1}{x-1} + 4\tan^{-1} x + C.$

3. $\int \dfrac{8x-3}{x^3-8}\, dx.$

   Ans. $\dfrac{1}{24}\log \dfrac{(x-2)^{26}}{(x^2+2x+4)^{13}} + \dfrac{57\sqrt{3}}{36}\tan^{-1}\!\left(\dfrac{x+1}{\sqrt{3}}\right) + C.$

4. $\int \dfrac{7x^2\, dx}{(x-1)^2(x^2+x+2)}.$

   Ans. $-\dfrac{7}{4(x-1)} + \dfrac{1}{32}\log\dfrac{(x-1)^{70}}{(x^2+x+2)^{35}} + \dfrac{9\sqrt{7}}{16}\tan^{-1}\!\left(\dfrac{2x+1}{\sqrt{7}}\right) + C.$

5. $\int \dfrac{8\, dx}{x(x-1)(3x^2+2x+1)}.$

6. $\int \dfrac{4x+5}{(x^2+x+1)(x^2-x+1)}\, dx.$

148. CASE IV. The denominator contains prime factors of the second degree, some of which are repeated. Consider the following integral:
$$\int \dfrac{3x+7}{(x^2+1)^2 x}\, dx.$$

In this case the factor x^2+1 is repeated. When such is the case we write an equation of the form

$$\dfrac{3x+7}{x(x^2+1)^2} = \dfrac{Ax+B}{(x^2+1)^2} + \dfrac{Cx+D}{x^2+1} + \dfrac{E}{x}.$$

Clearing this equation of fractions, we have

$$3x+7 = (Ax+B)x + x(Cx+D)(x^2+1) + E(x^2+1)^2.$$

Equating coefficients of like powers of x, we have the following equations:

$E+C=0,\quad D=0,\quad A+C+2E=0,\quad B+D=3,\quad E=7.$

Therefore $A = -7$, $B = 3$, $C = -7$, $D = 0$, $E = 7$.
Then we have
$$\int \frac{3x+7}{x(x^2+1)^2}\,dx = \int \frac{-7x+3}{(x^2+1)^2}\,dx - \int \frac{7x\,dx}{x^2+1} + \int \frac{7\,dx}{x}$$
$$= -7\int \frac{x\,dx}{(x^2+1)^2} + 3\int \frac{dx}{(x^2+1)^2} - 7\int \frac{x\,dx}{x^2+1}$$
$$+ 7\int \frac{dx}{x}$$
$$= \frac{7}{2(x^2+1)} + 3\int \frac{dx}{(x^2+1)^2} - \frac{7}{2}\log(x^2+1)$$
$$+ 7\log x + C.$$

The integral
$$\int \frac{dx}{(x^2+1)^2},$$
which is of the form
$$\int \frac{dx}{(x^2+px+q)^2},$$
may be integrated by the formula developed below.

Differentiate the function
$$\frac{x+\dfrac{p}{2}}{x^2+px+q}.$$

Then we have
$$\frac{d}{dx}\left(\frac{x+\dfrac{p}{2}}{x^2+px+q}\right) = \frac{1}{x^2+px+q} - \frac{2\left(x+\dfrac{p}{2}\right)^2}{(x^2+px+q)^2}$$
$$= \frac{-1}{x^2+px+q} + \frac{2\left(q-\dfrac{p^2}{4}\right)}{(x^2+px+q)^2},$$

or
$$d\left(\frac{x+\dfrac{p}{2}}{x^2+px+q}\right) = \left[\frac{-1}{x^2+px+q} + \frac{2\left(q-\dfrac{p^2}{4}\right)}{(x^2+px+q)^2}\right]dx. \quad (1)$$

On integrating both sides of (1), we have
$$\frac{x+\dfrac{p}{2}}{x^2+px+q} = -\int \frac{dx}{x^2+px+q} + 2\left(q-\dfrac{p^2}{4}\right)\int \frac{dx}{(x^2+px+q)^2}.$$

Consequently

$$\int \frac{dx}{(x^2+px+q)^2} = \frac{x+\frac{p}{2}}{2\left(q-\frac{p^2}{4}\right)(x^2+px+q)} \quad (2)$$

$$+ \frac{1}{2\left(q-\frac{p^2}{4}\right)} \int \frac{dx}{x^2+px+q}.$$

Formula (2) shows that an integral of the form

$$\int \frac{dx}{(x^2+px+q)^2}$$

can be made to depend on a known integral. Hence, by formula (2), we have

$$3\int \frac{dx}{(x^2+1)^2} = \frac{3x}{2(x^2+1)} + \frac{3}{2}\int \frac{dx}{x^2+1}$$

$$= \frac{3x}{2(x^2+1)} + \frac{3}{2}\tan^{-1} x + C.$$

Therefore the complete solution for the foregoing exercise is

$$\int \frac{3x+7}{x(x^2+1)^2} dx = \frac{7}{2(x^2+1)} + \frac{3x}{2(x^2+1)}$$

$$+ \frac{3}{2}\tan^{-1} x - \frac{7}{2}\log(x^2+1) + 7\log x + C.$$

REMARK. Any rational function of x may be integrated by means of the methods explained in the above four cases. For a rational function $R(x)$ is defined by the relation $R(x) = g(x)/G(x)$, where $g(x)$ and $G(x)$ are polynomials in x. Also, from algebra, we know that $G(x)$ may always be factored into real linear or quadratic factors (since we are assuming that the coefficients of $G(x)$ are real). If $g(x)$ is of degree greater than the degree of $G(x)$, then $R(x)$ may be reduced to a mixed expression consisting of a polynomial and a proper fraction. This proper fraction may be separated into partial fractions which will belong to one of the above four cases. Although this is always theoretically possible, in practice it is not always a simple matter to obtain the factors of $G(x)$.

EXERCISES

Evaluate the following integrals:

1. $\int \dfrac{dx}{(x-1)(x^2+1)^2}$.

   Ans. $\dfrac{1}{8}\log\dfrac{(x-1)^2}{x^2+1} + \dfrac{1-x}{4(1+x^2)} - \dfrac{1}{2}\tan^{-1} x + C.$

2. $\int \dfrac{x+7}{x(x^2+x+1)}\,dx.$

   Ans. $7\log x - \dfrac{7}{2}\log(x^2+x+1) - \dfrac{5}{\sqrt{3}}\tan^{-1}\left(\dfrac{2x+1}{\sqrt{3}}\right) + C.$

3. $\int \dfrac{dx}{(x^3+1)^2}$.

   Ans. $\dfrac{-x}{3(x^3+1)} + \dfrac{1}{9}\log\dfrac{(x+1)^2}{x^2-x+1} + \dfrac{1}{3\sqrt{3}}\tan^{-1}\left(\dfrac{2x-1}{\sqrt{3}}\right) + C.$

4. $\int \dfrac{7x+11}{x(x^2+2)^2}\,dx.$

   Ans. $\dfrac{1}{8}\log\dfrac{x^{22}}{(x^2+2)^{11}} + \dfrac{7x+11}{4(x^2+2)} + \dfrac{7}{4\sqrt{2}}\tan\dfrac{x}{\sqrt{2}} + C.$

5. $\int \dfrac{dx}{(x^2+a^2)^2}$.

6. $\int \dfrac{x^2+7}{(x-1)^2(x^2+4)^2}\,dx.$

149. Integrals containing $(ax+b)^{\frac{p}{q}}$. If the integrand of an integral is algebraic and contains the radical $(ax+b)^{\frac{1}{q}}$ but no other radical, it may be rationalized by the substitution

$$(ax+b) = z^q,$$

giving an expression in z which may be integrated by the methods already discussed. The following example will illustrate the method:

Example. Evaluate the integral

$$\int \dfrac{x\sqrt{x-1}}{x-2}\,dx.$$

Let $\qquad x - 1 = z^2.$

Then $\qquad dx = 2z\,dz.$

Substituting these values of x and dx in the integral, we obtain

$$\int \frac{x\sqrt{x-1}}{x-2}\,dx = \int \frac{(z^2+1)\cdot z \cdot 2z\,dz}{z^2-1}$$

$$= 2\int \left(z^2 + 2 + \frac{1}{z-1} - \frac{1}{z+1}\right) dz$$

$$= \frac{2}{3}z^3 + 4z + 2\log\left(\frac{z-1}{z+1}\right) + C.$$

Whence, substituting for z its value in terms of x, the final result is

$$\int \frac{x\sqrt{x-1}}{x-2}\,dx = \frac{2}{3}(x-1)^{\frac{3}{2}} + 4(x-1)^{\frac{1}{2}} + 2\log\left(\frac{\sqrt{x-1}-1}{\sqrt{x-1}+1}\right) + C.$$

150. Integrals containing the radical $\sqrt{x^2+px+q}$. If the integrand of an integral is algebraic and contains the radical $\sqrt{x^2+px+q}$ but no other radical, it may be freed from radicals by the substitution

$$\sqrt{x^2+px+q} = z - x.$$

Example. Evaluate the integral

$$\int \frac{dx}{\sqrt{x^2+x}}.$$

Let $\sqrt{x^2+x} = z - x.$

Then we have $\quad x = \dfrac{z^2}{2z+1}, \quad dx = \dfrac{2(z^2+z)}{(2z+1)^2}\,dz.$

Therefore $\quad \displaystyle\int \frac{dx}{\sqrt{x^2+x}} = \int \left(\frac{2(z^2+z)}{(2z+1)^2}\right)\left(\frac{2z+1}{z^2+z}\right) dz$

$$= 2\int \frac{dz}{2z+1} = \log(2z+1) + C.$$

Consequently

$$\int \frac{dx}{\sqrt{x^2+x}} = \log(2\sqrt{x^2+x} + 2x + 1) + C.$$

151. Integrals containing the radical $\sqrt{-x^2+px+q}$. If the integrand of an integral is algebraic and contains the radical $\sqrt{-x^2+px+q}$ but no other radical, the integral may be rationalized by either of the substitutions

$$\sqrt{-x^2+px+q} = (a-x)z$$

or
$$\sqrt{-x^2+px+q} = (x-b)z,$$

where $a-x$ and $x-b$ are the factors of $-x^2+px+q$. In this discussion the factors $a-x$ and $x-b$ are real.

Example. Evaluate the integral
$$\int \frac{dx}{\sqrt{2-x-x^2}}.$$

Let $\sqrt{2-x-x^2} = (1-x)z$.

Then $2-x-x^2 = (1-x)^2 z^2$, or $2+x = (1-x)z^2$.

Therefore $x = \dfrac{z^2-2}{z^2+1}$ and $dx = \dfrac{6z\,dz}{(z^2+1)^2}$.

Also $\sqrt{(2+x)(1-x)} = \dfrac{3z}{z^2+1}$.

Therefore
$$\int \frac{dx}{\sqrt{2-x-x^2}} = \int \left(\frac{z^2+1}{3z}\right)\left(\frac{6z\,dz}{(z^2+1)^2}\right) = 2\int \frac{dz}{z^2+1}$$
$$= 2\tan^{-1} z + C.$$

Substituting for z its value in terms of x, we have
$$\int \frac{dx}{\sqrt{2-x-x^2}} = 2\tan^{-1}\sqrt{\frac{2+x}{1-x}} + C.$$

Since a quadratic function may always be reduced to the sum or the difference of two squares, the integrals in this and the preceding article may be obtained by reducing them to the sum or the difference of two squares and then making a trigonometric substitution as in § 96.

The examples of this and the preceding section will illustrate the method.

Example 1. Evaluate the integral
$$\int \frac{dx}{\sqrt{x^2+x}}.$$

Now $\sqrt{x^2+x} = \sqrt{x^2+x+\tfrac{1}{4}-\tfrac{1}{4}} = \sqrt{(x+\tfrac{1}{2})^2 - (\tfrac{1}{2})^2}$.

Let $x + \tfrac{1}{2} = \tfrac{1}{2}\sec\theta$.

Then $dx = \tfrac{1}{2}\sec\theta\tan\theta\,d\theta$.

Also $\sqrt{x^2+x} = \tfrac{1}{2}\tan\theta$.

Therefore
$$\int \frac{dx}{\sqrt{x^2+x}} = \int \sec\theta\,d\theta = \log(\sec\theta + \tan\theta) + C.$$

Substituting for θ its value in terms of x, we obtain

$$\int \frac{dx}{\sqrt{x^2+x}} = \log\left(2\sqrt{x^2+x} + 2x + 1\right) + C.$$

Example 2. Evaluate the integral

$$\int \frac{dx}{\sqrt{2-x-x^2}}.$$

We have
$$\sqrt{2-x-x^2} = \sqrt{\tfrac{9}{4} - \tfrac{1}{4} - x - x^2} = \sqrt{\left(\tfrac{3}{2}\right)^2 - \left(x+\tfrac{1}{2}\right)^2}.$$

Let $\qquad x + \tfrac{1}{2} = \tfrac{3}{2}\sin\theta.$

Then $\qquad dx = \tfrac{3}{2}\cos\theta\, d\theta \quad$ and $\quad \sqrt{2-x-x^2} = \tfrac{3}{2}\cos\theta.$

Therefore $\qquad \int \dfrac{dx}{\sqrt{2-x-x^2}} = \int d\theta = \theta + C.$

Hence $\qquad \int \dfrac{dx}{\sqrt{2-x-x^2}} = \sin^{-1}\left(\dfrac{2x+1}{3}\right) + C.$

This result is apparently different from the one obtained above for the same example. The student may show that the two results are identical.

REMARK. In §§ 149–151 we have discussed substitutions whereby an algebraic expression containing linear or quadratic radicals may be transformed into a rational function of a new variable. Hence any integral of the form $\int R(x, \sqrt{ax^2 + bx + c})\,dx$, where R denotes a rational function of its two arguments, may be reduced to a rational algebraic integral which may be completely integrated. (See remark at end of § 148.) Likewise any integral of the form $\int R(x, \sqrt[p]{ax+b})\,dx$ may be completely integrated.

EXERCISES

Evaluate the following integrals:

1. $\int \dfrac{dx}{\sqrt{1+x}}.$ $\qquad\qquad\qquad$ Ans. $2\sqrt{1+x} + C.$

2. $\int \dfrac{dx}{(1+x)^{\frac{1}{3}}}.$ $\qquad\qquad\qquad$ Ans. $\tfrac{3}{2}(1+x)^{\frac{2}{3}} + C.$

3. $\int \dfrac{dx}{(1+x)^{\frac{1}{2}} + (1+x)^{\frac{1}{3}}}$.

Ans. $2(1+x)^{\frac{1}{2}} - 3(1+x)^{\frac{1}{3}} + (1+x)^{\frac{1}{6}} - 6\log\left[(1+x)^{\frac{1}{6}} + 1\right] + C$.

HINT. Let $1 + x = z^6$.

4. $\int \dfrac{x\,dx}{\sqrt{x^2 - x + 1}}$.

Ans. $\sqrt{x^2 - x + 1} + \tfrac{1}{2}\log\left(\dfrac{2x - 1 + 2\sqrt{x^2 - x + 1}}{\sqrt{3}}\right) + C$.

5. $\int \sqrt{5 - 4x - x^2}\,dx$.

Ans. $\left(\dfrac{2+x}{2}\right)\sqrt{5 - 4x - x^2} + \tfrac{9}{2}\sin^{-1}\left(\dfrac{2+x}{3}\right) + C$.

6. $\int \dfrac{x\,dx}{\sqrt{1 - x^2}}$.   Ans. $-\sqrt{1 - x^2} + C$.

7. $\int \dfrac{x\,dx}{x^{\frac{1}{2}} + 1}$.   Ans. $\tfrac{2}{3}x^{\frac{3}{2}} - x + 2x^{\frac{1}{2}} - 2\log\left(x^{\frac{1}{2}} + 1\right) + C$.

8. $\int \dfrac{\sqrt{3x + 2}}{x^2}\,dx$.

9. $\int \sqrt{x^2 + 2x + 3}\,dx$.

10. $\int_0^{\sqrt{2}} \sqrt{2 - x^2}\,dx$.   Ans. $\dfrac{\pi}{2}$.

11. $\int_0^4 \dfrac{dx}{\sqrt{7x + 5}}$.   Ans. $\tfrac{2}{7}(\sqrt{33} - \sqrt{5})$.

12. $\int_0^{\frac{1}{2}} \dfrac{\sqrt{x - x^2}}{1 + x}\,dx$.

152. Integration of $\int \sin^m x \cos^n x\,dx$. CASE I. If, in

$$\int \sin^m x \cos^n x\,dx,$$

m is an odd positive integer, the form can be readily integrated. This case has been treated in § 96. Another example will be worked here to illustrate the method.

Example. Evaluate the integral
$$\int \sin^5 x \cos^5 x \, dx.$$

This integral may be written
$$\int (1 - \cos^2 x)^2 \cos^5 x \sin x \, dx = \int (\cos^5 x - 2 \cos^7 x + \cos^9 x) \sin x \, dx$$
$$= -\frac{\cos^6 x}{6} + \frac{\cos^8 x}{4} - \frac{\cos^{10} x}{10} + C.$$

The same method applies if n is an odd positive integer.

CASE II. If, in
$$\int \sin^m x \cos^n x \, dx,$$

m and n are both even positive integers, the integral may be reduced to Case I by the use of the following trigonometric identities:
$$2 \sin x \cos x = \sin 2x,$$
$$2 \sin^2 x = 1 - \cos 2x,$$
$$2 \cos^2 x = 1 + \cos 2x.$$

The following example will illustrate their use:

Example. Show that
$$\int \sin^2 x \cos^2 x \, dx = \tfrac{1}{8}\left(x - \frac{\sin 4x}{4}\right) + C.$$

We may write
$$\int \sin^2 x \cos^2 x \, dx = \tfrac{1}{4}\int 4 \sin^2 x \cos^2 x \, dx$$
$$= \tfrac{1}{4}\int \sin^2 2x \, dx$$
$$= \tfrac{1}{4}\int \frac{1 - \cos 4x}{2} \, dx$$
$$= \tfrac{1}{8}\left(x - \frac{\sin 4x}{4}\right) + C.$$

153. Reduction formulas. If the integrand is an even power of $\sin x$ or $\cos x$, we may proceed as in § 152 or we may proceed by integration by parts.

We may write
$$\int \sin^n x \, dx = \int \sin^{n-1} x \cdot \sin x \, dx.$$

Integrating by parts (§ 95) by taking
$$u = \sin^{n-1} x, \quad dv = \sin x \, dx,$$
$$du = (n-1) \sin^{n-2} x \cos x \, dx, \quad v = -\cos x,$$
we obtain
$$\int \sin^n x \, dx = -\cos x \sin^{n-1} x + (n-1) \int \sin^{n-2} x \cos^2 x \, dx.$$
Substituting in the last integral $1 - \sin^2 x$ for $\cos^2 x$, we obtain
$$\int \sin^n x \, dx = -\cos x \sin^{n-1} x + (n-1) \int \sin^{n-2} x \, dx$$
$$- (n-1) \int \sin^n x \, dx.$$
Transposing the last integral and dividing by n, we have
$$\int \sin^n x \, dx = -\frac{\cos x \sin^{n-1} x}{n} + \frac{n-1}{n} \int \sin^{n-2} x \, dx. \quad (1)$$

By repeated applications of this process the integrand on the right-hand side of the above equation becomes $\sin^{n-4} x \, dx$, $\sin^{n-6} x \, dx$, etc., the last integral merely involving dx.

This process is also valid if n is an odd positive integer. The last integral in this case will be $\int \sin x \, dx$, which is a standard form.

In a similar manner we may develop the formula
$$\int \cos^n x \, dx = \frac{\cos^{n-1} x \sin x}{n} + \frac{n-1}{n} \int \cos^{n-2} x \, dx. \quad (2)$$

In the integral $\int \sin^m x \cos^n x \, dx$, if m and n are even positive integers, the integral may be reduced to several integrals containing even powers of $\sin x$ or of $\cos x$, so that formulas (1) and (2) may be used to evaluate such integrals.

Formulas (1) and (2) may also be written in the following forms:
$$\int \sin^{n-2} x \, dx = \frac{\cos x \sin^{n-1} x}{n-1} + \frac{n}{n-1} \int \sin^n x \, dx; \quad (3)$$
and
$$\int \cos^{n-2} x \, dx = -\frac{\sin x \cos^{n-1} x}{n-1} + \frac{n}{n-1} \int \cos^n x \, dx. \quad (4)$$

Formulas (3) and (4) raise the power of the function in the integrand from $n-2$ to n, and hence may be used for negative powers of $\sin x$ and $\cos x$. And since negative powers of $\sin x$ and $\cos x$ may be written as positive powers of $\csc x$ and $\sec x$, the formulas (3) and (4) may be used to integrate powers of $\sec x$ and $\csc x$.

The following examples will illustrate the use of the above formulas.

Example 1. Evaluate the integral

$$\int \sin^6 x \, dx.$$

Applying formula (1), we have

$$\int \sin^6 x \, dx = -\frac{\cos x \sin^5 x}{6} + \frac{5}{6} \int \sin^4 x \, dx$$

$$= -\frac{\cos x \sin^5 x}{6} + \frac{5}{6}\left[-\frac{\cos x \sin^3 x}{4} + \frac{3}{4} \int \sin^2 x \, dx\right]$$

$$= -\frac{\cos x \sin^5 x}{6} - \frac{5 \cos x \sin^3 x}{24} + \frac{5}{8} \int \sin^2 x \, dx$$

$$= -\frac{\cos x \sin^5 x}{6} - \frac{5 \cos x \sin^3 x}{24} - \frac{5 \cos x \sin x}{16} + \frac{5x}{16} + C.$$

Example 2. Evaluate the integral

$$\int \sec^3 x \, dx.$$

This may be written

$$\int \cos^{-3} x \, dx.$$

Applying formula (4), we obtain

$$\int \cos^{-3} x \, dx = -\frac{\sin x \cos^{-2} x}{-2} + \frac{-1}{-2} \int (\cos x)^{-1} \, dx$$

$$= \frac{\sin x \sec^2 x}{2} + \frac{1}{2} \int \sec x \, dx$$

$$= \frac{\sin x \sec^2 x}{2} + \frac{1}{2} \log (\sec x + \tan x) + C.$$

Therefore

$$\int \sec^3 x \, dx = \frac{\sin x \sec^2 x}{2} + \frac{1}{2} \log (\sec x + \tan x) + C.$$

If in the integral $\int \sec^n x \, dx$, n is an even positive integer, the integral may be evaluated by the following method more readily than by formula (4).

Example 3. Evaluate the integral
$$\int \sec^6 x \, dx.$$

This integral may be written
$$\int \sec^4 x \cdot \sec^2 x \, dx = \int (1 + \tan^2 x)^2 \sec^2 x \, dx$$
$$= \int (1 + 2 \tan^2 x + \tan^4 x) \sec^2 x \, dx$$
$$= \tan x + \frac{2 \tan^3 x}{3} + \frac{\tan^5 x}{5} + C.$$

Therefore $\int \sec^6 x \, dx = \tan x + \dfrac{2 \tan^3 x}{3} + \dfrac{\tan^5 x}{5} + C.$

EXERCISES

Evaluate the following integrals:

1. $\int \sin^3 x \, dx.$
 Ans. $-\cos x + \dfrac{\cos^3 x}{3} + C.$

2. $\int \cos^4 \theta \, d\theta.$
 Ans. $\frac{3}{8} \theta + \frac{1}{4} \sin 2\theta + \frac{1}{32} \sin 4\theta + C.$

3. $\int \sin^2 x \, dx.$
 Ans. $\dfrac{x}{2} - \dfrac{\sin 2x}{4} + C.$

4. $\int \dfrac{\sin^3 x \, dx}{1 - \cos 2x}.$
 Ans. $-\frac{1}{2} \cos x + C.$

5. $\int \sin^3 2x \cos^3 2x \, dx.$
 Ans. $-\dfrac{\cos^4 2x}{8} + \dfrac{\cos^6 2x}{12} + C.$

6. $\int \dfrac{1 - \cos x}{1 + \cos x} \, dx.$
 Ans. $2 \csc x - 2 \cot x - x + C.$

7. $\int \sin^{\frac{2}{3}} 3x \cos^3 3x \, dx.$
 Ans. $\frac{1}{5} \sin^{\frac{5}{3}} 3x - \frac{1}{11} \sin^{\frac{11}{3}} 3x + C.$

8. $\int \sin^7 x \, dx.$
 Ans. $-\cos x + \cos^3 x - \frac{3}{5} \cos^5 x + \dfrac{\cos^7 x}{7} + C.$

9. $\int \sec^4 5x \, dx.$
 Ans. $\frac{1}{5} \tan 5x + \frac{1}{15} \tan^3 5x + C.$

10. $\int \dfrac{dx}{(1 + \cos 4x)^3}.$
 Ans. $\frac{1}{16}(\tan 2x + \frac{2}{3} \tan^3 2x + \frac{1}{5} \tan^5 2x) + C.$

11. $\int \sec^5 3x\, dx$.

Ans. $\dfrac{1}{12}\dfrac{\sin 3x}{\cos^4 3x} + \dfrac{1}{8}\dfrac{\sin 3x}{\cos^2 3x} + \dfrac{1}{8}\log(\sec 3x + \tan 3x) + C$.

12. $\int \csc^3 2x\, dx$. 13. $\int \csc^4 3x\, dx$. 14. $\int \dfrac{7\sin x\, dx}{\cos^3 x}$.

154. Integration of $\int \tan^n x\, dx$ or of $\int \cot^n x\, dx$. The integration of $\int \tan^n x\, dx$ or of $\int \cot^n x\, dx$ may be accomplished when n is any integer, positive or negative. The identities

$$\tan^2 x = \sec^2 x - 1 \quad \text{and} \quad \cot^2 x = \csc^2 x - 1$$

are used to transform the integral.

Consider the integral $\int \tan^n x\, dx$.

This may be written

$$\int \tan^{n-2} x\, (\sec^2 x - 1)\, dx.$$

Therefore

$$\int \tan^n x\, dx = \int \tan^{n-2} x \sec^2 x\, dx - \int \tan^{n-2} x\, dx$$

$$= \dfrac{\tan^{n-1} x}{n-1} - \int \tan^{n-2} x\, dx.$$

This process may be continued until the original integral is made to depend on $\int \tan x\, dx$ or on $\int dx$, either of which is a standard integral. The integral $\int \cot^n x\, dx$ may be handled in a similar way.

Example. Show that

$$\int \tan^4 x\, dx = \dfrac{\tan^3 x}{3} - \tan x + x + C.$$

This may be written in the form

$$\int \tan^2 x\,(\sec^2 x - 1)\, dx = \int \tan^2 x \sec^2 x\, dx - \int \tan^2 x\, dx$$

$$= \dfrac{\tan^3 x}{3} - \int (\sec^2 x - 1)\, dx$$

$$= \dfrac{\tan^3 x}{3} - \tan x + x + C.$$

155. Integration of $\int \tan^n x \sec^m x \, dx$ **or of** $\int \cot^n x \csc^m x \, dx$.

The integration of $\int \tan^n x \sec^m x \, dx$ or of $\int \cot^n x \csc^m x \, dx$ may be easily accomplished, provided

       1. *m is an even positive integer.*
       2. *n is an odd positive integer.*

In case m is an even positive integer, $\sec^{m-2} x$ may be reduced to powers of $\tan x$ by means of the substitution

$$\sec^2 x = \tan^2 x + 1.$$

The integral may then be so written that each term of the integrand is an exact integral.

Example. Show that

$$\int \tan x \sec^4 x \, dx = \frac{\tan^4 x}{4} + \frac{\tan^2 x}{2} + C.$$

This may be written in the form

$$\int \tan x \sec^4 x \, dx = \int \tan x \, (1 + \tan^2 x) \sec^2 x \, dx$$
$$= \frac{\tan^4 x}{4} + \frac{\tan^2 x}{2} + C.$$

The following example will illustrate the procedure for the case when n is odd.

Example. Show that

$$\int \tan^3 2x \sec^3 2x \, dx = \frac{\sec^5 2x}{10} - \frac{\sec^3 2x}{6} + C.$$

We may write

$$\int \tan^3 2x \sec^3 2x \, dx = \int (\tan^2 2x \sec^2 2x) \tan 2x \sec 2x \, dx$$
$$= \int (\sec^2 2x - 1) \sec^2 2x \tan 2x \sec 2x \, dx$$
$$= \int (\sec^4 2x - \sec^2 2x) \tan 2x \sec 2x \, dx$$
$$= \frac{\sec^5 2x}{10} - \frac{\sec^3 2x}{6} + C.$$

REMARK. The integration of $\int \sec^n x \, dx$ or of $\int \csc^n x \, dx$ has already been discussed in § 153.

156. The substitution tan $x/2 = t$. In certain forms involving only rational trigonometric functions the substitution $\tan x/2 = t$ may be used to simplify the integral. The simplification is due to the fact that if $\tan x/2 = t$, then dx/dt and all the trigonometric functions of x are rational functions of t.

In fact, if $\tan \dfrac{x}{2} = t$, then

$$\sin x = \frac{2t}{1+t^2}; \quad \cos x = \frac{1-t^2}{1+t^2}; \quad \tan x = \frac{2t}{1-t^2}; \quad dx = \frac{2\,dt}{1+t^2}.$$

The following examples will illustrate the use of the substitution.

Example 1. Evaluate the integral

$$\int \csc x \, dx.$$

Here
$$\csc x = \frac{1+t^2}{2t}, \quad dx = \frac{2\,dt}{1+t^2}.$$

Therefore
$$\int \csc x \, dx = \int \frac{dt}{t} = \log t + C$$
$$= \log \tan \frac{x}{2} + C.$$

Example 2. Evaluate

$$\int \frac{dx}{1 - 2 \sin x}.$$

Here
$$dx = \frac{2\,dt}{1+t^2}, \quad 2 \sin x = \frac{4t}{1+t^2}.$$

Hence
$$\int \frac{dx}{1 - 2 \sin x} = 2 \int \frac{dt}{1 - 4t + t^2}$$
$$= \frac{1}{\sqrt{3}} \log \frac{\sqrt{3} - t + 2}{\sqrt{3} + t - 2} + C.$$

REMARK. Since the substitution $\tan \dfrac{x}{2} = t$ reduces all the trigonometric functions of x to rational functions of t and reduces dx to a rational function of t multiplied by dt, it follows that any rational function of the trigonometric functions may be reduced to a rational algebraic function; that is,

$$\int R(\sin x, \cos x) dx = \int S(t) dt,$$

where R and S are rational functions of the arguments indicated.

In the remark at the end of § 148 we have shown that $\int S(t)dt$ is completely integrable. Hence any rational function of the trigonometric functions is always integrable in terms of the elementary functions. There are, however, certain irrational functions of $\sin x$ and $\cos x$ which cannot be integrated in terms of the elementary functions. A discussion of these functions is beyond the scope of this book.

EXERCISES

Evaluate the following integrals:

1. $\int \tan^5 x \, dx.$ Ans. $\dfrac{\sec^4 x}{4} - \sec^2 x + \log \sec x + C.$

2. $\int \tan^3 2x \, dx.$ Ans. $\tfrac{1}{4} \sec^2 2x - \tfrac{1}{2} \log \sec 2x + C.$

3. $\int \tan^2 3x \cot 3x \, dx.$ Ans. $\tfrac{1}{3} \log \sec 3x + C.$

4. $\int \cot^7 2x \, dx.$

   Ans. $-\dfrac{\csc^6 2x}{12} + \dfrac{3}{8} \csc^4 2x - \dfrac{3}{4} \csc^2 2x + \dfrac{1}{2} \log \csc 2x + C.$

5. $\int \tan^2 x \sec^4 x \, dx.$ Ans. $\dfrac{\tan^5 x}{5} + \dfrac{\tan^3 x}{3} + C.$

6. $\int \cot^3 x \csc^7 x \, dx.$

7. $\int \tan^3 x \sec x \, dx.$

8. $\int \tan^3 2x \cos^2 2x \, dx.$

9. $\int \sec^5 2x \, dx.$

10. $\int \dfrac{7 \sin 2x}{\cos^3 2x} dx.$

11. $\int \dfrac{\tan x \sec x}{(2 + \sec x)^5} dx.$

12. $\int \csc^4 3x \cot 3x \, dx.$

13. $\int \dfrac{dx}{\sin^3 x}.$

14. $\int \dfrac{dx}{1 + \cos x}.$

15. $\int \dfrac{dx}{2 + 3 \sin x}.$

16. $\int \dfrac{dx}{2 + \tan x}.$

157. Use of a table of integrals. In the preceding sections of this chapter, methods have been developed for the integration of certain types of integrals. By the successive use of the method of integration by parts many other forms can be integrated. A further discussion of the integration of such forms is not necessary here, for no new theory is involved, and the results obtained by these methods have been compiled in tables of integrals, such as B. O. Peirce's "Short Table of Integrals." In practical work where integration is involved, it is desirable to use such a table rather than to work out each integration as it arises.

158. Approximate integration. Although we have succeeded in integrating a large number of forms, there are many integrals occurring in the practical applications which cannot be integrated in terms of the elementary functions. Some of these are so important that various methods have been devised for integrating them approximately. In what follows three methods for approximate integration will be explained.

159. Trapezoidal rule. Since the value of the definite integral $\int_a^b f(x)dx$ is a measure of the area under the curve $y = f(x)$ from $x = a$ to $x = b$, an approximation to the area under the curve will give an approximate value for the definite integral.

Suppose, then, that we have a curve A_0A_n (see Fig. 68) between the ordinates $x = a$ and $x = b$. Divide the interval from $x = a$ to $x = b$ into n equal parts and denote each part by Δx. At the points of division erect ordinates $y_i (i = 0, \cdots, n)$ cutting the curve in the points A_i. Draw the lines $A_j A_{j+1} (j = 0, \cdots, n-1)$. There are then formed n trapezoids.

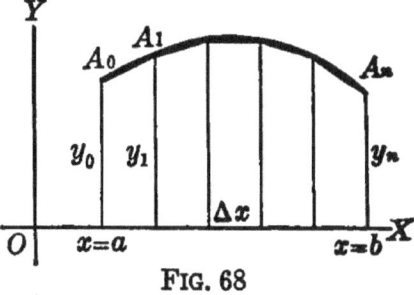

FIG. 68

The area of the first trapezoid $= \frac{1}{2}(y_0 + y_1)\Delta x$.
The area of the second trapezoid $= \frac{1}{2}(y_1 + y_2)\Delta x$.
. . . .
The area of the nth trapezoid $= \frac{1}{2}(y_{n-1} + y_n)\Delta x$.
The sum of the areas of the n trapezoids is equal to

$$\tfrac{1}{2}(y_0 + 2y_1 + 2y_2 + \cdots + 2y_{n-1} + y_n)\Delta x.$$

This result is known as the *trapezoidal rule* for area and is written
$$\text{area} = \tfrac{1}{2}(y_0 + 2y_1 + \cdots + 2y_{n-1} + y_n)\Delta x.$$

It is clear that in general the greater the number of parts into which the interval from $x = a$ to $x = b$ is divided, the closer will the sum of the trapezoids approach the area under the curve.

Example. Find $\log_e 5$ by the trapezoidal rule.

Now $\log_e 5$ is given by the integral

$$\int_1^5 \frac{dx}{x}.$$

This integral represents the area under the curve $y = 1/x$ from $x = 1$ to $x = 5$. Let us divide the interval from $x = 1$ to $x = 5$ into 8 equal parts. Then $\Delta x = .5$. Substituting the abscissas

$$x = 1, \ 1.5, \ 2, \ 2.5, \ 3, \ 3.5, \ 4, \ 4.5, \ 5$$

in the equation $y = 1/x$, we obtain for y the corresponding values

$$y = 1, \ \tfrac{2}{3}, \ \tfrac{1}{2}, \ \tfrac{2}{5}, \ \tfrac{1}{3}, \ \tfrac{2}{7}, \ \tfrac{1}{4}, \ \tfrac{2}{9}, \ \tfrac{1}{5}.$$

From the trapezoidal rule we then have the approximate result

$$\text{area} = \log_e 5 = \tfrac{1}{2}(1 + \tfrac{4}{3} + 1 + \tfrac{4}{5} + \tfrac{2}{3} + \tfrac{4}{7} + \tfrac{1}{2} + \tfrac{4}{9} + \tfrac{1}{5})\tfrac{1}{2}$$
$$= 1.63.$$

The value of $\log_e 5$ is 1.609 correct to three decimals.

160. Simpson's rule. Suppose we have a curve $A_0 A_n$ between the ordinates $x = a$ and $x = b$. Divide the interval between $x = a$ and $x = b$ into n equal parts (n being an even number) and denote each part by Δx. At the points of division erect ordinates $y_i (i = 0, \cdots, n)$ cutting the curve in the points A_i.

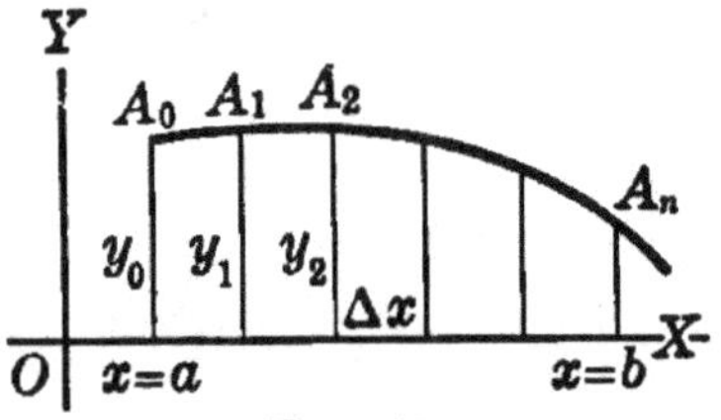

Now a parabola with its axis parallel to y_i may be made to pass through any three points. Let the equation of the parabola passing through $A_0 A_1 A_2$ be

$$y = mx^2 + bx + c.$$

FIG. 69

The area under this parabola between the ordinates y_0 and y_2 is given by the integral

$$\int_a^{a+2\Delta x} (mx^2 + bx + c)dx = (6a^2 m + 6ab + 6c + 12\,am\,\Delta x$$
$$+ 6b\,\Delta x + 8\,m\,\overline{\Delta x}^2)\frac{\Delta x}{3}.$$

Now since $x = a$ when $y = y_0$, $x = a + \Delta x$ when $y = y_1$ etc., we have
$$y_0 = ma^2 + ab + c.$$
$$y_1 = m(a + \Delta x)^2 + b(a + \Delta x) + c.$$
$$y_2 = m(a + 2\Delta x)^2 + b(a + 2\Delta x) + c.$$

Hence $y_0 + 4y_1 + y_2 = 6a^2m + 6ab + 6c + 12am\Delta x$
$$+ 6b\Delta x + 8m\Delta x^2.$$

Therefore
$$\int_a^{a+2\Delta x} (mx^2 + bx + c)dx = (y_0 + 4y_1 + y_2)\frac{\Delta x}{3}.$$

Similarly,

parabolic area under $A_2 A_3 A_4 = (y_2 + 4y_3 + y_4)\dfrac{\Delta x}{3}$,

. . . .

parabolic area under $A_{n-2} A_{n-1} A_n = (y_{n-2} + 4y_{n-1} + y_n)\dfrac{\Delta x}{3}.$

Adding all such areas, we obtain

total area $= (y_0 + 4y_1 + 2y_2 + 4y_3 + 2y_4 + \cdots + 2y_{n-2}$
$$+ 4y_{n-1} + y_n)\frac{\Delta x}{3}.$$

This is *Simpson's rule*, sometimes called Simpson's one-third rule.

Example. Calculate by Simpson's rule $\log_e 5$.

Let $\Delta x = \frac{1}{2}$, as in § 159. Then the corresponding values for x and y will be

$x = 1,\ 1.5,\ 2,\ 2.5,\ 3,\ 3.5,\ 4,\ 4.5,\ 5.$
$y = 1,\ \frac{2}{3},\ \frac{1}{2},\ \frac{2}{5},\ \frac{1}{3},\ \frac{2}{7},\ \frac{1}{4},\ \frac{2}{9},\ \frac{1}{5}.$

Simpson's rule then gives

area $= \log_e 5 = (1 + \frac{8}{3} + 1 + \frac{8}{5} + \frac{2}{3} + \frac{8}{7} + \frac{1}{2} + \frac{8}{9} + \frac{1}{5})\frac{1}{6} = 1.610.$

This is a very much better approximation than the one given by the trapezoidal rule in § 159.

161. Integration in series. In calculating integrals it is sometimes convenient to expand a function in infinite series and then integrate a few terms of the expansion. The student should apply this process only to convergent series.

Example. Find the length of a quadrant of the ellipse $x^2 + 4y^2 = 4$. The length of the arc is given by the integral

$$S = \int_0^2 \sqrt{1 + \left(\frac{dy}{dx}\right)^2}\, dx.$$

Introduce a parameter θ by means of the equations
$$x = 2\sin\theta, \quad y = \cos\theta.$$
Then we have
$$S = \int_0^{\frac{\pi}{2}} \sqrt{4 - 3\sin^2\theta}\, d\theta. \tag{1}$$

Expanding $\sqrt{4 - 3\sin^2\theta}$ by the binomial theorem, we get
$$\sqrt{4 - 3\sin^2\theta} = 2(1 - \tfrac{3}{8}\sin^2\theta - \tfrac{9}{128}\sin^4\theta + \cdots).$$
On inserting this series the integral (1) becomes
$$S = 2\int_0^{\frac{\pi}{2}}(1 - \tfrac{3}{8}\sin^2\theta - \tfrac{9}{128}\sin^4\theta + \cdots)\,d\theta.$$
Integrating the first three terms of the series, we have, $S = 2.47$, approximately.

162. Integrating machines. The areas hitherto computed have been areas bounded by curves whose equations could be written down. It happens, however, that sometimes in practice one desires to find an area bounded by an empirical curve. This area can be determined mechanically by the use of an integrating machine. The description and instructions for the use of such machines may be found in books on engineering practice.

163. Summary. In the present chapter we have discussed methods for integrating rational fractions, functions containing linear and quadratic radicals, various types of trigonometric functions, and three methods of approximate integration.

EXERCISES

By Simpson's rule and by the trapezoidal rule evaluate approximately the following integrals, setting in each case $\Delta x = \tfrac{1}{2}$.

1. $\int_1^3 x^2\, dx.$ *Ans.* $\tfrac{26}{3}, \tfrac{35}{4}.$

2. $\int_0^2 e^{-x^2}\, dx.$ *Ans.* .882, .881.

3. $\int_1^7 \dfrac{dx}{x}.$ *Ans.* 1.948, 1.946.

4. A field as shown in the figure is bounded by the three straight fences A_0B_0, B_0B_6, and A_6B_6 and the river A_0A_6. A surveyor finds that $B_0B_6 = 24$ rd. He therefore lays off $B_0B_1 = 4$ rd., $B_1B_2 = 4$ rd., etc. He also finds that $A_0B_0 = 8$ rd., $A_1B_1 = 9$ rd., $A_2B_2 = 9$ rd., $A_3B_3 = 7$ rd., $A_4B_4 = 10$ rd., $A_5B_5 = 11$ rd., $A_6B_6 = 12.5$ rd. By the use of Simpson's rule find the approximate area of the field.

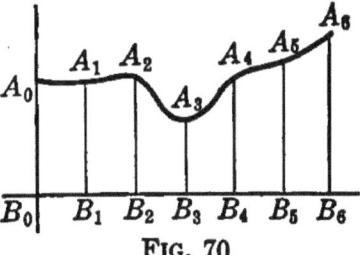

FIG. 70

5. Let y_1, y_2, y_3 be ordinates of the curve $y = ax^3 + bx^2 + cx + d$, and let y_2 be midway between y_1 and y_3. Show that the area under the curve between the ordinates y_1 and y_3 is

$$A = \frac{h}{3}(y_1 + 4y_2 + y_3),$$

where h is the distance between y_1 and y_2. *This is known as the prismoidal formula.*

6. Show that the prismoidal formula gives a correct volume in the following cases: (1) sphere, (2) cone, (3) cylinder, (4) pyramid, (5) segment of a sphere, (6) frustum of a cone or pyramid.

7. Using four intervals, compute by Simpson's rule the approximate value of $\pi/4$, starting from the formula

$$\frac{\pi}{4} = \int_0^1 \frac{dx}{1+x^2}.$$

8. Find the approximate value of $\int_0^1 \cos x^3 \, dx$ by expanding in series. *Ans.* .932.

9. Find the approximate value of $\int_0^1 e^{-x^2} \, dx$ by expanding in series. *Ans.* .746.

MISCELLANEOUS EXERCISES

Integrate the following:

1. $\int \dfrac{x^2 - 2}{x^3 - 3x^2} \, dx.$ *Ans.* $-\dfrac{2}{3x} + \dfrac{1}{9} \log x^2(x-3)^7 + C.$

2. $\int \dfrac{dx}{x^4 - 1}.$ *Ans.* $\dfrac{1}{4} \log \dfrac{x-1}{x+1} - \dfrac{1}{2} \tan^{-1} x + C.$

3. $\int \dfrac{x \, dx}{(x^2 + 1)^{\frac{3}{2}}}.$ *Ans.* $\dfrac{-1}{(x^2+1)^{\frac{1}{2}}} + C.$

4. $\int \dfrac{dx}{(x^2+1)^{\frac{3}{2}}}.$ Ans. $\dfrac{x}{(x^2+1)^{\frac{1}{2}}} + C.$

5. $\int \dfrac{\sqrt{x+1}}{x} dx.$ Ans. $2\sqrt{x+1} + \log \dfrac{\sqrt{x+1}-1}{\sqrt{x+1}+1} + C.$

6. $\int \dfrac{dx}{(1-x^2)^{\frac{3}{2}}}.$ Ans. $\dfrac{x}{\sqrt{1-x^2}} + C.$

7. $\int \dfrac{dx}{(2-x)^{\frac{1}{2}}}.$ Ans. $-2\sqrt{2-x} + C.$

8. $\int \dfrac{dx}{x^2(4-x^2)^{\frac{1}{2}}}.$ Ans. $-\dfrac{\sqrt{4-x^2}}{4x} + C.$

9. $\int \tan^4 5x\, dx.$

10. $\int x\sqrt{3+x}\, dx.$

11. $\int \dfrac{\sin 3x}{1+\cos 3x}\, dx.$

12. $\int \dfrac{dx}{\cos^8 x}.$

13. $\int \dfrac{dx}{1-\cos x}.$

14. $\int \dfrac{dx}{1-\sin 2x}.$

15. $\int \sin 2x \cos x\, dx.$

16. $\int \dfrac{dx}{\cos 4x}.$

17. $\int_0^{\frac{\pi}{2}} \sec^4 \dfrac{\theta}{2}\, d\theta.$

18. $\int \dfrac{dz}{z(1+\log z)}.$

19. $\int \dfrac{dx}{e^x + e^{-x}}.$ (HINT. Let $e^x = z$.)

20. $\int \dfrac{e^{3x}}{\sqrt{1-e^{6x}}}\, dx.$

21. By integrating in series find approximately the volume generated by revolving the curve
$$y = \dfrac{1}{1+x^2}$$
about the x-axis between the limits $x = -\tfrac{1}{2}$ and $x = \tfrac{1}{2}$.

22. Find approximately the volume in problem 21, using Simpson's rule.

23. Find approximately the length of the ellipse $x^2 + 9y^2 = 4$.

CHAPTER XIX

FUNCTIONS OF TWO OR MORE VARIABLES. DIFFERENTIATION

164. Continuous functions of two variables. A function $f(x, y)$ of two independent variables x and y is said to be continuous at $x = x_0$, $y = y_0$ if
$$\lim_{\substack{x \to x_0 \\ y \to y_0}} f(x, y) = f(x_0, y_0)$$

regardless of how x and y approach x_0 and y_0 respectively.

This may be illustrated geometrically by means of a figure as follows:

Let $z = f(x, y)$ be the equation of a surface (see Fig. 71). Let $P(x, y, z)$ be a point of this surface, and let $Q(x + \Delta x, y + \Delta y, z + \Delta z)$ be another point of this surface. Then if $f(x, y)$ is continuous, the point Q will approach the point P when Δx and Δy approach zero as a limit, regardless of how they approach this limit.

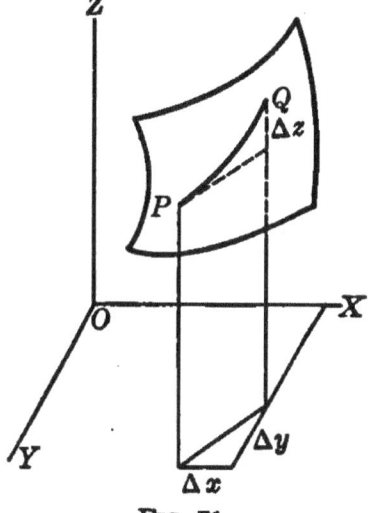

FIG. 71

In the following discussion only functions which are continuous will be considered.

165. Partial derivatives. Since x and y are independent variables, x may be allowed to vary while y remains constant. If x varies while y remains constant, then

$$\lim_{\Delta x \to 0} \frac{f(x + \Delta x, y) - f(x, y)}{\Delta x},$$

when it exists, is defined as the partial derivative of $f(x, y)$ with respect to x.

Similarly $\lim_{\Delta y \to 0} \dfrac{f(x, y + \Delta y) - f(x, y)}{\Delta y}$,

when it exists, is defined as the partial derivative of $f(x, y)$ with respect to y.

The partial derivative of $f(x, y)$ with respect to x is denoted by the symbol $\dfrac{\partial f}{\partial x}$ or by $f'_x(x, y)$. Similarly the partial derivative of $f(x, y)$ with respect to y is denoted by $\dfrac{\partial f}{\partial y}$ or $f'_y(x, y)$. The round ∂ is used instead of d to indicate partial differentiation as distinguished from total differentiation.

Partial derivatives of functions of three or more independent variables are defined in a similar way.

Example. If $z = 4 x^2 + 7 xy + 9 y^2$, find $\dfrac{\partial z}{\partial x}$ and $\dfrac{\partial z}{\partial y}$.

Differentiating with respect to x, treating y as a constant, we obtain
$$\frac{\partial z}{\partial x} = 8 x + 7 y.$$

Differentiating with respect to y, treating x as a constant, we obtain
$$\frac{\partial z}{\partial y} = 7 x + 18 y.$$

166. Higher partial derivatives. Partial derivatives may be differentiated again, giving higher partial derivatives. Thus we may have

$$\frac{\partial}{\partial x}\left(\frac{\partial f}{\partial x}\right), \quad \frac{\partial}{\partial x}\left(\frac{\partial f}{\partial y}\right), \quad \frac{\partial}{\partial y}\left(\frac{\partial f}{\partial x}\right), \quad \frac{\partial}{\partial y}\left(\frac{\partial f}{\partial y}\right), \text{ etc.}$$

These higher partial derivatives are represented symbolically by

$$\frac{\partial^2 f}{\partial x^2}, \quad \frac{\partial^2 f}{\partial x\, \partial y}, \quad \frac{\partial^2 f}{\partial y\, \partial x}, \quad \frac{\partial^2 f}{\partial y^2}, \text{ etc.,}$$

or $\quad f''_{xx}(x, y) \quad f''_{xy}(x, y), \quad f''_{yx}(x, y) \quad f''_{yy}(x, y)$, etc.

These considerations may be extended to partial derivatives of higher order when the function contains any number of independent variables. The notation indicates how many times the function has been differentiated partially and the order in which the differentiations have been performed. Thus $\dfrac{\partial^3 f}{\partial x\,\partial y^2}$ indicates three differentiations, the first two with respect to y and the third with respect to x.

It can be shown that
$$\frac{\partial^2 f}{\partial x\,\partial y} = \frac{\partial^2 f}{\partial y\,\partial x}$$
provided both derivatives are continuous;* that is, partial derivatives are independent of the order in which the differentiations are performed, provided all derivatives are continuous.

167. Partial derivatives interpreted geometrically. Let $z = f(x, y)$ be the equation of the surface in the figure. Cut the surface with the plane $x = a$. Then the equation of the trace APB of $z = f(x, y)$ on $x = a$ is $z = f(a, y)$. If the plane $x = a$ cuts the ZX-plane in $O'Z'$ and the XY-plane in $O'Y'$, and we consider $O'Z'$ as the Z-axis and $O'Y'$ as the Y-axis in the plane $Z'O'Y'$, then $z = f(a, y)$ is the equation of the curve APB. Hence

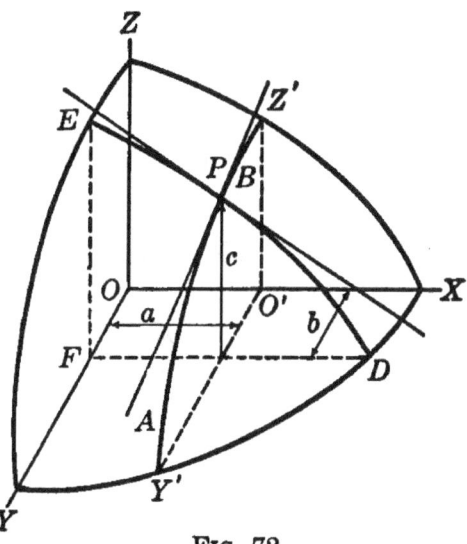

FIG. 72

$$\frac{dz}{dy} = \text{slope of } z = f(a, y) \text{ at } P \text{ on } z = f(x, y) \text{ where } x = a.$$

But $\dfrac{dz}{dy} = \dfrac{\partial z}{\partial y}$ at any point P on $z = f(x, y)$ where $x = a$.

Therefore

$$\frac{\partial z}{\partial y} = \text{slope of } z = f(a, y) \text{ at any point } P \text{ on } f(x, y) \text{ where } x = a.$$

Similarly, if a plane $y = b$ is passed through P, then

$$\frac{\partial z}{\partial x} = \text{slope of } z = f(x, b) \text{ at } P.$$

*See Wilson's "Advanced Calculus," § 50.

Example. Find the slope of the section which is cut from the ellipsoid $x^2 + 2y^2 + 3z^2 = 6$ by the plane $x = 1$, where $y = 1$ and $z = 1$.

Here $$\text{slope} = \frac{\partial z}{\partial y} = -\frac{4y}{6z}.$$

For $y = 1$ and $z = 1$ this has the value $-\frac{2}{3}$.

EXERCISES

In the first six exercises find $\dfrac{\partial}{\partial x}$, $\dfrac{\partial}{\partial y}$, $\dfrac{\partial^2}{\partial x \partial y}$, $\dfrac{\partial^2}{\partial y \partial x}$, $\dfrac{\partial}{\partial z}$:

1. $u = x^3 + 3xy^2$.
2. $S = x^2 + 2y^2$.
3. $t = e^x \sin y$.
4. $v = e^{x+y}$.
5. $u = \sin(xyz)$.
6. $v = \sin \dfrac{z}{x+y}$.

7. When $z = x^2 + 4y^2$, show that $x\dfrac{\partial z}{\partial x} + y\dfrac{\partial z}{\partial y} = 2z$.

8. When $z = \log(x^2 + y^2)$, show that $x\dfrac{\partial z}{\partial x} + y\dfrac{\partial z}{\partial y} = 2$.

9. When $u = \sin x \cos y$, show that $\dfrac{\partial u}{\partial x} + \dfrac{\partial u}{\partial y} = \cos(x+y)$.

10. When $u = \log(\tan x + \tan y)$, show that
$$\frac{\partial u}{\partial x} - \frac{\partial u}{\partial y} = \tan x - \tan y.$$

11. (1) The altitude of a right circular cone remains constant. Find the rate of change of the volume when the radius of the base changes. (2) The radius of the base remains constant. Find the rate of change of the volume when the altitude changes.

168. Total derivatives. The derivative of a function of a single variable x with respect to another variable t, where x is a function of t, has already been treated (see § 26). In fact, if $y = f(x)$ and $x = \phi(t)$, then
$$\frac{dy}{dt} = \frac{dy}{dx} \cdot \frac{dx}{dt}.$$

It is now proposed to find the derivative of a function of two variables x and y with respect to a variable t, when x and y are functions of t.

Let $z = f(x, y)$, where x and y are functions of t which together with their first derivatives are continuous over a region S. Let t assume the increment Δt. This will cause x to assume the increment Δx, y to assume the increment Δy, and z to assume the increment Δz. We then have

$$z + \Delta z = f(x + \Delta x, y + \Delta y).$$

Hence
$$\Delta z = f(x + \Delta x, y + \Delta y) - f(x, y). \tag{1}$$

This is called the total increment of z. To the right member of the equation (1) add and subtract the expression $f(x, y + \Delta y)$. The equation then becomes

$$\Delta z = f(x + \Delta x, y + \Delta y) - f(x, y + \Delta y) + f(x, y + \Delta y) - f(x, y). \tag{2}$$

Applying the theorem of the mean to equation (2), we obtain

$$\Delta z = f'_x(x + \theta_1 \Delta x, y + \Delta y)\Delta x + f'_y(x, y + \theta_2 \Delta y)\Delta y,$$

where $0 < \theta_1 < 1$, $0 < \theta_2 < 1$. Dividing by Δt, we get

$$\frac{\Delta z}{\Delta t} = f'_x(x + \theta_1 \Delta x, y + \Delta y)\frac{\Delta x}{\Delta t} + f'_y(x, y + \theta_2 \Delta y)\frac{\Delta y}{\Delta t}.$$

Then
$$\lim_{\Delta t \to 0} \frac{\Delta z}{\Delta t} = \frac{dz}{dt} = f'_x(x, y)\frac{dx}{dt} + f'_y(x, y)\frac{dy}{dt},$$

or
$$\frac{dz}{dt} = \frac{\partial z}{\partial x} \cdot \frac{dx}{dt} + \frac{\partial z}{\partial y} \cdot \frac{dy}{dt}. \tag{3}$$

The expression dz/dt in (3) is called the total derivative of z with respect to t.

Similarly, if z is a function of any number of variables, as

$$z = f(x, y, r, \cdots),$$

then
$$\frac{dz}{dt} = \frac{\partial z}{\partial x} \cdot \frac{dx}{dt} + \frac{\partial z}{\partial y} \cdot \frac{dy}{dt} + \frac{\partial z}{\partial r} \cdot \frac{dr}{dt} + \cdots. \tag{4}$$

If in (3) $t = x$, then
$$\frac{dz}{dx} = \frac{\partial z}{\partial x} + \frac{\partial z}{\partial y} \cdot \frac{dy}{dx}. \tag{5}$$

Also, if $t = x$, (4) becomes

$$\frac{dz}{dx} = \frac{\partial z}{\partial x} + \frac{\partial z}{\partial y} \cdot \frac{dy}{dx} + \frac{\partial z}{\partial r} \cdot \frac{dr}{dx} + \cdots. \tag{6}$$

It should be noted here that $\partial z/\partial x$ and dz/dx have entirely different meanings. The derivative $\partial z/\partial x$ has a definite value for any point (x, y), and for this reason is called a point function. But dz/dx depends not only on the point (x, y) but also on the derivatives dy/dx etc.; that is, dz/dx depends on the point (x, y) and the direction from which the point (x, y) is approached.

Example 1. If $z = x^3 \sin y$, $x = e^t$, $y = t^2$, find dz/dt.

Here
$$\frac{\partial z}{\partial x} = 3 x^2 \sin y, \quad \frac{dx}{dt} = e^t,$$

$$\frac{\partial z}{\partial y} = x^3 \cos y, \quad \frac{dy}{dt} = 2 t.$$

Therefore substitution in (3) gives

$$\frac{dz}{dt} = 3 x^2 e^t \sin y + 2 t x^3 \cos y,$$

which becomes
$$\frac{dz}{dt} = e^{3t} (3 \sin t^2 + 2 t \cos t^2)$$

after substituting for x and y their values in terms of t.

Example 2. The altitude of a right circular cylinder is 20 in. and is increasing at the rate of 10 in. per second. The radius of the base is 25 in. and is decreasing at the rate of 5 in. per second. Find the rate at which the volume is changing.

Let r = radius of base, h = altitude, and v = volume of cylinder.

Then
$$v = \pi r^2 h.$$

Hence
$$\frac{\partial v}{\partial r} = 2 \pi r h, \quad \frac{\partial v}{\partial h} = \pi r^2.$$

Therefore, from (3),
$$\frac{dv}{dt} = 2 \pi r h \cdot \frac{dr}{dt} + \pi r^2 \cdot \frac{dh}{dt}.$$

But
$$r = 25, \quad h = 20, \quad \frac{dr}{dt} = -5, \quad \frac{dh}{dt} = 10.$$

Consequently
$$\frac{dv}{dt} = 2 \pi \cdot 25 \cdot 20(-5) + \pi \cdot 625 \cdot 10$$
$$= 1250 \pi \text{ cu. in. per second, increase.}$$

169. Differentiation of implicit functions. An equation of the form $f(x, y) = 0$ defines y as an implicit function of x and defines x as an implicit function of y. Suppose now that

$$z = f(x, y) = 0.$$

Then, by (5), § 168, we have

$$\frac{dz}{dx} = \frac{\partial z}{\partial x} + \frac{\partial z}{\partial y} \cdot \frac{dy}{dx} = 0,$$

which gives

$$\frac{dy}{dx} = -\frac{\dfrac{\partial z}{\partial x}}{\dfrac{\partial z}{\partial y}},$$

a formula for differentiating implicit functions.

Example. Given $x^2y + \sin x = 0$. Find dy/dx.

Setting $z = x^2y + \sin x$, we have

$$\frac{\partial z}{\partial x} = 2\,xy + \cos x, \quad \frac{\partial z}{\partial y} = x^2.$$

Therefore

$$\frac{dy}{dx} = -\frac{2\,xy + \cos x}{x^2}.$$

170. Differentials. Considering in (3), § 168, the quantities dz, dx, dy, dt as differentials, we may multiply through by dt. We then have

$$dz = \frac{\partial z}{\partial x} \cdot dx + \frac{\partial z}{\partial y} \cdot dy. \tag{1}$$

The quantity dz is called the total differential of $f(x, y)$. The quantities $\partial z/\partial x \cdot dx$ and $\partial z/\partial y \cdot dy$ are called the partial differentials of $f(x, y)$. They are sometimes represented by the symbols $d_x z$ and $d_y z$ respectively. In this notation equation (1) becomes

$$dz = d_x z + d_y z. \tag{2}$$

This type of notation and the resulting formula may obviously be extended to functions of any number of variables.

Equation (2) expresses the fact that the total differential of a function is the sum of its partial differentials.

The difference between the total differential and the total increment of a function may be illustrated by the following example.

Let a rectangle have a base x and altitude y. The area is then

$$A = xy.$$

Let x be increased by dx and y by dy. Then the total increment of A is
$$\Delta A = x\,dy + y\,dx + dx\,dy.$$

But from (2) the differential of A is
$$dA = x\,dy + y\,dx.$$

Consequently the total increment of A differs from the total differential of A by the rectangle $dx\,dy$. Thus we see that the total increment of a function is not in general equal to its total differential. However, if dx and dy are taken sufficiently small, the formula for the total differential will give a good approximation to the total increment.

Example 1. Find approximately the change produced in the function $z = x^3 - y^3$ by changing x from $x = 2$ to $x = 2.01$, and changing y from $y = 1$ to $y = 1.01$.

From (2) we have $\quad dz = 3x^2\,dx - 3y^2\,dy$.

Now $x = 2$, $dx = .01$, $y = 1$, $dy = .01$. Therefore $dz = .09$.

Hence the increment of z is approximately .09. The precise total increment is .0903.

Example 2. Find approximately the change produced in a right circular cylinder by changing the height from 10 in. to 10.02 in. and the radius from 5 in. to 5.01 in.

We have $\quad v = \pi r^2 h.$
Also $\quad dv = 2\pi r h\,dr + \pi r^2\,dh.$

Now $r = 5$, $dr = .01$, $h = 10$, $dh = .02$. Therefore
$$dv = 2\pi \cdot 5 \cdot 10 \cdot .01 + \pi \cdot 25 \cdot .02 = 1.5\,\pi.$$

This is the approximate increment required.

EXERCISES

1. Given $z = x^3 + x^2 y + y^2$. Find $\dfrac{dz}{dx}$.

    Ans. $3x^2 + 2xy + x^2 \dfrac{dy}{dx} + 2y \dfrac{dy}{dx}$.

2. Given $s = e^x \sin y$. Find $\dfrac{ds}{dt}$.   Ans. $e^x \sin y \dfrac{dx}{dt} + e^x \cos y \dfrac{dy}{dt}$.

3. Given $v = e^{xyz}$. Find $\dfrac{dv}{dy}$.   Ans. $yze^{xyz}\dfrac{dx}{dy} + xze^{xyz} + xye^{xyz}\dfrac{dz}{dy}$.

4. Given $u = \dfrac{x^2}{x^2 + y^2}$. Find $\dfrac{du}{dx}$.

5. Given $z = \dfrac{x+y}{x-y}$. Find $\dfrac{dz}{dy}$.

6. A parallelepiped has for edges $a = 3$, $b = 4$, $c = 5$; a is increasing at the rate of 4 units per second, b is decreasing at the rate of 2 units per second, and c is decreasing at the rate of 5 units per second. Find the rate at which the volume is changing.

$\qquad\qquad\qquad\qquad\qquad\qquad$ Ans. -10 cubic units.

7. Suppose the characteristic equation of a gas to be $T = .4\,pv$. If at any instant $p = 15$ lb. and $v = 10$ cu. ft., the value of T is increasing at the rate of $.05°$ per second, and p is increasing at the rate of 1 lb. per second, find how fast the volume is changing.

$\qquad\qquad\qquad\qquad\qquad\qquad$ Ans. $-.66$ cu. ft. per second.

8. At a distance r in space the potential due to an electric charge is $v = e/r$. If $e = 2$ amperes and is increasing at the rate of 5 amperes per second, and $r = 8$ ft. and is increasing at the rate of 2 ft. per second, find how fast v is changing. $\qquad\qquad$ Ans. $\tfrac{9}{16}$.

Differentiate the following implicit functions by the method of § 169:

9. $x^3 y + xy^3 = 6$. $\qquad\qquad$ Ans. $\dfrac{dy}{dx} = -\dfrac{y^3 + 3x^2 y}{x^3 + 3xy^2}$.

10. $e^x \sin y = 7$. $\qquad\qquad$ Ans. $\dfrac{dy}{dx} = -\tan y$.

11. $\dfrac{xy}{x^2 + y^2} = 3$.

12. $\sin x^2 y^2 = x + y$.

13. $x^2 + y^2 = 4\,ax$.

14. $e^x + \log y = 6$.

15. $7\,ax + \log xy = 7$. (a is constant.)

16. $x^3 + y^3 = 3\,axy$.

17. $(x^2 + y^2)^2 = a^2 x^2$.

18. If $z = f(ax + by)$, show that $b\,\dfrac{\partial z}{\partial x} = a\,\dfrac{\partial z}{\partial y}$.

19. One side of a right triangle increases from 2 in. to 2.01 in. The other side decreases from 2 in. to 1.99 in. Find the total increment and the total differential (1) of the hypotenuse, (2) of the area. $\qquad\qquad$ Ans. (1) $.0001, 0$; (2) $.0001, 0$.

20. The period of a pendulum is
$$T = 2\pi\sqrt{\frac{l}{g}}.$$

If, in measuring l and g, errors of 1 per cent are made, find the greatest possible percentage of error in T.

21. The area of a triangle when two sides and the included angle are known is given by the formula
$$K = \tfrac{1}{2} bc \sin A.$$

If $b = 15$, $c = 12$, $A = 30°$, find the approximate change in area when b increases 1 unit, c decreases $\tfrac{1}{2}$ unit, and A decreases $1°$.

171. Exact differentials. It very frequently happens in practice that a form $M\,dx + N\,dy$ occurs, where M and N are functions of x and y. The question then arises whether there is a function $z = f(x, y)$ such that

$$dz = M\,dx + N\,dy.$$

If such a function $z = f(x, y)$ exists, then $M\,dx + N\,dy$ is called an exact differential; if not, it is called an inexact differential.

No complete discussion of the differential form $M\,dx + N\,dy$ will be attempted here. For such a discussion the reader is referred to works on differential equations, Wilson's "Advanced Calculus" or Goursat-Hedrick's "Mathematical Analysis," Vol. I. Only the necessary condition that $M\,dx + N\,dy$ be an exact differential will be developed here.

Suppose $M\,dx + N\,dy$ is an exact differential. Then there exists a function
$$z = f(x, y)$$
such that
$$dz = M\,dx + N\,dy.$$

But
$$dz = \frac{\partial f}{\partial x} \cdot dx + \frac{\partial f}{\partial y} \cdot dy.$$

Therefore
$$\frac{\partial f}{\partial x} \cdot dx + \frac{\partial f}{\partial y} \cdot dy = M\,dx + N\,dy.$$

Hence, since dx and dy are independent increments, it follows that
$$\frac{\partial f}{\partial x} = M, \quad \frac{\partial f}{\partial y} = N. \tag{1}$$

Differentiating (1), we obtain

$$\frac{\partial^2 f}{\partial y\, \partial x} = \frac{\partial M}{\partial y}, \quad \frac{\partial^2 f}{\partial x\, \partial y} = \frac{\partial N}{\partial x}.$$

But
$$\frac{\partial^2 f}{\partial x\, \partial y} = \frac{\partial^2 f}{\partial y\, \partial x}.$$

Therefore
$$\frac{\partial M}{\partial y} = \frac{\partial N}{\partial x}. \tag{2}$$

Equation (2) states a necessary condition that $M\, dx + N\, dy$ be an exact differential. In the references just cited above it is proved that it is also a sufficient condition.

If $\int (M\, dx + N\, dy) = z = f(x, y)$, the following rule may be stated for finding $f(x, y)$, provided that M and N are polynomials in x and y.

RULE. *Integrate $M\, dx$ considering y as constant, then integrate the terms in N which do not contain x, and add the results.*

Example. Show that the differential $3\, x^2 y\, dx + (x^3 + 2\, y)dy$ is exact, and find the function z which gives rise to it.

We have
$$M = 3\, x^2 y, \quad \frac{\partial M}{\partial y} = 3\, x^2,$$

$$N = x^3 + 2\, y, \quad \frac{\partial N}{\partial x} = 3\, x^2.$$

Therefore $3\, x^2 y\, dx + (x^3 + 2\, y)dy$ is an exact differential. Applying the above rule to this differential, we obtain

$$z = x^3 y + y^2 + c.$$

The student may check the result by finding dz.

The rule, though given only for the case when M and N are polynomials, may be tentatively employed in other cases also. In these other cases the result should always be checked by forming the differential of the resulting function.

The difficulty in applying the rule to expressions of general form, as well as to polynomials, lies in the fact that the word *term* is then not well defined.

EXERCISES

Determine which of the first ten differentials are exact, and, where exact, determine in each case the function from which the differential arises.

1. $x\,dy + y\,dx$.   Ans. $z = xy + c$.
2. $x^3 y\,dx + (x+y)dy$.   Ans. Not exact.
3. $\cos x \cos y\,dx - \sin x \sin y\,dy$.   Ans. $z = \sin x \cos y + c$.
4. $\tan y\,dx + \sec x\,dy$.   Ans. Not exact.
5. $(3x^2 y + e^x)dx + x^3\,dy$.   Ans. $z = x^3 y + e^x + c$.
6. $(2x - y)dx + (y - x)dy$.   Ans. $z = x^2 - xy + \dfrac{y^2}{2} + c$.
7. $(8x^3 y + 1)dx + (3x^2 - y)dy$.   Ans. Not exact.
8. $(x + \log y)dx + \left(y + \dfrac{1}{x}\right)dy$.   Ans. Not exact.
9. $(3x^2 + 2xy)dx + (x^3 + xy^2)dy$.   Ans. Not exact.
10. $\log y\,dx + \left(\dfrac{x}{y} + 2y\right)dy$.   Ans. $z = x \log y + y^2 + c$.

172. Envelopes. The equation $f(x, y, a) = 0$ represents a curve if a is constant. If, however, a is allowed to assume a series of values, the equation represents a series of curves called a family, and a is called the parameter of the family. Thus

$$(x - a)^2 + y^2 = 4 \qquad (1)$$

represents all the circles with radius 2 whose centers lie on the x-axis. Likewise the equation

$$ax + 2 = y \qquad (2)$$

represents all the lines through the point $(0, 2)$ except the one parallel to the y-axis.

It may happen that all the curves of a family of curves are tangent to another curve. If such is the case, the curve to which each curve of the family is tangent is called the envelope of the family. The circles (1) are all tangent to the lines $y = \pm 2$. Hence the lines $y = \pm 2$ constitute the envelope of the family of circles (1). The family (2) has no envelope.

173. Method of determining the envelope. Let

$$f(x, y, a) = 0 \tag{1}$$

be the equation of a family of curves, and suppose that this family has an envelope. Let P_a be the point of tangency of the envelope with the family. The coördinates of P_a evidently depend on the value of a. The parametric equations of the envelope may therefore be written

$$x = \phi(a), \quad y = \psi(a). \tag{2}$$

The problem now is to determine the functions $\phi(a)$ and $\psi(a)$.

Since (1) and (2) are tangent at P, we may write

$$-\frac{\dfrac{\partial f}{\partial x}}{\dfrac{\partial f}{\partial y}} = \frac{\psi'(a)}{\phi'(a)},$$

or $\quad \phi'(a)\dfrac{\partial f}{\partial x} + \psi'(a)\dfrac{\partial f}{\partial y} = 0. \tag{3}$

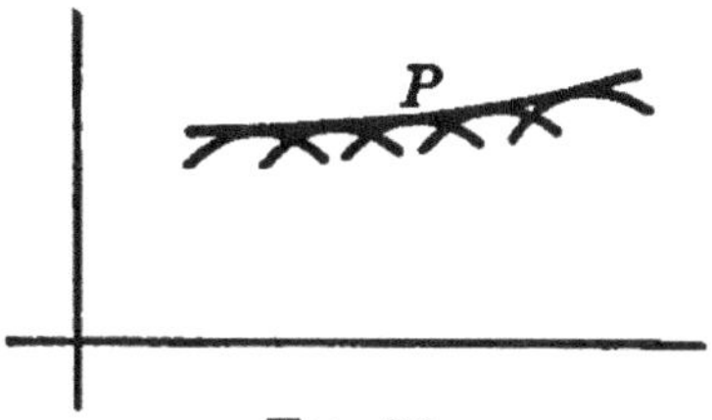

FIG. 73

The total derivative of (1) is

$$\frac{\partial f}{\partial x} \cdot \frac{dx}{da} + \frac{\partial f}{\partial y} \cdot \frac{dy}{da} + \frac{\partial f}{\partial a} = 0. \tag{4}$$

Also, from (2), $\quad \dfrac{dx}{da} = \phi'(a); \quad \dfrac{dy}{da} = \psi'(a).$

Therefore (4) may be written

$$\frac{\partial f}{\partial x} \cdot \phi'(a) + \frac{\partial f}{\partial y} \cdot \psi'(a) + \frac{\partial f}{\partial a} = 0. \tag{5}$$

Equations (3) and (5) then give

$$\frac{\partial f}{\partial a} = 0. \tag{6}$$

The simultaneous solution of (1) and (6) for x and y determines the required functions $\phi(a)$ and $\psi(a)$. The elimination of a from equations (2) then gives the Cartesian equation of the envelope of the curve.

It may also be noted that the elimination of a from equations (1) and (6) will also give the equation of the envelope of the family. The student is advised to work several problems through by the method just discussed before using this suggestion.

Example. Find the envelope of the family of lines
$$y = ax - \frac{1}{a}.$$

Here
$$f(x, y, a) = y - ax + \frac{1}{a} = 0. \tag{1'}$$

Let
$$x = \phi(a), \quad y = \psi(a). \tag{2'}$$

Then, substituting the values of x and y from (2') in (1'), we obtain
$$\psi(a) - a\phi(a) + \frac{1}{a} = 0. \tag{3'}$$

The slope of (3') is a, and the slope of (2') is $\psi'(a)/\phi'(a)$. Therefore
$$a = \frac{\psi'(a)}{\phi'(a)}. \tag{7'}$$

Differentiating (3') with respect to a, we obtain
$$\psi'(a) - a\phi'(a) - \phi(a) - \frac{1}{a^2} = 0. \tag{4'}$$

Substituting (7') in (4'), we obtain
$$\phi(a) = -\frac{1}{a^2} = x. \tag{8'}$$

Substituting the value of $\phi(a)$ from (8') in (3'), we have
$$\psi(a) = -\frac{2}{a} = y.$$

Therefore
$$x = -\frac{1}{a^2} \quad \text{and} \quad y = -\frac{2}{a}$$

are the parametric equations of the envelope. The elimination of a from these equations gives
$$y^2 + 4x = 0,$$

which is the Cartesian equation of the envelope.

The student should observe that every curve may be considered as the envelope of its tangents.

EXERCISES

1. Find the equation of the envelope of the family of straight lines $y = ax + a^2$, a being the variable parameter.   Ans. $y = -\dfrac{x^2}{4}$.

2. Find the envelope of the family of lines $y = x/a + a^2$.
   Ans. $4 y^3 = 27 x^2$.

3. Find the envelope of the family of parabolas $y = a(x - a)^2$, a being the variable parameter.   Ans. $y = \dfrac{4 x^3}{27}$.

4. Find the envelope of the straight line which makes with the coördinate axes a triangle of constant area.
   Ans. $4 xy = k$, where $\dfrac{k}{2} =$ area.

5. Find the envelope of the line $x/a + y/b = 1$, when $a^2 + b^2 = k^2$, k being a constant.

   HINT. $b = \sqrt{k^2 - a^2}$.

Therefore the equation of the line may be written

$$\frac{x}{a} + \frac{y}{\sqrt{k^2 - a^2}} = 1,$$

a being then the variable parameter.

6. Find the envelope of the family of lines $y = mx + a\sqrt{1 + m^2}$, m being the variable parameter.   Ans. $x^2 + y^2 = a^2$.

7. Find the envelope of the family of ellipses $x^2/a^2 + y^2/b^2 = 1$, where $ab = k$.   Ans. $2 xy = k$.

8. Find the envelope of the family of circles which pass through the origin and have their centers on the parabola $y^2 = 4 x$.

9. Find the envelope of the line $x \cos \alpha + y \sin \alpha = p$, α being the variable parameter.

10. Find the envelope of the line $x/a + y/b = 1$, where $a + b = k$.

11. Find the envelope of the family of circles $(x - a)^2 + y^2 = 2 a^2$, a being the variable parameter.   Ans. No envelope.

174. The evolute as the envelope of the normals. In § 135 the evolute of a curve was defined as the locus of the center of curvature of a curve. It may now be shown that the evolute of a curve is the envelope of its normals.

The equation of the normal to a curve at the point (x', y') is

$$(x - x') + \frac{dy'}{dx'}(y - y') = 0, \qquad (1)$$

in which x' is the variable parameter, since y' and dy'/dx' depend on x'. Let the parametric equations of the envelope be

$$x = \phi(x'), \quad y = \psi(x'). \qquad (2)$$

The slope of (1) is $-dx'/dy'$ and the slope of (2) is $\psi'(x')/\phi'(x')$. Therefore, since these slopes are equal, it follows that

$$\frac{dy'}{dx'}\psi'(x') = -\phi'(x'). \qquad (3)$$

Substituting in (1) the values of x and y from (2), we obtain

$$\phi(x') - x' + \frac{dy'}{dx'}[\psi(x') - y'] = 0. \qquad (4)$$

Differentiating (4) with respect to x', we obtain

$$\phi'(x') - 1 + \frac{d^2y'}{dx'^2}\left[\psi(x') - y'\right] + \frac{dy'}{dx'}\psi'(x') - \left(\frac{dy'}{dx'}\right)^2 = 0. \qquad (5)$$

A solution of (3) and (5) gives

$$\psi(x') = y = y' + \frac{1 + \left(\dfrac{dy'}{dx'}\right)^2}{\dfrac{d^2y'}{dx'^2}}. \qquad (6)$$

Substituting the value of $\psi(x')$ from (6) into (4), we have

$$\phi(x') = x = x' - \frac{\dfrac{dy'}{dx'}\left[1 + \left(\dfrac{dy'}{dx'}\right)^2\right]}{\dfrac{d^2y'}{dx'^2}}. \qquad (7)$$

Since the expressions (6) and (7) are identical with those developed in § 135 for the coördinates of the center of curvature, it follows that the envelope of the normals is the evolute of the curve.

Example. Find the evolute of the parabola $y^2 = 4\,ax$.

The equation of the normal to the parabola at the point (x', y') is

$$y - y' + \frac{y'}{2\,a}(x - x') = 0,$$

or $\qquad\qquad 2\,ay - 2\,ay' + xy' - x'y' = 0. \qquad\qquad (1)$

Let $\qquad\qquad x = \phi(x'), \quad y = \psi(x') \qquad\qquad\qquad (2)$

be the parametric equations of the envelope of the normals. The slope of the normal is $-y'/2\,a$ and the slope of the envelope is $\psi'(x')/\phi'(x')$. Therefore

$$-\frac{y'}{2\,a} = \frac{\psi'(x')}{\phi'(x')},$$

or $\qquad\qquad y'\phi'(x') + 2\,a\psi'(x') = 0. \qquad\qquad (3)$

Differentiating (1), we get

$$2\,a\psi'(x') - 2\,a\left(\frac{dy'}{dx'}\right) + y'\phi'(x') + \phi(x')\frac{dy'}{dx'} - y' - x'\frac{dy'}{dx'} = 0. \quad (4)$$

A solution of (3) and (4) gives

$$\phi(x') = \frac{(2\,a + x')\dfrac{dy'}{dx'} + y'}{\dfrac{dy'}{dx'}}. \qquad\qquad (5)$$

Now $\qquad\dfrac{dy'}{dx'} = \dfrac{2\,a}{y'} = \dfrac{a^{\frac{1}{2}}}{x'^{\frac{1}{2}}} \quad\text{and}\quad y' = 2\,a^{\frac{1}{2}}x'^{\frac{1}{2}}.$

Substituting these values in (5), we obtain

$$\phi(x') = 2\,a + 3\,x' = x. \qquad\qquad (6)$$

A substitution of this value of x in (1) gives

$$y = \psi(x') = -2\,a^{-\frac{1}{2}}x'^{\frac{3}{2}}. \qquad\qquad (7)$$

Elimination of x' from (6) and (7) gives

$$4(x - 2\,a)^3 = 27\,ay^2,$$

which is the equation of the evolute of the parabola.

EXERCISES

1. Find the evolute of the hyperbola $2\,xy = a^2$.

   Ans. $(x+y)^{\frac{2}{3}} - (x-y)^{\frac{2}{3}} = 2\,a^{\frac{2}{3}}$.

2. Find the evolute of the hypocycloid $x^{\frac{2}{3}} + y^{\frac{2}{3}} = a^{\frac{2}{3}}$.

   Ans. $(x+y)^{\frac{2}{3}} + (x-y)^{\frac{2}{3}} = 2\,a^{\frac{2}{3}}$.

3. Find the evolute of the ellipse $x^2/a^2 + y^2/b^2 = 1$.

   Ans. $(ax)^{\frac{2}{3}} + (by)^{\frac{2}{3}} = (a^2 - b^2)^{\frac{2}{3}}$.

175. Tangent plane to a surface. In the adjoining figure consider the point $P(a, b, c)$ on the surface $z = f(x, y)$. The tangent to the curve APB at $P(a, b, c)$ satisfies the relations

$$z - c = \frac{\partial z}{\partial y}\bigg]_P (y - b); \quad x = a. \tag{1}$$

The symbol $\dfrac{\partial z}{\partial y}\bigg]_P$ denotes the slope of the curve APB at the point P. Also the tangent to the curve EPD satisfies the relations

$$z - c = \frac{\partial z}{\partial x}\bigg]_P (x - a); \quad y = b. \tag{2}$$

The relations (1) and (2) evidently satisfy the equation

$$z - c = \frac{\partial z}{\partial x}\bigg]_P (x - a) + \frac{\partial z}{\partial y}\bigg]_P (y - b). \tag{3}$$

Equation (3) is of the first degree in x, y, and z and is therefore the equation of a plane. This plane contains the tangent lines determined by the relations (1) and (2). It is therefore the plane tangent to the surface $z = f(x, y)$ at P.

Suppose now that the equation of the surface is given in the form

$$F(x, y, z) = 0. \tag{4}$$

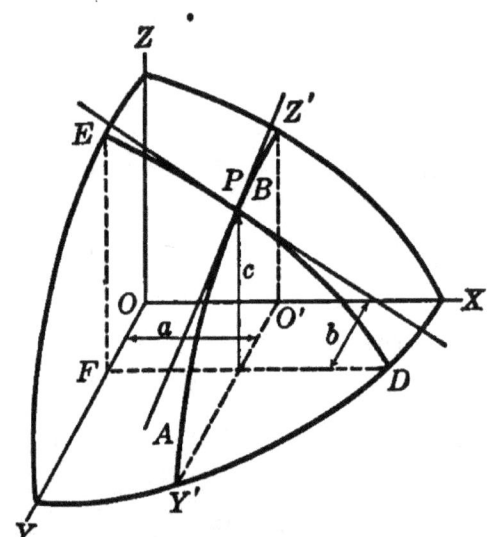

FIG. 74

From equation (4) we obtain by differentiation

$$\frac{\partial F}{\partial x} \cdot dx + \frac{\partial F}{\partial y} \cdot dy + \frac{\partial F}{\partial z} \cdot dz = 0. \tag{5}$$

Also, since in $F(x, y, z) = 0$, z is a function of x and y, we may write

$$dz = \frac{\partial z}{\partial x} \cdot dx + \frac{\partial z}{\partial y} \cdot dy. \tag{6}$$

Substitution of the value of dz from (6) in (5) gives

$$\left(\frac{\partial F}{\partial x} + \frac{\partial F}{\partial z} \cdot \frac{\partial z}{\partial x}\right) dx + \left(\frac{\partial F}{\partial y} + \frac{\partial F}{\partial z} \cdot \frac{\partial z}{\partial y}\right) dy = 0. \quad (7)$$

But dx and dy are arbitrary independent increments. Therefore (7) can be satisfied only if the coefficients of dx and dy vanish separately. This gives

$$\frac{\partial z}{\partial x} = -\frac{\dfrac{\partial F}{\partial x}}{\dfrac{\partial F}{\partial z}}; \quad \frac{\partial z}{\partial y} = -\frac{\dfrac{\partial F}{\partial y}}{\dfrac{\partial F}{\partial z}}. \quad (8)$$

A substitution in (3) of the values of $\partial z/\partial x$ and $\partial z/\partial y$ from (8) gives

$$\frac{\partial F}{\partial x}\bigg]_P (x - a) + \frac{\partial F}{\partial y}\bigg]_P (y - b) + \frac{\partial F}{\partial z}\bigg]_P (z - c) = 0, \quad (9)$$

which is the equation of the tangent plane to the surface $F(x, y, z) = 0$ at the point $P(a, b, c)$.

176. Line normal to a surface. The equations of the normal to the surface $z = f(x, y)$ at $P(a, b, c)$ are the equations of the line through $P(a, b, c)$ perpendicular to the plane (1) of §175.

Now the angles made by the tangent plane with the coördinate planes are equal to the angles made by the normal with the coördinate axes. Consequently the direction cosines of a line perpendicular to the plane (1) of §175 are proportional to

$$\frac{\partial z}{\partial x}, \quad \frac{\partial z}{\partial y}, \quad -1.$$

Hence the equations of the normal to the surface $z = f(x, y)$ at $P(a, b, c)$ are

$$\frac{x - a}{\dfrac{\partial z}{\partial x}\bigg]_P} = \frac{y - b}{\dfrac{\partial z}{\partial y}\bigg]_P} = \frac{z - c}{-1}. \quad (1)$$

Similarly, if the equation of the surface is $F(x, y, z) = 0$, the equations of the normal to the surface at $P(a, b, c)$ are

$$\frac{x - a}{\dfrac{\partial F}{\partial x}\bigg]_P} = \frac{y - b}{\dfrac{\partial F}{\partial y}\bigg]_P} = \frac{z - c}{\dfrac{\partial F}{\partial z}\bigg]_P}$$

Example. Find the equations of the tangent plane and the normal to the surface $x^2 + 2y^2 + z^2 = 18$ at the point $(1, 2, 3)$.

Here
$$\frac{\partial F}{\partial x} = 2x; \quad \frac{\partial F}{\partial y} = 4y; \quad \frac{\partial F}{\partial z} = 2z,$$

and
$$\left.\frac{\partial F}{\partial x}\right]_P = 2; \quad \left.\frac{\partial F}{\partial y}\right]_P = 8; \quad \left.\frac{\partial F}{\partial z}\right]_P = 6.$$

Therefore the equation of the tangent plane is
$$2(x - 1) + 8(y - 2) + 6(z - 3) = 0,$$

which reduces to $\quad x + 4y + 3z = 18.$

The equations of the normal are
$$\frac{x-1}{2} = \frac{y-2}{8} = \frac{z-3}{6}.$$

EXERCISES

Find the equations of the tangent planes and normals to the following surfaces at the points specified:

1. $x^2 + y^2 + z^2 = 9$, at $(2, 2, 1)$.
   Ans. $2x + 2y + z = 9$; $x - 2 = y - 2 = 2z - 2$.

2. $z = x - y^2$, at $(4, 1, 3)$.
   Ans. $2y - x + z = 1$; $4 - 2x = y - 1 = 2z - 6$.

3. $z = xy$, at origin.
   Ans. $z = 0$; $y = x = 0$.

4. $z = x^2 - y^2$, at $(1, 1, 0)$.
   Ans. $2y - x + z = 0$; $1 - x = y - 1 = 2z$.

5. $x^2 + y^2 + z^2 = a^2$, at (x', y', z').
   Ans. $xx' + yy' + zz' = a^2$; $\dfrac{x-x'}{x'} = \dfrac{y-y'}{y'} = \dfrac{z-z'}{z'}$.

6. $3x^2 + y^2 - 2z = 0$, at $(1, 1, 2)$.
   Ans. $3x + y - z = 2$; $x - 1 = 3y - 3 = 6 - 3z$.

7. $x^2 y^2 + 2x + z^3 = 16$, at $(2, 1, 2)$.

8. Determine the equation of the tangent plane to the surface $x^{\frac{2}{3}} + y^{\frac{2}{3}} + z^{\frac{2}{3}} = a^{\frac{2}{3}}$ at (x', y', z'), and show that the sum of the squares of its intercepts on the coördinate axes is constant.

9. Prove that the tetrahedron formed by the coördinate planes and a tangent to the surface $xyz = a^3$ has a constant volume.

10. Find the equations of the projections on the coördinate planes of the normal to the surface $x = y + z^2$ at $(2, 1, 1)$.

177. Summary. In this chapter we have developed the formulas for partial and total differentiation of functions of two or more independent variables. We have applied this theory to problems of rates which depend on more than one variable, to the differentiation of implicit functions of two variables, to the finding of approximate values of functions, to the determination of envelopes of families of curves, and to the determination of the equations of tangent planes and normals to a surface.

CHAPTER XX

FUNCTIONS OF TWO OR MORE VARIABLES. INTEGRATION

178. Definitions and processes. The symbol

$$\iint f(x, y) dx\, dy$$

is called a double integral. It means that $f(x, y)$ is to be integrated with respect to x, y remaining constant, and the result is to be integrated with respect to y, x being considered constant.

Let $\quad f(x, y) = x^2 + y^2.$

Then write $\quad \iint (x^2 + y^2) dx\, dy.$

Integrating with respect to x, considering y constant, we have

$$\iint (x^2 + y^2) dx\, dy = \int \left(\frac{x^3}{3} + xy^2 + c_1\right) dy. \tag{1}$$

Integrating this result with respect to y, treating x as constant, we obtain

$$\int \left(\frac{x^3}{3} + xy^2 + c_1\right) dy = \frac{x^3 y}{3} + \frac{xy^3}{3} + \int c_1 dy + c_2. \tag{2}$$

This integration differs from the successive integration of functions of a single variable in one respect, namely, in the form of the constant of integration.

In equation (1) c_1 is the constant of integration. However, in this integration y was treated as a constant. Hence c_1 may involve y. We may therefore write

$$c_1 = \phi(y),$$

and equation (1) may be written as follows:

$$\iint (x^2 + y^2) dx\, dy = \int \left[\frac{x^3}{3} + xy^2 + \phi(y)\right] dy. \tag{3}$$

347

Integrating the right-hand member of (3), we obtain

$$\int \left[\frac{x^3}{3} + xy^2 + \phi(y)\right] dy = \frac{x^3 y}{3} + \frac{xy^3}{3} + \int \phi(y) dy + c_2.$$

The constant c_2 may involve x, since x was treated as a constant in this integration. We may therefore write

$$c_2 = \psi(x).$$

We then have finally

$$\iint (x^2 + y^2) dx\, dy = \frac{x^3 y}{3} + \frac{xy^3}{3} + \int \phi(y) dy + \psi(x).$$

The two constants of integration, $\phi(y)$ and $\psi(x)$, are arbitrary functions of y and x respectively.

The symbol $\iiint \cdots f(x, y, z \cdots) dx\, dy\, dz \cdots$

is called a multiple integral and means that $f(x, y, z, \cdots)$ is to be integrated with respect to x, the remaining variables being treated as constants, this result to be integrated with respect to y, with the remaining variables treated as constants, etc.*

The symbol
$$\int_c^d \int_a^b f(x, y) dx\, dy$$

is called a definite double integral. The symbol means that $f(x, y)$ is integrated between the limits $x = a$ and $x = b$, y being treated as a constant, this result to be integrated with respect to y between the limits $y = c$ and $y = d$. The limits a and b may be functions of y. The limits c and d are usually constants. Definite multiple integrals with a greater number of variables are defined in a similar manner. The following example will illustrate the definition:

Example. Evaluate the integral

$$\int_1^2 \int_y^{y^2} xy\, dx\, dy.$$

---

* Textbooks differ in the convention adopted concerning the order in which the integrations are to be performed. Some use the above order, and others use the reverse order. In reading a text one must be sure to find out which convention has been adopted.

Integration with respect to x gives

$$\int_1^2 \frac{x^2 y}{2}\bigg]_y^{y^2} dy = \int_1^2 \left(\frac{y^5}{2} - \frac{y^3}{2}\right) dy.$$

Integration of this result with respect to y gives

$$\frac{y^6}{12} - \frac{y^4}{8}\bigg]_{y=1}^{y=2} = \frac{27}{8}.$$

EXERCISES

Integrate the following:

1. $\iint xy\, dx\, dy.$
   Ans. $\dfrac{x^2 y^2}{4} + \int \phi(y)dy + \psi(x).$

2. $\iiint xyz\, dx\, dy\, dz.$
   Ans. $\dfrac{x^2 y^2 z^2}{8} + \iint \psi(yz)dy\, dz + \int \phi(xz)dz + f(xy).$

3. $\int_1^2 \int_1^2 (x+y)\, dx\, dy.$ Ans. 3.

4. $\int_0^4 \int_x^{2x} xy\, dy\, dx.$ Ans. 96.

5. $\int_1^2 \int_0^x (x^2 + y^2)\, dy\, dx.$ Ans. 5.

6. $\int_0^{\frac{\pi}{2}} \int_0^{\sin\theta} r\cos\theta\, dr\, d\theta.$ Ans. $\frac{1}{6}.$

7. $\int_3^4 \int_0^{\pi} r\theta\, d\theta\, dr.$ Ans. $\frac{7}{4}\pi^2.$

8. $\int_1^2 \int_y^{4y} (x^2 + y^2)\, dx\, dy.$ Ans. 90.

9. $\int_0^a \int_0^{\sqrt{a^2 - x^2}} dy\, dx.$ Ans. $\dfrac{\pi a^2}{4}.$

10. $\int_1^2 \int_1^2 \int_0^1 (xyz)\, dx\, dz\, dy.$ Ans. $\frac{9}{8}.$

11. $\int_0^\pi \int_0^\pi \int_{\sin\theta}^{\sin\phi} dr\, d\theta\, d\phi.$

12. $\int_{-1}^2 \int_{-2}^0 \dfrac{xy}{4+y^2}\, dx\, dy.$

13. $\int_0^a \int_0^x \int_0^{x+y} (xyz)\, dz\, dy\, dx.$

14. $\int_{-1}^1 \int_0^\phi \int_{-\theta}^{\theta-\phi} r\, dr\, d\theta\, d\phi.$

15. $\int_1^4 \int_0^{\log 2} ye^x\, dx\, dy.$

16. $\int_0^a \int_0^{\sqrt{a^2-x^2}} (x+y)\, dy\, dx.$

CHAPTER XXI

FURTHER APPLICATIONS TO GEOMETRY

179. Plane area by means of a double integration; rectangular coördinates. Let $y = f(x)$ and $y = \phi(x)$ be the equations of two curves. And let it be proposed to find the area bounded by the curves $y = f(x)$, $y = \phi(x)$ and the ordinates $x = a$ and $x = b$ (see Fig. 75), assuming that $f(x)$ and $\phi(x)$ are continuous and have no common points in the interval from $x = a$ to $x = b$.

From § 99, the area $ABB'A'$ is given by

$$\int_a^b \phi(x)dx,$$

and the area $ABB''A''$ is given by

$$\int_a^b f(x)dx.$$

Therefore the area $A'B'B''A''$ is given by

$$\int_a^b [f(x) - \phi(x)]dx.$$

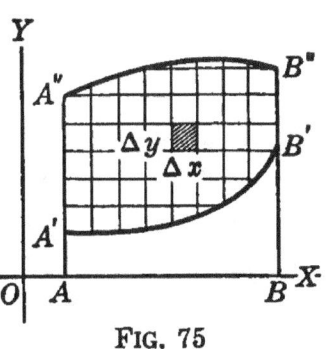

Fig. 75

But, from the definition of a definite double integral,

$$\int_a^b \int_{\phi(x)}^{f(x)} dy\, dx = \int_a^b [f(x) - \phi(x)]dx.$$

Hence the area $A'B'B''A''$ is given by the formula

$$A = \int_a^b \int_{\phi(x)}^{f(x)} dy\, dx.$$

Similarly, if the equations of the curves are $x = \psi(y)$ and $x = \zeta(y)$, and the area to be found is bounded by the curves $x = \psi(y)$, $x = \zeta(y)$ and the lines $y = c$ and $y = d$, the formula for area is

$$A = \int_c^d \int_{\zeta(y)}^{\psi(y)} dx\, dy.$$

The area $A'B'B''A''$ may also be considered as the limit of a double summation. For, suppose the area $A'B'B''A''$ to be divided by lines drawn parallel to the x-axis and y-axis into rectangles with bases Δx and altitudes Δy, plus some irregular areas as shown in the figure. An approximation to the area will be obtained by adding all the rectangles in each column and then adding the columns. Let this addition be denoted by the symbol

$$\sum_{a}^{b}\left[\sum_{\phi(x)}^{f(x)}\Delta y\right]\Delta x.$$

It then follows that

$$\lim_{\substack{\Delta y=0 \\ \Delta x=0}}\sum_{a}^{b}\left[\sum_{\phi(x)}^{f(x)}\Delta y\right]\Delta x = \text{area } A'B'B''A''.$$

Therefore $\quad A = \lim\limits_{\substack{\Delta y=0 \\ \Delta x=0}}\sum_{a}^{b}\left[\sum_{\phi(x)}^{f(x)}\Delta y\right]\Delta x = \int_{a}^{b}\int_{\phi(x)}^{f(x)}dy\,dx.$ \hfill (1)

If Δx and Δy are considered as differentials, a rectangle with Δx as base and Δy as altitude may be called a differential element of area. The summation in the left-hand member of (1) then means that:

First. All elements of area in a column are added giving the area of a strip, such as the shaded strip in Fig. 76, parallel to the y-axis, and extending from $y = \phi(x)$ to $y = f(x)$ except for an irregular area at either end.

Second. All such strips from $x = a$ to $x = b$ are added.

This notion is very useful in setting the limits on the integrals in the right-hand member of (1). Since the first summation extends from $y = \phi(x)$ to $y = f(x)$, the first set of limits on the integral will be from $y = \phi(x)$ to $y = f(x)$. And since the second summation calls for the addition of all strips from $x = a$ to $x = b$, the second set of limits on the integral will be from $x = a$ to $x = b$.

Example. Find the area bounded by the parabola $y = x^2$ and the line $y = 2x$.

Let us consider the area as divided into strips parallel to the y-axis. The solution may then be thought of as, first, finding the area of a strip, such as the shaded strip in the figure, running from $y = x^2$ to $y = 2x$, and then adding all such strips from $x = 0$ to $x = 2$, 0 and 2 being the two values of x where the curves intersect. Therefore

$$A = \int_0^2 \int_{x^2}^{2x} dy\, dx = \int_0^2 (2x - x^2) dx = \tfrac{4}{3}.$$

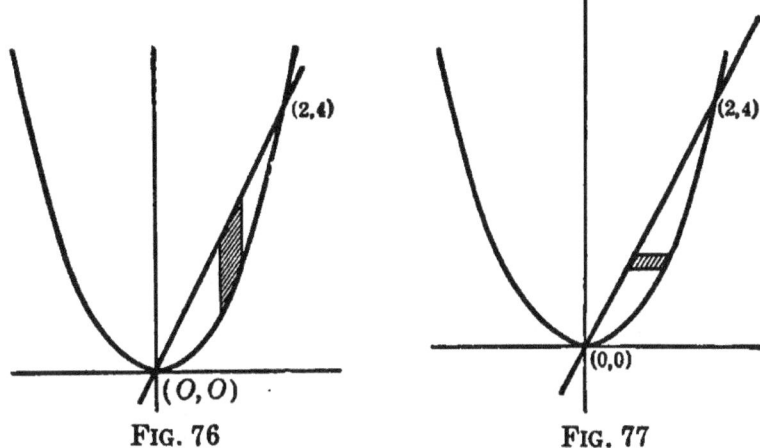

FIG. 76 FIG. 77

Suppose now that the area is divided into strips parallel to the x-axis. The solution may then be thought of as finding the area of a strip, such as the shaded strip in the figure, running from $x = y/2$ to $x = \sqrt{y}$ and then adding all such strips from $y = 0$ to $y = 4$. Therefore

$$A = \int_0^4 \int_{\frac{y}{2}}^{\sqrt{y}} dx\, dy = \int_0^4 \left(\sqrt{y} - \tfrac{y}{2}\right) dy = \tfrac{4}{3}.$$

In the above example either method of summation is equally advantageous. However, in some problems a proper choice of the order of summation is important. The following example will illustrate this point:

Example. Find the area bounded by the line $x = 0$, the parabola $y = 2x^2$, and the circle $x^2 + y^2 = 5$.

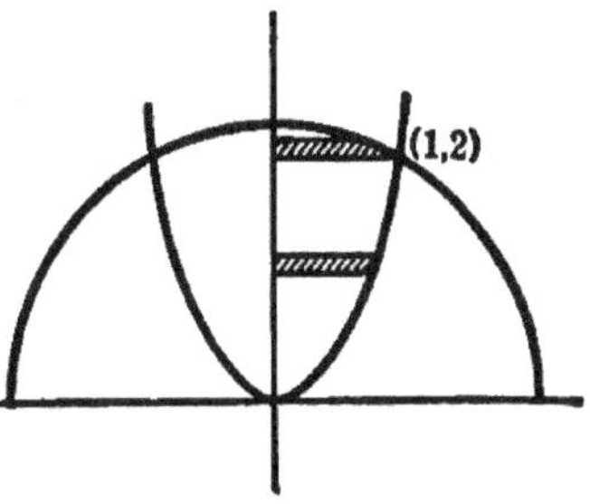

In this problem, if the strips are considered vertical, they will run from $y = 2x^2$ to $y = \sqrt{5 - x^2}$, and these strips will extend from $x = 0$ to $x = 1$. The formula for area will therefore be

FIG. 78

$$A = \int_0^1 \int_{2x^2}^{\sqrt{5-x^2}} dy\, dx = \int_0^1 \left[\sqrt{5 - x^2} - 2x^2\right] dx = \frac{1}{3} + \frac{5}{2} \sin^{-1} \frac{1}{\sqrt{5}}.$$

If, however, the strips are taken horizontal, some of them will run from $x=0$ to $x=\sqrt{y/2}$ and others will run from $x=0$ to $x=\sqrt{5-y^2}$. The first set of strips will extend from $y=0$ to $y=2$ and the second set from $y=2$ to $y=\sqrt{5}$. In this case the area must be considered in two portions, and will be given by the formula

$$A = \int_0^2 \int_0^{\sqrt{y/2}} dx\,dy + \int_2^{\sqrt{5}} \int_0^{\sqrt{5-y^2}} dx\,dy = \frac{1}{3} + \frac{5}{2}\sin^{-1}\frac{1}{\sqrt{5}}.$$

The student will observe that the first solution of this problem is simpler than the second. He should be constantly on the lookout for similar situations.

EXERCISES

In the first four exercises find the areas indicated by double integration, assuming first that the strips run parallel to the y-axis and second that the strips run parallel to the x-axis:

1. Area bounded by the parabola $y = 4x^2$ and the lines $x = 4$ and $y = 0$.   Ans. $2\frac{5 \cdot 6}{3}$.

2. Area bounded by the line $y = 0$, the curve $y = x^3$, and the line $x = 2$.   Ans. 4.

3. Area bounded by the parabola $y - 2 = -4x^2$ and the x-axis.   Ans. $\dfrac{4\sqrt{2}}{3}$.

4. Area bounded by the circle $x^2 + y^2 = 25$ and the line $x + y = 5$.   Ans. $\dfrac{25}{2}\left[\dfrac{\pi}{2} - 1\right]$.

5. Find by double integration the area bounded by the curve $x + y = a$ and the coördinate axes.   Ans. $\frac{1}{2}a^2$.

6. Find by double integration the area of one arch of the sine curve.   Ans. 2.

7. Find by double integration the area bounded by the curves $4y^2 = x^3$ and $y = x$.   Ans. $\frac{8}{5}$.

8. Find by double integration the area in the first quadrant bounded by the x-axis, the parabola $y = x^2$, and the circle $x^2 + y^2 = 25$.

180. Plane area by the means of double integration; polar coördinates. The area bounded by the curves $\rho = f_1(\theta)$ and $\rho = f_2(\theta)$ and the radii vectores $\theta = \theta_1$ and $\theta = \theta_2$, where $\rho = f_1(\theta)$

and $\rho = f_2(\theta)$ are continuous and have no common points between $\theta = \theta_1$ and $\theta = \theta_2$, may be calculated by the method of § 136, giving
$$A = \tfrac{1}{2}\int_{\theta_1}^{\theta_2}\{[f_2(\theta)]^2 - [f_1(\theta)]^2\}\,d\theta.$$

But by the definition of the definite double integral we have
$$\int_{\theta_1}^{\theta_2}\int_{f_1(\theta)}^{f_2(\theta)}\rho\,d\rho\,d\theta = \tfrac{1}{2}\int_{\theta_1}^{\theta_2}\{[f_2(\theta)]^2 - [f_1(\theta)]^2\}\,d\theta.$$

Hence the area required is given by

$$A = \int_{\theta_1}^{\theta_2}\int_{f_1(\theta)}^{f_2(\theta)}\rho\,d\rho\,d\theta. \quad (1)$$

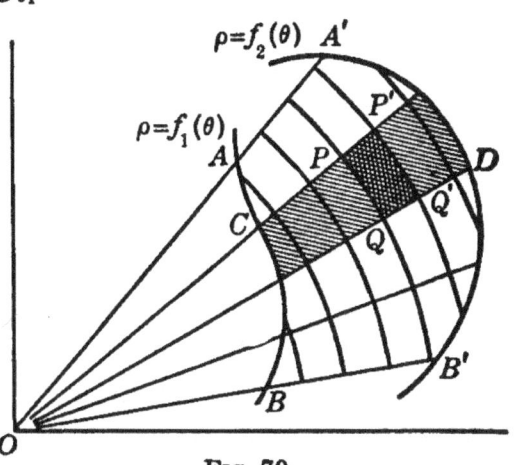

Fig. 79

The area may also be considered as the limit of a double summation, as follows. Let the area $AA'B'B$ (see Fig. 79) be divided by radii drawn from O and by arcs of circles drawn with O as center into curvilinear quadrilaterals plus some irregular figures as shown. Let $PP'Q'Q$ be one such quadrilateral. Let

$$\angle POQ = \Delta\theta \text{ and } PP' = \Delta\rho. \quad (OP = \rho)$$

Then area of sector $POQ = \tfrac{1}{2}\rho^2\Delta\theta$.
area of sector $P'OQ' = \tfrac{1}{2}(\rho + \Delta\rho)^2\Delta\theta$.

Hence area of $PP'Q'Q = \rho\Delta\rho\Delta\theta + \tfrac{1}{2}\overline{\Delta\rho}^2\Delta\theta$.

Now keeping θ and $\Delta\theta$ constant and adding such areas as $PP'Q'Q$ from $\rho = f_1(\theta)$ to $\rho = f_2(\theta)$ gives approximately the area of the shaded strip CD. The sum of all such strips gives approximately the area $AA'B'B$.

Let this sum be denoted by the symbol
$$\sum_{\theta_1}^{\theta_2}\left[\sum_{f_1(\theta)}^{f_2(\theta)}(\rho\Delta\rho + \tfrac{1}{2}\overline{\Delta\rho}^2)\right]\Delta\theta.$$

Then $\lim\limits_{\substack{\Delta\rho=0 \\ \Delta\theta=0}} \sum\limits_{\theta_1}^{\theta_2}\left[\sum\limits_{f_1(\theta)}^{f_2(\theta)}(\rho\Delta\rho + \tfrac{1}{2}\overline{\Delta\rho}^2)\right]\Delta\theta = \text{area } AA'BB'$.

Therefore
$$A = \lim_{\substack{\Delta\rho=0\\ \Delta\theta=0}} \sum_{\theta_1}^{\theta_2}\left[\sum_{f_1(\theta)}^{f_2(\theta)}(\rho\,\Delta\rho + \tfrac{1}{2}\overline{\Delta\rho}^2)\right]\Delta\theta = \int_{\theta_1}^{\theta_2}\int_{f_1(\theta)}^{f_2(\theta)} \rho\,d\rho\,d\theta.$$

If in the above summation ρ and $\Delta\rho$ are kept constant, we will obtain a segment of a circular ring. The limit of the sum of all such rings as $\Delta\rho$ and $\Delta\theta$ approach zero as a limit will also give the required area. This summation gives

$$A = \int_{\rho_1}^{\rho_2}\int_{\phi_1(\rho)}^{\phi_2(\rho)} \rho\,d\theta\,d\rho.$$

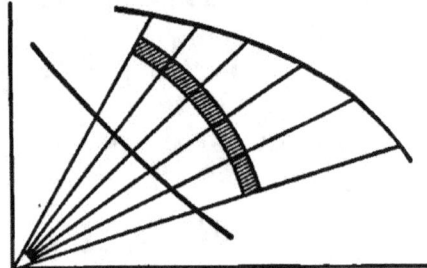

FIG. 80

Example. Find the area bounded by the circles $\rho = a$ and $\rho = a\cos\theta$.

This area is to be considered in two parts (see Fig. 81). The limits on the first part will be for ρ from $\rho = a\cos\theta$ to $\rho = a$, and for θ from $\theta = 0$ to $\theta = \pi/2$. The limits on the second part will be for ρ from $\rho = 0$ to $\rho = a$, and for θ from $\theta = \pi/2$ to $\theta = \pi$. The formula for the area is therefore

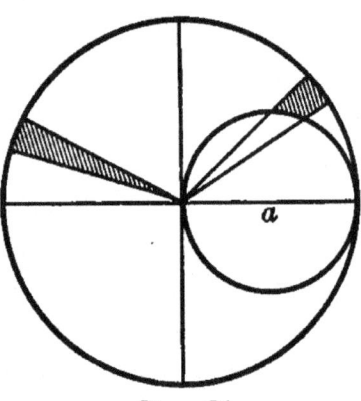

FIG. 81

$$A = 2\int_0^{\frac{\pi}{2}}\int_{a\cos\theta}^{a} \rho\,d\rho\,d\theta + 2\int_{\frac{\pi}{2}}^{\pi}\int_0^a \rho\,d\rho\,d\theta = \frac{3\,\pi a^2}{4}.$$

EXERCISES

1. Find by double integration the area of the circle $\rho = a$.
   Ans. πa^2.

2. Find the area bounded by the cardioid $\rho = a(1 - \cos\theta)$.
   Ans. $\tfrac{3}{2}\pi a^2$.

3. Find the area of one loop of the curve $\rho = a\sin 2\theta$.
   Ans. $\tfrac{1}{8}\pi a^2$.

4. The circle $\rho = a$ and the curve $\rho = a\sin 2\theta$ are tangent at the point $(a, \pi/4)$. Find by double integration the area bounded by the circle $\rho = a$ and the curve $\rho = a\sin 2\theta$ between the initial line and their point of tangency.
   Ans. $\tfrac{1}{16}\pi a^2$.

5. Find the entire area bounded by the lemniscate $\rho^2 = a^2 \cos 2\theta$.
Ans. a^2.

6. Find the area bounded by the parabola $\rho = a \sec^2 (\theta/2)$ and the circle $\rho = 2a$ from $\theta = 0$ to $\theta = \pi/2$. *Ans.* $a^2(\pi - \frac{4}{3})$.

7. Find the area of the circle with center at the point (2, 0) and passing through the origin, using (1) rectangular, (2) polar coördinates. *Ans.* 4π.

8. Find the area of an ellipse with major axis 6 and minor axis 4 by double integration, using (1) rectangular coördinates, (2) polar coördinates. *Ans.* 6π.

9. Find the area bounded by $\rho = 10$ and $\rho \sin \theta = 5$.

10. Find the area bounded by $\rho \cos \theta = 5$ and $\rho = 10 \cos \theta$.

181. Volume by double integration. Let $z = f(x, y)$ be the equation of a surface. Let T be a portion of this surface with a boundary PQ, and let RS be the projection of PQ on the xy-plane. Let it be required to find the volume bounded by T, the xy-plane, and the cylindrical surface determined by PQ and RS.

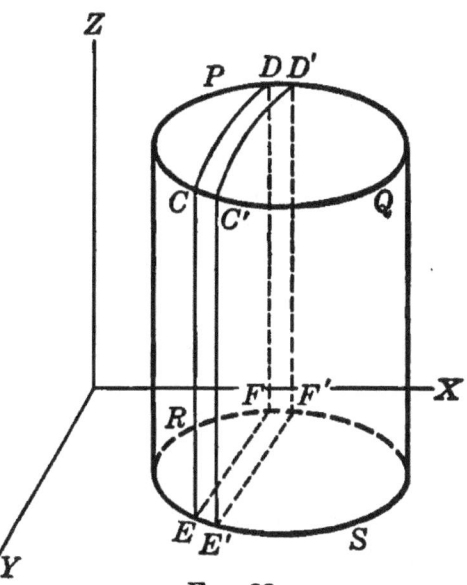

FIG. 82

Let the plane $x = x_1$ cut T in the curve CD and the projection of T in the line EF. The coördinates of E are $x = x_1$, $y = y_1$, and of F are $x = x_1$, $y = y_2$. The equation of the curve CD is $z = f(x_1, y)$, and since for the present x_1 is considered fixed, z is a function of y alone. Therefore the area of $CDFE$ is given by the formula

$$\text{area } CDFE = \int_{y_1}^{y_2} z\, dy = \int_{y_1}^{y_2} f(x_1, y)\, dy.$$

Now suppose the plane $x = x_1 + \Delta x$ cuts T in the curve $C'D'$ and the projection of T in the line $E'F'$. Then the volume CF'

is given approximately by the area $CDFE$ times Δx, that is, by the expression
$$\left[\int_{y_1}^{y_2} f(x_1, y)\,dy\right]\Delta x.$$

The sum of all such slices over the area bounded by RS is approximately the required volume V'. An approximate formula for V' is therefore
$$V' = \sum_a^b \left[\int_{y_1}^{y_2} f(x_1, y)\,dy\right]\Delta x.$$

Hence the exact volume is given by
$$V = \lim_{\Delta x \to 0} \sum_a^b \left[\int_{y_1}^{y_2} f(x_1, y)\,dy\right]\Delta x = \int_a^b \int_{y_1}^{y_2} f(x, y)\,dy\,dx.$$

It should be noted by the student that y_1 and y_2 are functions of x alone, and are determined by the projection of T on the xy-plane.

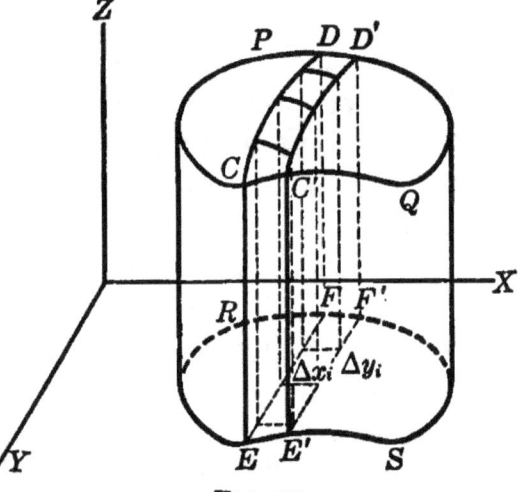

Fig. 83

The volume $PQRS$ may be considered as the limit of a double summation as follows. Draw lines in the xy-plane parallel to the x-axis at a distance Δy_i apart. Through these lines pass planes perpendicular to the xy-plane. Also draw lines parallel to the y-axis at a distance Δx_i apart, and pass through these lines planes perpendicular to the xy-plane. These two sets of parallel planes divide the volume into prismatic elements plus some irregular portions as shown in the figure.

The volume of one of these prismatic elements is approximately
$$z_i\,\Delta x_i\,\Delta y_i,$$
where z_i is a value of z within the element considered. The addition of all such elements in the slice CF' will give approxi-

mately the volume of CF'. Let this summation be denoted by the symbol

$$\left[\sum_{E}^{F} z\,\Delta y\right]\Delta x.$$

The sum of all such slices will be an approximation to the required volume. Let this sum be denoted by the symbol

$$\sum_{a}^{b}\left[\sum_{E}^{F} z\,\Delta y\right]\Delta x.$$

The exact volume will then be given by the formula

$$V = \lim_{\substack{\Delta y = 0 \\ \Delta x = 0}} \sum_{a}^{b}\left[\sum_{E}^{F} z\,\Delta y\right]\Delta x = \int_{a}^{b}\int_{y_1}^{y_2} f(x, y)\,dy\,dx.$$

The student will observe from this expression for V that integration with respect to y may be thought of as a summation of elements of a thin slice of volume parallel to the y-axis, and that the second integration with respect to x may be considered as the addition of these slices.

A change in the order of the summation above will give the following formula for volume:

$$V = \int_{c}^{d}\int_{\phi(y)}^{\psi(y)} f(x, y)\,dx\,dy.$$

In a similar way it may be shown that the volume between the surface and the xz-plane is given by the formula

$$V = \int_{a}^{b}\int_{f_1(x)}^{f_2(x)} f(x, z)\,dz\,dx.$$

Also the volume between the surface and the yz-plane is given by

$$V = \int_{c}^{d}\int_{f_1(y)}^{f_2(y)} f(y, z)\,dz\,dy.$$

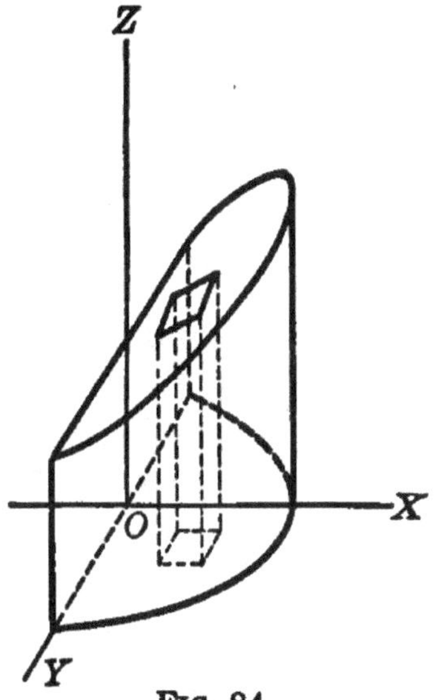

FIG. 84

Example. Find the volume bounded by the cylinder $x^2 + y^2 = a^2$ and the planes $z = 2x + a$, $z = 0$, and $x = 0$.

The element of volume is a prismatic element extending from the plane $z = 0$ to the plane $z = 2x + a$. Integration first with respect to x and then with respect to y requires the following integral:

$$V = 2\int_0^a \int_0^{\sqrt{a^2-y^2}} (2x + a)dx\,dy = 2\int_0^a (a^2 - y^2 + a\sqrt{a^2-y^2})dy$$
$$= \frac{4a^3}{3} + \frac{\pi a^3}{2}.$$

EXERCISES

1. Find the volume bounded by the plane $z = 6 - 3x - 2y$ and the coördinate planes. *Ans.* 6.

2. Find the volume common to the cylinders $x^2 + y^2 = a^2$ and $x^2 + z^2 = a^2$. *Ans.* $\frac{16}{3} a^3$.

3. Find the volume bounded by the surface $z = xy$ and the planes $z = 0$, $x = 1$, $x = 4$, $y = 0$, and $y = 6$. *Ans.* 135.

4. Find the volume of a sphere by double integration. *Ans.* $\frac{4}{3}\pi a^3$.

5. Find the volume bounded by the cylinder $x + 2 = y^2$ and the planes $z = 0$ and $z = x$. *Ans.* $\dfrac{32\sqrt{2}}{15}$.

6. Find the volume bounded by the surface $x^{\frac{1}{2}} + y^{\frac{1}{2}} + z^{\frac{1}{2}} = a^{\frac{1}{2}}$ and the coördinate planes. *Ans.* $a^3/90$.

7. Find the volume bounded by the paraboloid $z = x^2 + y^2 - 1$ and the plane $z = 0$. *Ans.* $\dfrac{\pi}{2}$.

8. Find the volume bounded by the cone $x^2 + y^2 = 4z^2$ and the plane $z = 4$.

182. Volumes by triple integration. Let $z = f_1(x, y)$ and $z = f_2(x, y)$ be the equations of two surfaces. Let T_1 and T_2 be portions of these surfaces which have the same projection on the xy-plane. Then, by § 181, it has been shown that the volume bounded by T_1, T_2, and the cylindrical surface determined by the boundaries of T_1 and T_2 is given by

$$V = \int_a^b \int_{y_1}^{y_2} f_2(x, y)dy\,dx - \int_a^b \int_{y_1}^{y_2} f_1(x, y)dy\,dx$$
$$= \int_a^b \int_{y_1}^{y_2} [f_2(x, y) - f_1(x, y)]dy\,dx.$$

But, by the definition of multiple integrals,

$$\int_a^b \int_{y_1}^{y_2} \int_{f_1(x,\,y)}^{f_2(x,\,y)} dz\,dy\,dx = \int_a^b \int_{y_1}^{y_2} [f_2(x,\,y) - f_1(x,\,y)]\,dy\,dx.$$

Hence
$$V = \int_a^b \int_{y_1}^{y_2} \int_{f_1(x,\,y)}^{f_2(x,\,y)} dz\,dy\,dx. \tag{1}$$

We may also think of volume as the limit of a triple summation. For let $P(x_1, y_1, z_1)$ be any point in the volume considered, and let $Q(x_1 + \Delta x, y_1 + \Delta y, z_1 + \Delta z)$ be another point of the volume so situated that the parallelepiped bounded by the planes $x = x_1$, $x = x_1 + \Delta x$, $y = y_1$, $y = y_1 + \Delta y$, $z = z_1$, $z = z_1 + \Delta z$ lies entirely in the volume. The volume V' of this parallelepiped is

$$V' = \Delta x\,\Delta y\,\Delta z.$$

The entire volume will then be given by

$$V = \lim_{\substack{\Delta z = 0 \\ \Delta y = 0 \\ \Delta x = 0}} \sum\sum\sum \Delta z\,\Delta y\,\Delta x. \tag{2}$$

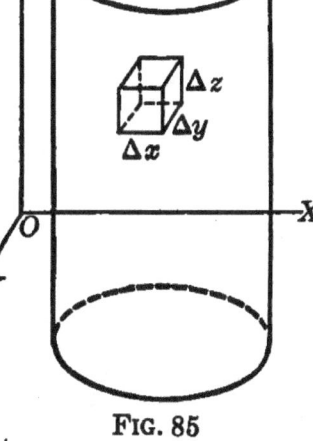

FIG. 85

The triple summation (2) may be evaluated by means of formula (1) by three successive integrations:

1. Integration with respect to z, the limits being from the value of z in terms of x and y on one surface to the value of z in terms of x and y on the other surface.

2. Integration with respect to y, the limits being in terms of x and being determined by the projection of the surfaces on the xy-plane.

3. Integration with respect to x, the limits being the extreme values of x in the projection of the surface on the xy-plane.

The following example will illustrate the meaning of these statements:

Example. Find the volume bounded by the paraboloid $z = x^2 + y^2$ and the plane $z = 9$.

In the figure consider one fourth the volume.

The prismatic element runs from the z of the paraboloid to the z of the plane. The first set of limits will therefore be from $z = x^2 + y^2$ to $z = 9$. The plane cuts the paraboloid in the circle $x^2 + y^2 = 9$, which lies in the plane $z = 9$. The projection of this circle on the xy-plane is the circle $x^2 + y^2 = 9$. The limits on the y will then be from $y = 0$ to $y = \sqrt{9 - x^2}$, as determined by this projection. And, lastly, the limits on the x will be from $x = 0$ to $x = 3$. The integral for the required volume will therefore be

$$V = 4 \int_0^3 \int_0^{\sqrt{9-x^2}} \int_{x^2+y^2}^9 dz\, dy\, dx$$

$$= 4 \int_0^3 \int_0^{\sqrt{9-x^2}} (9 - x^2 - y^2) dy\, dx$$

$$= \tfrac{8}{3} \int_0^3 (9 - x^2)^{\frac{3}{2}} dx = \frac{81\,\pi}{2}.$$

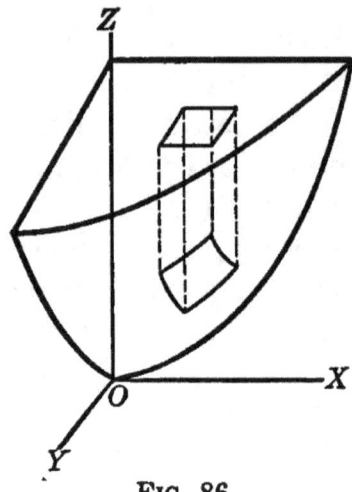

FIG. 86

183. Area of a surface. Let $z = f(x, y)$ be the equation of a surface, and let A be a portion of this surface whose projection on the xy-plane is a rectangle B with sides parallel to the x- and y-axes and of dimensions Δx and Δy. Let P be a point of A, and let the tangent plane at P be drawn. Let A' be the portion of the tangent plane which projects into B. Let γ be the angle which the normal to A at the point P makes with the z-axis. Then

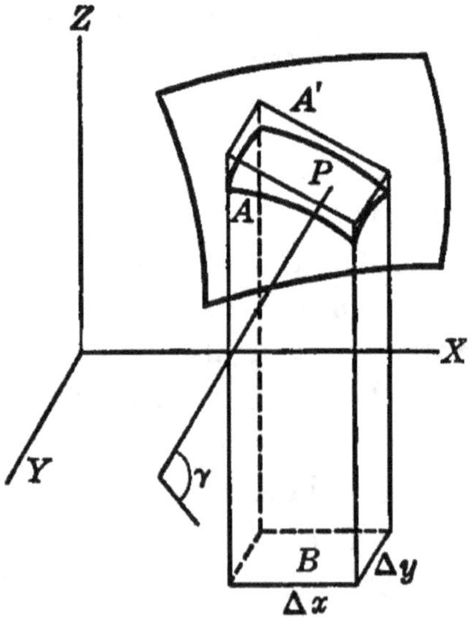

$$A' = \frac{\Delta y \cdot \Delta x}{\cos \gamma}.$$

FIG. 87

From § 176, $\quad \cos \alpha : \cos \beta : \cos \gamma = \dfrac{\partial z}{\partial x} : \dfrac{\partial z}{\partial y} : -1.$ \hfill (1)

Also $\quad\quad\quad\quad \cos^2 \alpha + \cos^2 \beta + \cos^2 \gamma = 1.$ \hfill (2)

From equations (1) and (2) we obtain

$$\frac{1}{\cos \gamma} = \sqrt{\left(\frac{\partial z}{\partial x}\right)^2 + \left(\frac{\partial z}{\partial y}\right)^2 + 1}.$$

Therefore $\quad A' = \sqrt{\left(\frac{\partial z}{\partial x}\right)^2 + \left(\frac{\partial z}{\partial y}\right)^2 + 1} \, \Delta y \cdot \Delta x.$

The sum of all surfaces A' over the required surface will give

$$S' = \sum \sum \sqrt{\left(\frac{\partial z}{\partial x}\right)^2 + \left(\frac{\partial z}{\partial y}\right)^2 + 1} \, \Delta y \cdot \Delta x,$$

which is an approximation to the required surface S. Hence if we pass to the limit, we have

$$S = \lim_{\substack{\Delta y \to 0 \\ \Delta x \to 0}} \sum \sum \sqrt{\left(\frac{\partial z}{\partial x}\right)^2 + \left(\frac{\partial z}{\partial y}\right)^2 + 1} \, \Delta y \cdot \Delta x \quad (3)$$

$$= \iint \sqrt{\left(\frac{\partial z}{\partial x}\right)^2 + \left(\frac{\partial z}{\partial y}\right)^2 + 1} \, dy \, dx.$$

Corresponding formulas for the surface may be written if the projection is on the xz-plane or the yz-plane.

Example. Find the area of that portion of the surface $z = y + x^2$ whose projection on the xy-plane is the rectangle formed by the lines $y = 0$, $y = 1$, $x = 0$, $x = 2$.

In this case
$$\frac{\partial z}{\partial x} = 2x, \qquad \frac{\partial z}{\partial y} = 1.$$

Hence the area of the required surface is

$$S = \int_0^2 \int_0^1 \sqrt{4x^2 + 1 + 1} \, dy \, dx$$
$$= \int_0^2 \sqrt{4x^2 + 2} \, dx$$
$$= 9\sqrt{2} + \tfrac{1}{2} \log (4 + 9\sqrt{2}) - \tfrac{1}{2} \log \sqrt{2}.$$

184. Summary. In the present chapter areas, volumes, and curved surfaces have been computed by means of double integration, and volumes have been computed by means of triple integration.

In working the problems in this chapter three things have been emphasized. They are:

1. The proper choice of the element of area, surface, or volume.
2. The order of the integrations.
3. The setting of the proper limits to correspond to the order of the integration. The student should make himself familiar with these three points and keep them in mind when solving problems.

EXERCISES

Find by triple integration the volumes of the following:

1. A sphere. *Ans.* $\frac{4}{3}\pi r^3$.

2. The solid bounded by the coördinate planes and the plane $x + y + z = a$. *Ans.* $\dfrac{a^3}{6}$.

3. The ellipsoid $\dfrac{x^2}{a^2} + \dfrac{y^2}{b^2} + \dfrac{z^2}{c^2} = 1$. *Ans.* $\dfrac{4\pi abc}{3}$.

4. The volume common to the two cylinders $x^2 + y^2 = a^2$ and $x^2 + z^2 = a^2$.

Find by the method of this section the surfaces:

5. Of a sphere.

6. Of a right circular cone.

7. Of the surface determined by the coördinate planes and the plane $x + y + z = a$.

8. Of the cylinder $x^2 + z^2 = a^2$ in the first octant cut out by the cylinder $x^2 + y^2 = a^2$.

CHAPTER XXII

FURTHER APPLICATIONS TO MECHANICS

185. Heterogeneous masses; density. If a homogeneous solid, that is, a solid whose density is the same at all points, has a mass m and a volume V, its density d is defined by the equation
$$d = \frac{m}{V}.$$

If the solid is heterogeneous, that is, if it has a density which varies from point to point, we may define density at a point in the following manner. Let ΔV be a small element of the solid and let Δm be its mass. Let P be a point lying within ΔV, and let ΔV approach zero as a limit in such a way that it always contains P. The density d at P is defined by the equation
$$d = \lim_{\Delta V \to 0} \frac{\Delta m}{\Delta V}$$
if the limit here written exists.

The mass of ΔV will be given approximately by the equation
$$\Delta m = d \, \Delta V.$$

The sum of all such masses as ΔV approaches zero as a limit will be exactly the mass m of the solid. This is expressed by the equation
$$m = \lim_{\Delta V \to 0} \sum d \cdot \Delta V = \int d \cdot dv. \tag{1}$$

The student should be careful to choose the element ΔV so that it will approach homogeneity as ΔV approaches zero as a limit.

Formulas similar to (1) apply to surfaces and lengths considered as masses. Example 3 will make clear the meaning of this statement.

Example 1. Find the mass of a straight wire of length l whose density varies as the distance from one end.

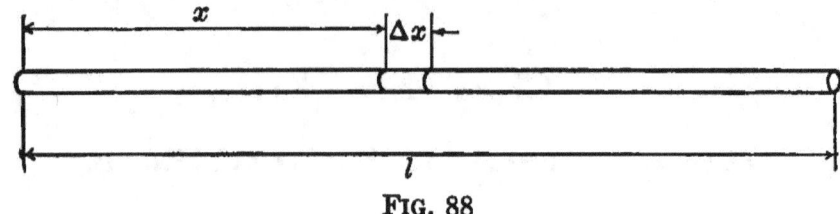

FIG. 88

Let x = distance of a cross section from one end,
   a = area of a cross section,
   $a\,\Delta x$ = element of volume = ΔV.
Then $\Delta m = kax\,\Delta x$.
The mass will then be given by
$$m = \int_0^l kax\,dx = \frac{kal^2}{2}.$$

Example 2. Find the mass of a cube of edge a when the density varies as the square of the distance from one edge.

Let the cube be situated as in the figure, and let the edge from which the density varies coincide with the x-axis. The mass will then be given by
$$m = k\int_0^a\int_0^a\int_0^a (y^2 + z^2)\,dx\,dy\,dz$$
$$= \frac{2\,ka^5}{3}.$$

Example 3. Find the mass of a rectangular surface if the density varies as the distance from the base.

Let b = length of base of rectangle,
   a = length of altitude of rectangle.

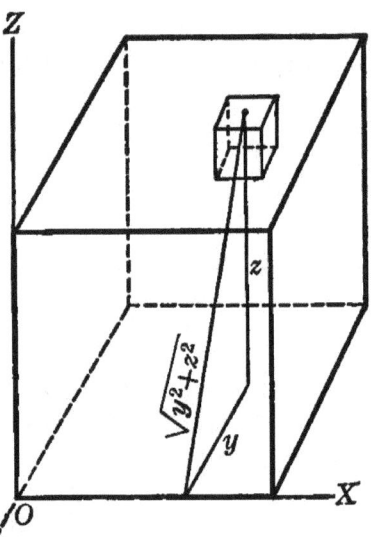

FIG. 89

Then an element of mass is
$$\Delta m = k \cdot bx\,\Delta x$$
and the entire mass will be
$$m = \int_0^a kbx\,dx = \frac{ka^2b}{2}.$$

EXERCISES

Determine the masses of the following bodies:

1. A straight rod of length l whose density varies as the nth power of its distance from one end.
$$\text{Ans. } \frac{kl^{n+1}}{n+1}.$$

2. A cube whose density varies as the square of the distance from one corner. $\quad$ Ans. ka^5.

3. A circular plate whose density varies (1) as the distance from the center, (2) as the distance from a fixed diameter.
$$\text{Ans. (1) } \frac{2\pi ka^3}{3}; \quad (2) \frac{4}{3}ka^3.$$

4. The surface of a hemisphere whose density varies as the distance from its base. $\quad$ Ans. πka^3.

5. A sphere whose density varies as the distance from the center.
$\quad$ Ans. πka^4.

6. A cone whose density varies as the distance from the base.
$$\text{Ans. } \frac{\pi ka^2 h^2}{12}.$$

7. A circular cylinder whose density varies as the distance from the axis.
$$\text{Ans. } \frac{2\pi ka^3 h}{3}.$$

8. A circular plate whose density varies as the distance from a point on the circumference.
$$\text{Ans. } \frac{32\, ka^3}{9}.$$

9. An isosceles right triangle whose density varies as the distance from the right angle.

186. Moments; centroids. Let any mass m (a line, surface, or solid may be considered as a mass, as we have seen) be divided into elements in any suitable manner. Let one of these elements Δm be at a distance x from a line l. Then $x \cdot \Delta m$ is defined as the moment of Δm about the line l. Similarly, if Δm is at a distance x from a plane α, $x \cdot \Delta m$ is defined as the moment of Δm about α. The limit of the sum of the moments of all such elements of m as Δm approaches zero is defined as the moment of m about l or α. Thus, if M_l denotes the moment of m about l, and M_α the moment of m about α, we have

$$M_l = \lim_{\Delta m \to 0} \sum x \cdot \Delta m, \quad M_\alpha = \lim_{\Delta m \to 0} \sum x \cdot \Delta m.$$

Let us now choose a point G and a set of rectangular axes so that the coördinates of G are $(\bar{x}, \bar{y}, \bar{z})$ (see Fig. 90).

Let Δm be an element of a mass m such that the coördinates of a point P in Δm are (x, y, z). The moment of Δm about a plane β through G parallel to the yz-plane is $(x - \bar{x})\Delta m$. The moment of m about β is therefore

$$M_\beta = \int (x - \bar{x}) dm.$$

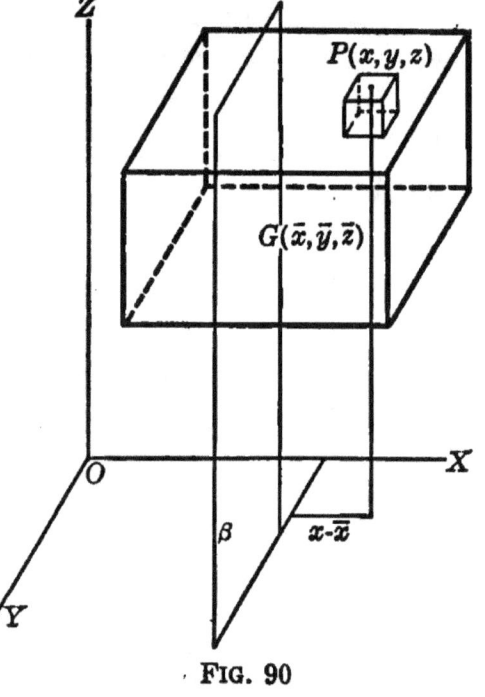

FIG. 90

Suppose now that G is such a point that $M_\beta = 0$.

Then $\int (x - \bar{x}) dm = 0.$

Hence $\quad \bar{x} = \dfrac{\int x\, dm}{\int dm}.$

Similarly, if a plane γ be drawn through G parallel to the xz-plane, the moment of m about γ is given by

$$M_\gamma = \int (y - \bar{y}) dm.$$

Now if G is a point such that $M_\gamma = 0$, then

$$\bar{y} = \frac{\int y\, dm}{\int dm}.$$

Also if a plane δ is drawn through G parallel to the xy-plane, the moment of m about δ is

$$M_\delta = \int (z - \bar{z}) dm,$$

and if $M_\delta = 0$, we have $\quad \bar{z} = \dfrac{\int z\, dm}{\int dm}.$

The point $G(\bar{x}, \bar{y}, \bar{z})$ defined above is called the centroid of the mass m.

187. Centroids of curves. Let $y = f(x)$ be the equation of a curve in the xy-plane, and let it be required to find the centroid of an arc of the curve $y = f(x)$. Since the curve is in the xy-plane, $\bar{z} = 0$. In actual calculation this coördinate is disregarded. In this case length of arc may be considered as mass. Then the formulas of § 186 for $\bar{x}$ and $\bar{y}$ become

$$\bar{x} = \frac{\int x\, ds}{\int ds} = \frac{\int x \sqrt{1 + \left(\dfrac{dy}{dx}\right)^2}\, dx}{\int \sqrt{1 + \left(\dfrac{dy}{dx}\right)^2}\, dx}. \tag{1}$$

$$\bar{y} = \frac{\int y\, ds}{\int ds} = \frac{\int y \sqrt{1 + \left(\dfrac{dy}{dx}\right)^2}\, dx}{\int \sqrt{1 + \left(\dfrac{dy}{dx}\right)^2}\, dx}. \tag{2}$$

These formulas determine the coördinates of the centroid of any plane curve.

Example. Find the centroid of the quadrant of a circle.

The equation of the circle may be written in the form $x^2 + y^2 = a^2$. We consider the arc lying in the first quadrant. Then

$$\frac{dy}{dx} = -\frac{x}{y},$$

and substitution in (1) gives

$$\bar{x} = \frac{a \int_0^a x(a^2 - x^2)^{-\frac{1}{2}}\, dx}{\dfrac{\pi a}{2}} = \frac{2\, a}{\pi}.$$

From the symmetry of the figure we see that
$$\bar{y} = \frac{2a}{\pi}.$$

If the equation of the curve is given in polar coördinates, the coördinates of the centroid are given as illustrated by the following example:

Example. Find the centroid of the arc of the circle $\rho = a \cos \theta$ between $\theta = 0$ and $\theta = \pi/2$.

If we consider the initial line as the x-axis, $\bar{x} = a/2$ on account of symmetry. Also, since
$$y = \rho \sin \theta$$
and
$$ds = \sqrt{\rho^2 + \left(\frac{d\rho}{d\theta}\right)^2}\, d\theta,$$
we have
$$\bar{y} = \frac{\int_0^{\frac{\pi}{2}} \rho \sin \theta \sqrt{\rho^2 + \left(\frac{d\rho}{d\theta}\right)^2}\, d\theta}{\frac{\pi a}{2}}$$

$$= \frac{a^2 \int_0^{\frac{\pi}{2}} \sin \theta \cos \theta \sqrt{\sin^2 \theta + \cos^2 \theta}\, d\theta}{\frac{\pi a}{2}}$$

$$= \frac{a}{\pi}.$$

EXERCISES

Find the centroids of the following arcs:

1. A straight line whose density varies as the distance from one end. Ans. $\frac{2l}{3}$.

2. A semicircle. Ans. $\bar{x} = \frac{2a}{\pi}$, $\bar{y} = 0$.

3. One quadrant of the curve $x^{\frac{2}{3}} + y^{\frac{2}{3}} = a^{\frac{2}{3}}$. Ans. $\bar{x} = \bar{y} = \frac{2}{5} a$.

4. One arch of the cycloid $x = a \operatorname{vers}^{-1} \frac{y}{a} - \sqrt{2ay - y^2}$.
Ans. $\bar{x} = \pi a$, $\bar{y} = \frac{4}{3} a$.

5. An arc of the parabola $y^2 = 4x$ from $x = 0$ to $x = 2$.

6. A semicircle whose density varies as the distance from the diameter.

188. Centroids of areas. If it is desired to find the centroid of a plane area, the formulas are

$$\bar{x} = \frac{\int x\, dA}{\int dA}, \quad \bar{y} = \frac{\int y\, dA}{\int dA}.$$

Example 1. Find the centroid of the area bounded by the parabola $y = x^2$ and the line $y = x$.

Here we have $dA = dy\, dx$.

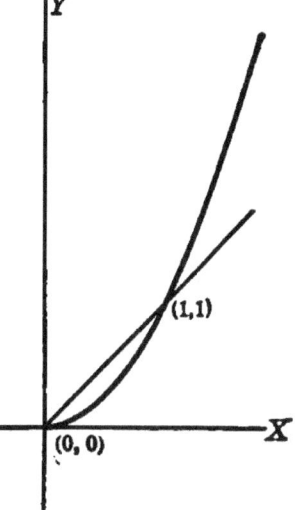

FIG. 91

Hence
$$\bar{x} = \frac{\int_0^1 \int_{x^2}^{x} x\, dy\, dx}{\int_0^1 \int_{x^2}^{x} dy\, dx} = \frac{1}{2}.$$

$$\bar{y} = \frac{\int_0^1 \int_{x^2}^{x} y\, dy\, dx}{\int_0^1 \int_{x^2}^{x} dy\, dx} = \frac{2}{5}.$$

Example 2. Find the centroid of the area bounded by the initial line and the circle $\rho = a \cos \theta$ between the limits $\theta = 0$ and $\theta = \pi/2$.

Here we have $dA = \rho\, d\rho\, d\theta$, $x = \rho \cos \theta$, $y = \rho \sin \theta$.

Hence
$$\bar{x} = \frac{\int_0^{\frac{\pi}{2}} \int_0^{a \cos \theta} \rho^2 \cos \theta\, d\rho\, d\theta}{\int_0^{\frac{\pi}{2}} \int_0^{a \cos \theta} \rho\, d\rho\, d\theta} = \frac{a}{2}.$$

$$\bar{y} = \frac{\int_0^{\frac{\pi}{2}} \int_0^{a \cos \theta} \rho^2 \sin \theta\, d\rho\, d\theta}{\int_0^{\frac{\pi}{2}} \int_0^{a \cos \theta} \rho\, d\rho\, d\theta} = \frac{2a}{3\pi}.$$

EXERCISES

Find the centroids of the following areas:

1. The area bounded by the curves $y = x^2$ and $y^2 = x$.

    Ans. $\bar{y} = \bar{x} = \frac{9}{20}$.

2. The area of a quadrant of a circle.

    Ans. $\bar{x} = \bar{y} = \dfrac{4a}{3\pi}$.

3. The area between the circle $x^2 + y^2 = a^2$ and the line $x + y = a$.

$$\text{Ans. } \bar{x} = \bar{y} = \frac{2a}{3\pi - 6}.$$

4. The area of a rectangle if the density varies as the distance from the base.

$$\text{Ans. } \bar{y} = \frac{2a}{3}.$$

5. The area of a semicircle if the density varies as the distance from the diameter.

$$\text{Ans. } \frac{3\pi a}{16}.$$

6. The area of one loop of the curve $\rho = a \sin 2\theta$.

$$\text{Ans. } \bar{x} = \bar{y} = \frac{128\, a}{105\, \pi}.$$

7. The area of a semicircle if the density varies as the distance from the center.

$$\text{Ans. } \bar{x} = 0, \bar{y} = \frac{3a}{2\pi}.$$

8. The area bounded by the circles $x^2 + y^2 = 4a^2$ and $x^2 + y^2 = 2ax$.

9. The area of one arch of the sine curve $y = \sin x$.

10. The area of one arch of the cycloid

$$x = a(\theta - \sin \theta), \quad y = a(1 - \cos \theta).$$

11. The area of a semi-ellipse determined by the major axis.

12. The area of one loop of the lemniscate $\rho^2 = a^2 \cos 2\theta$.

189. Cylindrical coördinates. Before proceeding to find the centroids of solids it will be well to define two more systems of coördinates. The first of these, termed cylindrical coördinates, is defined as follows:

Let P be a point of space (see Fig. 92) and let Q be its projection on the xy-plane. Let ρ, θ be the polar coördinates of Q in the xy-plane. Then the coördinates ρ, θ, z will locate P. They are called the cylindrical coördinates of P.

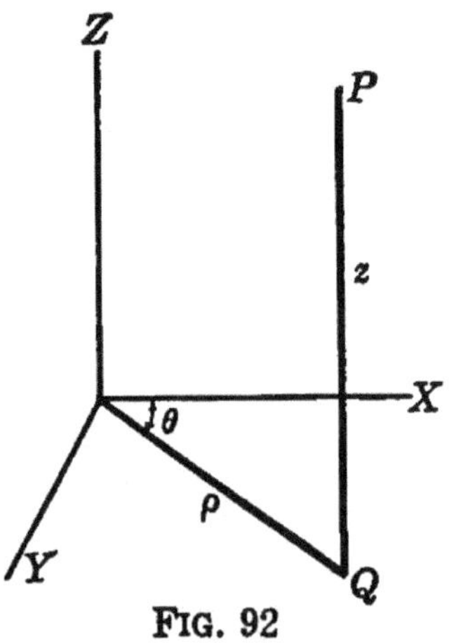

We now proceed to develop a formula for finding volumes by cylindrical coördinates. Let $P(\rho, \theta, z)$ and $Q(\rho + \Delta\rho, \theta + \Delta\theta, z + \Delta z)$ be

FIG. 92

two points in space. Pass planes through P and Q, whose common line is the z-axis. Pass planes through P and Q perpendicular to the z-axis. Pass circular cylindrical surfaces with the z-axis as axis through the points P and Q. The volume bounded by these four planes and two cylindrical surfaces will be given by the formula

$$\Delta V = \left[\int_{\theta}^{\theta + \Delta\theta} \int_{\rho}^{\rho + \Delta\rho} \rho \, d\rho \, d\theta \right] \Delta z.$$

Hence the entire volume may be calculated by the integral

$$V = \iiint \rho \, d\rho \, d\theta \, dz.$$

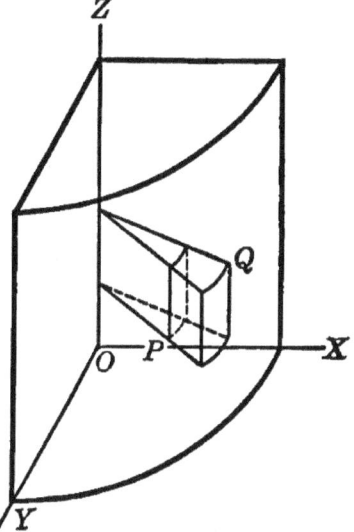

FIG. 93

Cylindrical coördinates can be used to good advantage in problems relating to cylinders, cones, and spheres.

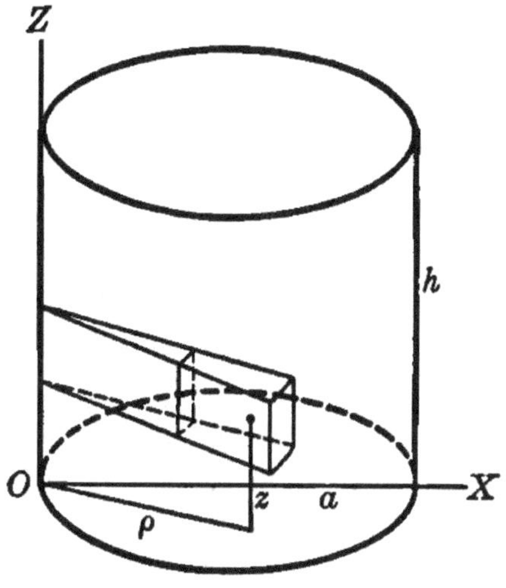

Example. Find the mass of a right circular cylinder whose density varies as the square of the distance from a point on the circumference of the base.

Let a = diameter of the base,
h = height.

Let O be the point from which density is calculated. Let the diameter through O be the initial line, and the element through O be the z-axis.

FIG. 94

Then $\quad dm = \text{density} \cdot dv = k(\rho^2 + z^2)\rho \, d\rho \, d\theta \, dz.$

Hence $\quad m = 2k \int_0^h \int_0^{\frac{\pi}{2}} \int_0^{a\cos\theta} (\rho^2 + z^2)\rho \, d\rho \, d\theta \, dz$

$$= \frac{2k}{9} (2 a^3 h + 3 a h^3).$$

190. Spherical coördinates. Another system of coördinates which is very useful is defined as follows. Let P be any point in space (see Fig. 95), and let a set of rectangular axes be set up with O as origin. Let the projection of P on the xy-plane be Q. Draw OQ and OP. Let the angle which OQ makes with the x-axis be θ, the angle which OP makes with the z-axis be ϕ, and the length of OP be ρ. The three quantities ρ, θ, ϕ are called the spherical coördinates of P. If ρ is kept constant the locus of P is a sphere with O as center. If ϕ is kept constant the locus of P is a cone with O as vertex and OZ as axis. If θ is kept constant the locus

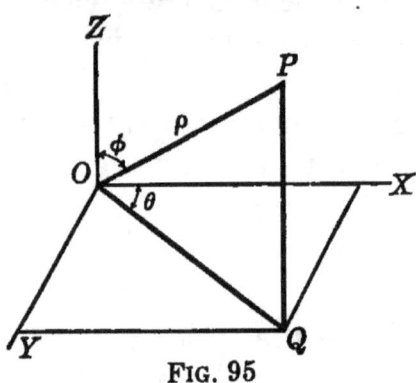

Fig. 95

of P is a plane passing through OZ. The element of volume is the volume bounded by the spheres with centers at O and radii ρ and $\rho + \Delta\rho$, the cones with O as vertices, OZ as axis, and with semivertical angles ϕ and $\phi + \Delta\phi$, and the planes $\theta = \theta_1$ and $\theta = \theta_1 + \Delta\theta$. If $\Delta\rho, \Delta\theta,$ and $\Delta\phi$ are small the element approximates a rectangular parallelepiped with PQ, QR, and RS as edges, as shown in the figure. Then we have approximately

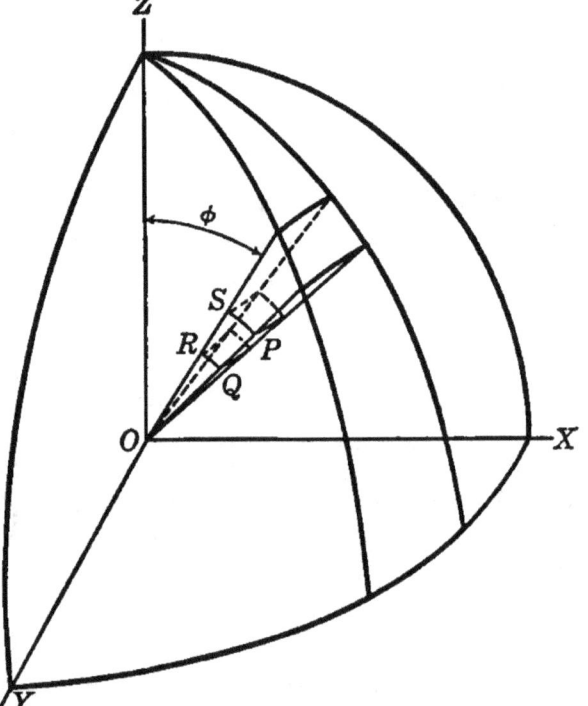

Fig. 96

$$RS = \Delta\rho, \quad RQ = \rho\Delta\phi, \quad QP = \rho \sin \phi \Delta\theta.$$

The approximate volume of this element is therefore

$$\Delta V' = \rho^2 \sin \phi \, \Delta\theta \, \Delta\phi \, \Delta\rho.$$

From these considerations it may be shown that the required volume is given by the limit

$$V = \lim_{\substack{\Delta\theta \to 0 \\ \Delta\phi \to 0 \\ \Delta\rho \to 0}} \sum\sum\sum \rho^2 \sin\phi\, \Delta\theta\, \Delta\phi\, \Delta\rho.$$

This sum may be evaluated by the triple integral

$$V = \iiint \rho^2 \sin\phi\, d\theta\, d\phi\, d\rho.$$

Example. Find the volume of a sphere, using spherical coördinates. Applying the above formula, we have

$$\begin{aligned}
V &= \int_0^a \int_0^\pi \int_0^{2\pi} \rho^2 \sin\phi\, d\theta\, d\phi\, d\rho \\
&= 2\pi \int_0^a \int_0^\pi \rho^2 \sin\phi\, d\phi\, d\rho \\
&= 4\pi \int_0^a \rho^2\, d\rho \\
&= \tfrac{4}{3}\pi a^3.
\end{aligned}$$

191. Centroids of solids. By § 186 the centroid of a solid is given by the three formulas

$$\bar{x} = \frac{\int x\, dm}{\int dm}, \qquad \bar{y} = \frac{\int y\, dm}{\int dm}, \qquad \bar{z} = \frac{\int z\, dm}{\int dm}.$$

The integrals in the above formulas may be written in rectangular, cylindrical or spherical coördinates. The student should observe the solid and use his judgment as to the best set of coördinates to use in a particular problem.

Example 1. Find the centroid of a right circular cylinder whose density varies as the square of the distance from a point on the circumference of the base.

Let
$$a = \text{diameter of the base},$$
$$h = \text{height}.$$

Let O be the point from which the density is calculated. Let the diameter through O be the x-axis and an element through O be the

z-axis. Then, from the symmetry of the figure, we see that $\bar{y} = 0$.
We have
$$\bar{x} = \frac{\int x \, dm}{\int dm}.$$

Using cylindrical coördinates, we have
$$x = \rho \cos \theta,$$
$$dm = \text{density} \cdot dv = k(\rho^2 + z^2)\rho \, d\rho \, d\theta \, dz.$$

Hence
$$\bar{x} = \frac{k \int_0^h \int_0^{\frac{\pi}{2}} \int_0^{a \cos \theta} (\rho^2 + z^2)\rho^2 \cos \theta \, d\rho \, d\theta \, dz}{k \int_0^h \int_0^{\frac{\pi}{2}} \int_0^{a \cos \theta} (\rho^2 + z^2)\rho \, d\rho \, d\theta \, dz}$$
$$= \frac{2 a(3 a^2 + 2 h^2)}{(9 a^2 + 8 h^2)}.$$

Similarly,
$$\bar{z} = \frac{\int_0^h \int_0^{\frac{\pi}{2}} \int_0^{a \cos \theta} z(z^2 + \rho^2)\rho \, d\rho \, d\theta \, dz}{\int_0^h \int_0^{\frac{\pi}{2}} \int_0^{a \cos \theta} (\rho^2 + z^2)\rho \, d\rho \, d\theta \, dz}$$
$$= \frac{3 h(4 h^2 + 3 a^2)}{2(8 h^2 + 9 a^2)}.$$

Example 2. Find the centroid of a hemisphere if the density varies as the distance from the center of the diametral plane.

Let the diametral plane be the xy-plane, the origin coinciding with the center of the circular boundary of the hemisphere. When this is the case we have
$$\bar{x} = 0, \quad \bar{y} = 0.$$
For $\bar{z}$ we have the equation
$$\bar{z} = \frac{\int z \, dm}{\int dm}.$$

Using spherical coördinates, we have
$$z = \rho \cos \phi,$$
$$dm = \text{density} \cdot dv = k\rho^3 \sin \phi \, d\theta \, d\phi \, d\rho.$$

Hence
$$\bar{z} = \frac{k \int_0^a \int_0^{\frac{\pi}{2}} \int_0^{2\pi} \rho^4 \cos \phi \sin \phi \, d\theta \, d\phi \, d\rho}{k \int_0^a \int_0^{\frac{\pi}{2}} \int_0^{2\pi} \rho^3 \sin \phi \, d\theta \, d\phi \, d\rho}$$
$$= .4 a.$$

EXERCISES

Find the centroids of the following solids:

1. A pyramid with a square base of side b and with an altitude h.

HINT. Use for element of volume a slice parallel to the base.

Ans. $\frac{1}{4} h$.

2. A right circular cone of altitude h. *Ans.* $\frac{1}{4} h$.

3. A right circular cone whose density varies as the distance from the base. *Ans.* $\frac{2}{5} h$.

4. A right circular cone whose density varies as the distance from the axis. *Ans.* $\frac{h}{5}$.

5. A right circular cone whose density varies as the distance from the center of the base. *Ans.* $\dfrac{a^2 h + h^3}{6 a^2 + 2 h^2}$.

6. A rectangular parallelepiped if the density varies as the distance from the base. *Ans.* $\frac{2}{3} h$.

7. A cube if the density varies as the square of the distance from an edge. *Ans.* $\bar{x} = \bar{y} = \dfrac{5 a}{8}, \ \bar{z} = \dfrac{a}{2}$.

8. A hemisphere. *Ans.* $\dfrac{3 a}{8}$.

9. A hemisphere if the density varies as the square of the distance from the base.

10. The volume generated by revolving the parabola $y^2 = 4x$ about the x-axis from $x = 0$ to $x = 4$.

11. The volume generated by revolving the cardioid $\rho = a(1 - \cos\theta)$ about the initial line.

192. Theorems of Pappus. The following two theorems due to the Greek geometer Pappus may be easily proved by means of the formulas for centroids.

I. The surface of a solid of revolution is equal to the length of the generating arc multiplied by the circumference of the circle described by the centroid.

Suppose the arc revolved about the x-axis. We have

$$\bar{y} = \frac{\int y\,ds}{\int ds} = \frac{\int y\,ds}{s}.$$

Hence
$$2\,\pi\bar{y}s = 2\,\pi \int y\,ds. \tag{1}$$

The right-hand member of (1) is the formula for the area of a surface of revolution formed by revolving a curve about the x-axis, and the left-hand member is the length of the circumference traveled by the centroid of the arc multiplied by the length of the arc. This proves the theorem.

II. The volume of a solid of revolution is equal to the product of the generating area and the circumference of the circle described by the centroid of the area.

Let the area be revolved about the x-axis. We have

$$\bar{y} = \frac{\int y\,dA}{A}.$$

Hence
$$2\,\pi\bar{y}A = 2\,\pi \int y\,dA. \tag{2}$$

This formula is a symbolic statement of the theorem.

These theorems may be used to find certain volumes or surfaces which it would be difficult otherwise to obtain. They may also be used to find the centroids of surfaces and curves.

Example 1. Find the volume generated by revolving a circle with radius a about a line at a distance b from the center of the circle when $b > a$.

Since the centroid of the circle is its center, we have

$$2\,\pi b = \text{circumference described by the centroid.}$$

The area of the circle is πa^2. Hence, by (2),

$$V = 2\,\pi b \pi a^2 = 2\,\pi^2 a^2 b.$$

Example 2. Find the centroid of the area of a quadrant of a circle. If the area be revolved about one of its radii, it will generate a hemisphere. Then, by theorem II,

volume = (area of surface) × (circumference described by the centroid of the area).

Therefore $\quad \dfrac{2\pi a^3}{3} = \dfrac{\pi a^2}{4} \cdot 2\pi \cdot \bar{y}, \quad \text{or} \quad \bar{y} = \dfrac{4a}{3\pi}.$

By considerations of symmetry we see that

$$\bar{x} = \frac{4a}{3\pi}.$$

EXERCISES

Find by the theorems of Pappus the following:

1. The lateral surface and volume of a right circular cone.
$\qquad$ *Ans.* $\pi a\sqrt{a^2+h^2}$; $\frac{1}{3}\pi a^2 h$.

2. The centroid of a semicircular arc. $\qquad$ *Ans.* $\dfrac{2a}{\pi}$.

3. The centroid of the area of a semicircle. $\qquad$ *Ans.* $\dfrac{4a}{3\pi}$.

4. The volume cut away by making a groove around a cylinder 4 in. in radius if the cross section of the groove is a right triangle of height $\frac{1}{2}$ in. and base 1 in. The base of the triangle coincides with an element of the cylinder. $\qquad$ *Ans.* $\dfrac{23\pi}{12}$.

5. The volume generated by revolving a semicircle of radius a about a line parallel to the base and at a distance b from the base on the side opposite the arc. $\qquad$ *Ans.* $\pi^2 a^2\left(\dfrac{4a}{3\pi} + b\right)$.

6. The surface generated by the revolving semicircle of problem 5. $\qquad$ *Ans.* $2\pi^2 a\left(\dfrac{2a}{\pi} + b\right)$.

7. The volume of a cylinder obtained by revolving a rectangle about one side. $\qquad$ *Ans.* $\pi a^2 h$.

193. Moments of inertia. Let Δm be a small mass and let P be a point in Δm. Let P be at a distance r from a given line l or plane α. Then
$$I = \lim_{\Delta m \to 0} \sum r^2 \Delta m = \int r^2 \, dm \tag{1}$$
is defined as the moment of inertia of a mass m about the line l or the plane α. The mass m is supposed to be continuous, and Δm approaches zero in such a way that it always includes P.

By moment of inertia of a surface or length is meant the value obtained by replacing the differential of mass in the integral (1) by the differential of surface or length.

Example 1. Find the moment of inertia of a right circular cylinder about its axis.

Let
$$h = \text{height of the cylinder,}$$
$$a = \text{radius of the base.}$$

Let the axis of the cylinder be the z-axis. Using cylindrical coördinates, the element of mass is $\rho \, d\rho \, d\theta \, dz$, and the distance of this element from the axis is ρ. Therefore
$$I = \int_0^h \int_0^{2\pi} \int_0^a \rho^3 \, d\rho \, d\theta \, dz$$
$$= \frac{\pi a^4 h}{2} = m \cdot \frac{a^2}{2},$$
where m denotes the mass.

Example 2. Find the moment of inertia of a right circular cone about its axis. Using cylindrical coördinates, we obtain
$$I = \int_0^h \int_0^{2\pi} \int_0^{\frac{a(h-z)}{h}} \rho^3 \, d\rho \, d\theta \, dz$$
$$= \frac{\pi a^4 h}{10} = m \cdot \tfrac{3}{10} a^2.$$

In the foregoing two examples the moment of inertia is equal to the mass of the solid (density being considered unity) multiplied by the square of a distance. This distance is called the radius of gyration. In the first example the radius of gyration is $\dfrac{a}{\sqrt{2}}$ and in the second $a\sqrt{\tfrac{3}{10}}$.

194. Summary. In the present chapter we have applied the principles of the integral calculus to the problem of finding heterogeneous masses and to the problems of finding the centroids and moments of inertia of volumes, surfaces, and curves.

EXERCISES

Find the moments of inertia and radii of gyration of the following:

1. A rectangle with base b and altitude h (1) about the base, (2) about a line through the center parallel to the base.

$$\text{Ans. (1) } \frac{h^2}{3} m; \quad \frac{h^2}{3} = r^2.$$

2. A rod whose density varies as the distance from one end, about an axis perpendicular to the rod and passing through this end.

$$\text{Ans. } \frac{kl^4}{4}; \quad \frac{l^2}{2} = r^2.$$

3. A square about a line through a vertex perpendicular to its plane. $\quad$ Ans. $\frac{2}{3} a^4$; $\frac{2}{3} a^2 = r^2$.

4. The area of a semicircle about its base.

$$\text{Ans. } \tfrac{1}{8} \pi a^4; \quad \tfrac{1}{4} a^2 = r^2.$$

5. The arc of a semicircle about its base. Ans. $\tfrac{1}{2} ma^2$; $\tfrac{1}{2} a^2 = r^2$.

6. A cylinder about a line intersecting its axis at right angles (1) in one base, (2) midway between the bases.

$$\text{Ans. } \pi a^2 l \left(\frac{a^2}{4} + \frac{l^2}{3} \right).$$

7. A cube about one edge. $\quad$ Ans. $\frac{2}{3} a^5$.

8. A hemisphere about a diameter in its base.

9. The surface of a sphere about a diameter.

10. A cylinder about an element.

11. A flywheel 36 in. in diameter, the rim being 4 in. wide and 4 in. thick, the hub being a cylinder 6 in. in diameter with a 3-inch hole through it, and the six spokes being cylinders 2 in. in diameter.

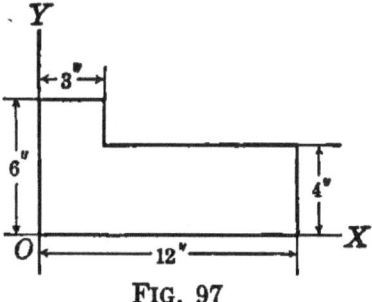

FIG. 97

12. The area shown in Fig. 97 (1) about OX, (2) about OY.

13. A right circular cone about a line intersecting its axis in the plane of the base.

Bibiography For <u>Further Reading In Basic Calculus</u>

As I've said several times, the purpose of this series is to revitalize the standard university calculus course to the mathematical level it had been before the American university system decided to monetize it and make it a requirement of far too many students for reasons entirely disjoint from their educational welfare. Calculus should be a revelation for true students seeing it for the first time, regardless of what their future career plans are. Having hopefully stimulated interest in calculus-and mathematics, by extension-it's natural to suggest further reading, particularly for those who wish to study the serious theory of calculus.

This bibliography will focus on sources to supplement our definition of "standard" textbooks in calculus i.e. books that present the main elements of calculus carefully but not completely rigorously, without sophisticated results from analysis. Honors calculus textbooks-of which we have several planned for publication in the near future-*do* give a completely rigorous presentation of calculus. These books will have their own recommended reading lists.

One of the reasons we began this Series was the insane cost of the average calculus textbook today. Hence a guiding principle to this recommendations list will be to suggest inexpensive sources. Fortunately, there are quite a few good ones. Even better, several of these are reprinted classic sources of the same vintage as the books in this series.

The ubiquitous *Schaum's Outlines* series from McGraw Hill are the modern student's preferred shortcut study aids. Many are excellent, but the calculus entries in the series are particularly good. The 2 main entries, (1) and (2) are good enough to function together as an inexpensive calculus textbook. They were both co-written by one of my former professors. Elliott Mendelson, a superior researcher in his day in mathematical logic and a fine teacher. He frequently taught calculus at all levels at my alma mater, CUNY. The first edition of the main book (1) appeared in the 1970's and it hasn't changed in overall style drastically since then. It covers all the major topics of a basic calculus course of one and several variables-algebra and analytic geometry, limits, derivatives, differentials, integration, coordinate systems, vectors, parametric equations, partial derivatives and the differential of several variables, multiple integrals and differential equations-clearly, succinctly and with an enormous number of solved problems in addition to additional exercises for the student to practice. (2) is exactly what the title says it is and supplies about as many solved problems as you need in any given course to act as either examples or as a makeshift solutions manual for your textbook. The books are soft compared to the level most of the books in the OSC series are, there are not many rigorous geometric or analytic problems. But they do provide excellent support for most of the standard subjects in a calculus course. Both are a must have for a basic calculus class.

These days, a wealth of free resources in calculus can be found online. All one needs to do is to go to one's favorite search engine and spend some time looking under "calculus". Many if not all can be found at my website, www.tuloomath.com, where I've collected just about all the major free resources on mathematics available at that time on the internet. In other words, to paraphrase an old detergent commercial, I worked hard so you don't have to. At the basic calculus page, one will find a legion of free lecture notes and textbooks-yes, entire free textbooks in calculus-that are available now. Not only do I have them listed with links (some of which may have to be updated-if the original links are broken and the notes have moved or been deleted, please email me at themathemagician369@gmail.com or bluecollarscholar2018@gmail.com), I've carefully read and reviewed each set of notes at the site so you don't waste your time with them. Of particular interest, in no specific order, for serious calculus students here are the lecture notes of Eleftherios Gkioulekas of the Pan American University, Arthur Mattuck at MIT, Joel Feldman at the University of British Columbia and Ken Kuniyuki of San Diego Mesa College. I'm quite proud of the site-particularly the calculus and analysis pages-and I hope to give the site a much needed updating by the end of 2018. But until then, I hope it remains a major site for math students to find help. The basic calculus page can be found at (3).

(I don't want to begin recommending online material in this bibliography because the purpose of BSC is to make hard copy inexpensive mathematics texts available. It's not that I have anything against online material, don't get me wrong. It's just if we begin listing those sources here, it'll defeat the entire purpose As I said, I have a separate website for that purpose.)

Several quite good and inexpensive calculus textbooks at this level are available in paperback. (4), (5), (6), and (7) are all pitched at about the same level, that is, about the same level as "standard" textbooks in the OSC series-which are a bit more sophisticated than today's books, but still accessible to the average student. They all have very different styles and each has unique characteristics.

(4) is about the same level as Carmicheal, et. al or (1). It was published in 1984 and sadly, it looks it, as the "watering down" of calculus at most American universities was already in full swing. Except for the absence of computer-generated graphs and programs, the book could have been written today. On the plus side, it does have many well-chosen and presented examples and exercises. It's solidly written but pretty run of the mill and pedestrian. Nothing striking about it. Still, with the cost of today's calculus texts, this conceivably could be used as an inexpensive replacement in a standard course.

Morris Kline's (5) is a classic textbook by one of the greatest applied mathematicians and educators of mathematics of the 20th century. It's more sophisticated than the previous texts, about the same level as Phillips. After 25 years of teaching calculus at many different levels to students at New York University, the author wrote this book to serve as a careful calculus textbook that deliberately emphasized applications to the physical sciences without disregarding rigorous proofs of important results. Kline believed that a serious presentation of calculus through its' applications in a detailed and broad presentation is the best way for beginning but serious students to understand it. Only in the last chapter is a brief but lucid discussion of precise proofs of calculus results given. Many applications and geometric explanations are given, but Kline is adamant that these are not proofs, but explanations. A number of these examples in the text are unusual for a first course, such as applications to gravitational and electromagnetic fields. It is masterfully presented and beautifully written by a master teacher and at no time does Kline pretend he's doing a rigorous treatment. It is an applied course in the very best sense-carefully presented arguments based on physics and geometry, while still keeping in mind that a prelude to a rigorous treatment is part of the intent of the course. Indeed, I would recommend all instructors of the subject to read Kline's comments in the prefaces in detail, as they are wonderfully enlightening on teaching calculus to a general audience. The following quote summarizes the method behind his madness:

Rigor has its place in mathematics education. It is a check on the creations and it permits an aesthetic (as well as an anaesthetic) presentation. But it is also to some extent gilt on the lily and an interdiction against the inclusion of functions which rarely occur in practice and which must even be invented with Weierstrassian ingenuity. A rigorous first course in calculus reminds one of the words of Samuel Johnson; "I have found you an argument but I am not obliged to find you an understanding."Even if the rigorous material is understood, its value is limited. As Henri Lebesgue pointed out:"Logic makes us reject certain arguments but it cannot make us believe any argument."

Kline is an absolute jewel and it's a book that should be in every teacher and student of calculus' library at this price. This the book that today's calculus books and courses should take as the model of a general purpose calculus course. A must have.

(6) is another unusual and ambitious book, it combines a relatively careful first course in calculus of one and several variables with a first course in probability and statistics. It's about the same level as (5), although a bit more rigorous. This perspective supplies many unorthodox examples and applications of calculus that aren't usually available to the usual first course in calculus, such as the use of integration in computing expectations and frequency distributions. It's quite well written and clear. An adventurous instructor or student with an interest in probability and statistics, a very popular and lucrative career path today, should be quite intrigued by it.

(7) , by the same author, is a much more standard course in calculus for beginning students that emphasizes geometric intuition, although it does give fairly rigorous definitions, as most books from this period did. It's about the same level as Carmicheal,et. all. With the exception of precision of the definitions, it's a pretty run of the mill calculus text, although the author explains things quite well. Worth a look for the beginner.

No recommendation list for introductory calculus would be complete without the incredible and inexpensive study guide for calculus by Adrian Banner (8). This terrific book gives a detailed, deeply conceptual and visual coverage of all the standard topics and techniques of calculus for beginning students. It can also be a very effective refresher for people who have forgotten calculus. The book is amazingly user friendly and comprehensive. It begins literally from scratch, with basic algebra and geometry and ends with a brief optional discussion of rigorous limits at the end. He presents limits, derivatives, integrals, etc. from all perspectives in a step by step manner and with hundreds of diverse solved examples. Banner writes in a very deliberate, wordy yet warm and conversational style and has a remarkable ability not to miss any steps or observations. This is clearly someone with considerable experience in teaching this subject to students at many levels and backgrounds and it shines through on every page. Indeed, if supplemented with a collection of exercises in calculus, it would make an excellent low priced single variable calculus textbook. This is a book any beginning student of calculus would love their teacher for handing them. My one quibble is that the book only covers single variable calculus. A new edition or sequel that does the same for multivariable calculus would be a godsend for students. An absolute must have for either students or teachers of calculus.

Bibliography

    1) Ayres Jr., Frank, Mendelson, Elliott, *Schaum's Outline of Calculus*, McGraw-Hill Education,6th edition,2012

    2) Mendelson, Elliott, *3000 Solved Problems In Calculus*, McGraw-Hill Education,3rd edition,2013

3) http://www.tuloomath.com/tuloomath-mathematics-site/basic-calculus-calculus-without-theory/ .

4) Gerstling, Judith L, *Technical Calculus with Analytic Geometry* , Dover Books, 1992

5) Kline, Morris, *Calculus: An Intuitive and Physical Approach*, Dover Books, 1998

6) Gemignani, Michael, *Calculus And Statistics*, Dover Books, 2006

7) Gemignani, Michael, *Calculus: A Short Course*, Dover Books, 2004

8) Banner, Adrian, *The Calculus Lifesaver: All the Tools You Need to Excel at Calculus*, Princeton University Press,2007

INDEX*

(The numbers refer to pages)

Acceleration, average, 86; instantaneous, 86; normal component of, 92; tangential component of, 92; angular, 98
Algebra, formulas from, 1
Alternating series, 243, 249
Amplitude of motion, 96
Analytics, formulas from, 3
Antiderivative, 49, 60, 75
Approach to a limit, 20
Approximate computation, by differentials, 109, 333; by series, 258; by trapezoidal rule, 320; by Simpson's rule, 321; by integration in series, 322
Approximate integration, 320
Arc length, 111, 190, 285
Area, derivative of an, 48
Areas, computation of, 51, 189, 192, 283; considered as limits of sums, 55; by double integration, 350; of surface, 361; centroids of, 370
Argument of a function, 5
Arithmetic mean, 198
Asymptotes, 269–273
Average value, 198

Barrow, Isaac, 83
Boyle's law for gases, 41, 295

Calculus, infinitesimal, 20; differential, 40; integral, 40; origin of, 83
Cavalieri, 83
Center of curvature, 276
Centroids, 366; of curves, 368; of areas, 370; of solids, 374
Circle of curvature, 276
Component accelerations, 92
Component velocities, 90

Composite functions, 47
Compound-interest law, 158
Computation, of areas, 51, 189, 192, 283; by differentials, 109, 333; by series, 258
Concavity, 263
Conjugate point, 265
Constant, 3, 4; of integration, 201, 204
Continuity, 25; theorems on, 27
Convergence, to a limit, 18; absolute, 244; conditional, 244; of series, 245; tests of, 245; interval of, 251
Coördinates, 11; cylindrical, 371; spherical, 373
Critical value, 123
Curvature, 275; center of, 276; circle of, 276; radius of, 276
Curve, slope of, 12, 42; tangent to, 98; normal to, 98, 280, 340
Curve tracing, 274
Curves, properties of, 263–282; centroids of, 368
Cusp, 265
Cycloid, 282

Decimals, repeating, 24
Decrease of functions, 120
Density, 364
Derivative, 38; of $\sin x$ and $\cos x$, 45; of a function of a function, 46; of an area, 48; of indefinite integral, 62; of inverse functions, 77; of higher order, 81; successive, 81; as a rate, 102; partial, 326; total, 329
Differentiable functions, 39
Differential calculus, 40

Differential coefficient, 38
Differential equation, 208; degree of, 208; order of, 208; integrable, 208; with variables separable, 209; homogeneous, 211, 214; linear, 212, 214; nonhomogeneous, 214
Differentials, 108, 332; of arc length, 113, 114; successive, 118; exact, 335
Differentiation, general principles, 38–48; of sine and cosine functions, 45; of a function of a function, 46; theorems on, 66; of inverse functions, 77; of implicit functions, 80, 331; applications of, 84–107; of trigonometric functions, 130–132; of inverse trigonometric functions, 139–141; of logarithmic functions, 146; of exponential functions, 147; of hyperbolic functions, 156; successive, 225; partial, 326

e, base of natural logarithms, 35–37, 258
End point, 265
Envelopes, 337, 338, 340
Equations, parametric, 79; solutions of, with multiple roots, 101; differential, 208–224
Evaluation of definite integrals, 58, 60
Evolutes, 281, 340
Exact differentials, 335
Expansions of functions, 253–262
Explicit functions, 80
Exponential functions, 147, 148, 157

Fermat, 83
Force, 292
Forms, indeterminate, 15, 30–37, 234
Formulas for reference, 1
Four-Step Rule, 40
Frequency, 96
Function, 4–7; of a function, 46
Functional notation, 7–10

Functions, double-valued, 6; continuous, 25; discontinuous, 25; differentiable, 39; composite, 47; inverse, 77; represented parametrically, 78; explicit, 80, 331; implicit, 80, 331; decreasing, 120; increasing, 120; maxima and minima of, 121, 232; hyperbolic, 152; integrals of discontinuous, 186; complementary, 217; expansion of, 253–262; rational, 298; of two or more variables, 326–349

Gases, Boyle's law for, 41, 295
Geometry, formulas from, 1
Graph, 10, 27
Gyration, radius of, 379

Harmonic motion, simple, 94, 96, 205
Harmonic series, 243
Historical note, 82
Hooke's law, 295
Hyperbolic functions, 152; relations between, 154; inverse, 155; differentiation of, 156

Implicit functions, 80, 331
Increase of functions, 120
Increment, 13, 38
Indeterminate forms, 15, 30–37, 234
Inertia, moment of, 379
Infinite limits of integration, 185
Infinite series, 19, 24, 240–252
Infinitesimal, 20, 82
Infinitesimal calculus, 20, 82
Infinity, 22
Inflection, points of, 263
Instantaneous speed, 90
Integrable differential equation, 208
Integral, definite, 57; indefinite, 62; particular, 217; double, 347; repeated, 347, 348; multiple, 348
Integral calculus, 40
Integrals, standard forms of, 163; table of, 319
Integrand, 57
Integrating machines, 323

Integration, 163–188; applications of, 84–98, 283–297, 350–380; by parts, 177; miscellaneous methods of, 181; successive, 227; of special classes of functions, 298–319; in series, 322; of functions of two or more variables, 347
Interval, 11, 251
Inverse functions, 77
Involute, 279
Isolated point, 265

Law, of the mean, 44, 230; compound-interest, 158; of organic growth, 158
Leibniz, 83
Length, 111, 190, 285
Limit, 15, 18–27; of $(\sin x)/x$ as x approaches zero, 33; of sequence, 54
Limiting value of a function, 20
Limits, theorems on, 21; special, 23; of integration, 57, 185
Logarithm, natural, 36; differentiation of, 146
Logarithmic differentiation, 149
Logarithms, 36, 259, 260

Machines, integrating, 323
Maclaurin's series, 254
Mass, 364
Maxima, 121, 232
Mean, arithmetic, 198
Mean value, 198
Method of undetermined coefficients, 222
Minima, 121, 232
Moments, 366, 379
Motion, 84–98; with assigned acceleration, 87; curvilinear, 90; simple harmonic, 94, 96, 205; amplitude of, 96; period of, 96; periodic, 96; sinusoidal, 96; state of, 96
Multiple integral, 348
Multiple point, 265
Multiple roots, 101
Multiple-valued function, 6

Newton, 83
Normal, to a curve, 98, 280, 340; length of, 99; to a surface, 344
Normals, envelope of, 340
Notation, functional, 7–10
Number, e, 35–37, 258; π, 259

Orthogonal trajectories, 202
Oscillating function, 21
Oscillation, 21

π, 259
Pappus, theorems of, 376
Parameter, 4
Parametric equations, 79
Parametric representation of functions, 78
Partial derivatives, 326
Parts, integration by, 177
Power series, 250–253
Prismoidal formula, 324
Problems of the calculus, 15
Projectile, 93, 204
p-series, 247

Radius, of curvature, 276; of gyration, 379
Range of variable, 4, 6, 7
Rates, 15, 102
Ratio test for convergence, 247
Rational functions, integration of, 298
Reduction formulas, 312
Relative maximum, 122
Relative minimum, 122
Remainder in Taylor's theorem, 255
Repeated integral, 347, 348
Revolution, surfaces of, 288, 290
Rolle's theorem, 229
Roots, multiple, 101
Rotation, 97
Roulettes, 278
Rule, Four-Step, 40; trapezoidal, 320; Simpson's, 321

Salient point, 265
Scale, 10, 11
Separation of variables, 209

Sequence, 53; partial sum, 240
Series, infinite, 19, 24, 240–252; terminating, 241; convergent, 242; divergent, 243; harmonic, 243; absolutely convergent, 244; conditionally convergent, 244; power, 250–253; Maclaurin's, 254; Taylor's, 254; integration in, 322
Simple harmonic motion, 94, 96, 205
Simpson's rule, 321
Single-valued function, 6
Singular points of curves, 265
Slope, 12, 42
Solids, centroids of, 374
Speed, 16, 84, 90, 97
Subnormal, 100
Subtangent, 100
Successive differentiation, 225
Successive integration, 227
Sum of infinite series, 19, 24, 242
Summation, fundamental process of, 55
Surfaces, 288, 290, 361, 376; tangent planes to, 343; normals to, 344

Table of integrals, 319
Tacnode, 265

Tangent, to a curve, 98; length of, 99
Tangent plane, 343
Tangential component of acceleration, 92
Taylor's expansion, 255
Taylor's series, 254
Taylor's theorem, 231, 255
Tests for convergence, 245; comparison, 247; ratio, 247
Total derivatives, 329
Trajectories, orthogonal, 202
Trapezoidal rule, 320
Trigonometric functions, 130–138; inverse, 139–145
Trigonometry, formulas from, 2

Value, critical, 123; mean or average, 198
Variable, 3, 4; range of, 4, 6, 7; dependent, 5; independent, 5
Variables, separation of, 209
Vector, 91
Velocity, 84; components of, 90
Volumes, 194, 286, 356, 359

Work, 294